Pelican Books
Asimov's Guide to Scie
The Physical Sciences

Dr Isaac Asimov was born in Russia in 1920 and received a Ph.D. from Columbia University. A Professor of Biochemistry at the School of Medicine, Boston University, he has published many works, both fiction and non-fiction. Asimov's chief contribution to science fiction is said to have been his use of robots, from the collection of stories *I, Robot* (1950) to *The Caves of Steel* (1954) and *The Naked Sun* (1957), both of which employ a robot detective, as does *Asimov's Mysteries* (1968), a collection which combines science fiction with the mystery story. Under the pseudonym Paul French, he has also written for children. Among his other works are *The Foundation Trilogy* (1963), *Nine Tomorrows* (1959), a collection of tales, and *The Universe* (1967).

Isaac Asimov

Asimov's Guide to Science

Volume 1
The Physical Sciences

Penguin Books

Penguin Books Ltd,
Harmondsworth, Middlesex, England
Penguin Books Australia Ltd,
Ringwood, Victoria, Australia
Penguin Books (N.Z.) Ltd,
182–190 Wairau Road, Auckland 10, New Zealand

First published by Basic Books, Inc., New York, 1972
Published in Pelican Books, 1975

Made and printed in Great Britain by
Hazell Watson & Viney Ltd, Aylesbury, Bucks
Set in Monotype Ehrhardt

To Janet Jeppson
who shares my interest in science

Contents

List of Plates

Preface

The rapid advance of science is exciting and exhilarating to anyone who is fascinated by the unconquerability of the human spirit and by the continuing efficacy of the scientific method as a tool for penetrating the complexities of the universe.

But what if one is also dedicated to keeping up with every phase of scientific advance for the deliberate purpose of interpreting that advance for the general public? For him, the excitement and exhilaration is tempered by a kind of despair.

Science will not stand still. It is a panorama that subtly dissolves and changes even while we watch. It cannot be caught in its every detail at any one moment of time without being left behind at once.

In 1960, *The Intelligent Man's Guide to Science* was published, and at once the advance of science flowed past it. In order to consider quasars and lasers, for instance (which were unknown in 1960 and household words a couple of years later), *The New Intelligent Man's Guide to Science* was published in 1965.

But still science drove on inexorably. Now there came the question of pulsars, of black holes, of continental drift, men on the moon, REM sleep, gravitational waves, holography, cyclic-AMP, and so on, and so on, and so on – all post-1965.

So it is time for a new edition, the third. But what do we call it now? *The New New Intelligent Man's Guide to Science*? Obviously not.

Since 1965, however, I have done, with my own name in the title, a two-volume guide to the Bible and also a two-volume guide to Shakespeare. Why not use the same system here? Enter, then, the 1972 edition of my guide to science, entitled, straightforwardly, *Asimov's Guide to Science.*

ISAAC ASIMOV

New York
1972

Note

The English forms of billion (a million million), trillion (a million billion), etc., are used in this edition.

1 What is Science?

Almost in the beginning was curiosity.

Curiosity, the overwhelming desire to know, is not characteristic of dead matter. It is also not characteristic of some forms of living organism, which, for that very reason, we can scarcely bring ourselves to consider alive.

A tree does not display curiosity about its environment in any way we can recognize; nor does a sponge or an oyster. The wind, the rain, the ocean currents bring them what is needful, and from it they take what they can. If the chance of events is such as to bring them fire, poison, predators, or parasites, they die as stoically and as undemonstratively as they lived.

Early in the scheme of life, however, independent motion was developed by some organisms. It meant a tremendous advance in their control of the environment. A moving organism no longer had to wait in stolid rigidity for food to come its way; it went out after it.

This meant that adventure had entered the world – and curiosity. The individual that hesitated in the competitive hunt for food, that was overly conservative in its investigation, starved. As early as that, curiosity concerning the environment was enforced as the price of survival.

The one-celled paramecium, moving about in a searching way, cannot have conscious volitions and desires in the sense that we do, but it has a drive, even if only a 'simple' physical–chemical one, which causes it to behave as if it were investigating its surroundings for food. And this 'act of curiosity' is what we most easily recognize as being inseparable from the kind of life that is most akin to ours.

As organisms grew more intricate, their sense organs multiplied and became both more complex and more delicate. More messages of greater variety were received from and about the external environment. Along with that (whether as cause or effect we cannot tell), there developed an increasing complexity of the nervous system, the living instrument that interpreted and stored the data collected by the sense organs.

There comes a point where the capacity to receive, store, and interpret messages from the outside world may outrun sheer necessity. An organism may for the moment be sated with food, and there may, at the moment, be no danger in sight. What does it do then?

It might lapse into an oyster-like stupor. But the higher organisms, at least, still show a strong instinct to explore the environment. Idle curiosity, we may call it. Yet, though we may sneer at it, we judge intelligence by it. The dog, in moments of leisure, will sniff idly here and there, pricking up its ears at sounds we cannot hear; and so we judge it to be more intelligent than the cat, which in its moments of leisure grooms itself or quietly and luxuriously stretches out and falls asleep. The more advanced the brain, the greater the drive to explore, the greater the 'curiosity surplus'. The monkey is a byword for curiosity. Its busy little brain must and will be kept going on whatever is handy. And in this respect, as in many others, man is but a super-monkey.

The human brain is the most magnificently organized lump of matter in the known universe, and its capacity to receive, organize, and store data is far in excess of the ordinary requirements of life. It has been estimated that in a lifetime a human being can learn up to 15 billion items of information.

It is to this excess that we owe our ability to be afflicted by that supremely painful disease, boredom. A human being forced into a situation where he has no opportunity to utilize his brain except for minimal survival will gradually experience a variety of unpleasant symptoms, up to and including serious mental disorganization.

What it amounts to, then, is that the normal human being has an intense and overwhelming curiosity. If he lacks the opportunity to satisfy it in ways immediately useful to him, he will satisfy it in other ways – even regrettably ways to which we have attached admonitions such as: 'Curiosity killed the cat', 'Mind your own business'.

The overriding power of curiosity, even with harm as the penalty, is reflected in the myths and legends of the human race. The Greeks had the tale of Pandora and her box. Pandora, the first woman, was given a box that she was forbidden to open. Quickly and naturally enough she opened it and found it full of the spirits of disease, famine, hate, and all kinds of evil – which escaped and have plagued the world ever since.

In the Biblical story of the temptation of Eve, it seems fairly certain (to me, at any rate) that the serpent had the world's easiest job. He might have saved his tempting words: Eve's curiosity would have driven her to taste the forbidden fruit even without temptation. If you are of a mind to interpret the Bible allegorically, you may think of the serpent as simply the representation of this inner compulsion; in the conventional cartoon picturing Eve standing under the tree with the forbidden fruit in her hand, the serpent coiled around the branch might be labelled 'Curiosity'.

If curiosity, like any other human drive, can be put to ignoble use – the prying invasion of privacy that has given the word its cheap and unpleasant connotation – it nevertheless remains one of the noblest properties of the human mind. For its simplest definition is 'the desire to know'.

This desire finds its first expression in answers to the practical needs of human life – how best to plant and cultivate crops, how best to fashion bows and arrows, how best to weave clothing – in short, the 'applied arts'. But after these comparatively limited skills have been mastered, or the practical needs fulfilled, what then? Inevitably the desire to know leads on to less limited and more complex activities.

It seems clear that the 'fine arts' (designed to satisfy inchoate and boundless and spiritual needs) were born in the agony of boredom. To be sure, one can easily find more mundane uses and excuses for the fine arts. Paintings and statuettes were used as fertility charms and as religious symbols, for instance. But one cannot help suspecting that the objects existed first and the use second.

To say that the fine arts arose out of a sense of the beautiful may also be putting the cart before the horse. Once the fine arts were developed, their extension and refinement in the direction of beauty would have followed inevitably, but even if this had not happened, the fine arts would have developed nevertheless. Surely the fine arts antedate any possible need or use for them, other than the elementary need to occupy the mind as fully as possible.

Not only does the production of a work of fine art occupy the mind satisfactorily; the contemplation or appreciation of the work supplies a similar service to the audience. A great work of art is great precisely because it offers a kind of stimulation that cannot readily be found elsewhere. It contains enough data of sufficient complexity to cajole the brain into exerting itself past the usual needs, and, unless a person is hopelessly ruined by routine or stultification, that exertion is pleasant.

But if the practice of the fine arts is a satisfactory solution to the problem of leisure, it has this disadvantage: it requires, in addition to an active and creative mind, a physical dexterity. It is just as interesting to pursue mental activities that involve only the mind, without the supplement of manual skill. And, of course, such an activity is available. It is the pursuit of knowledge itself, not in order to do something with it but for its own sake.

Thus the desire to know seems to lead into successive realms of greater etherealization and more efficient occupation of the mind – from knowledge of accomplishing the useful, to knowledge of accomplishing the aesthetic, to 'pure' knowledge.

Knowledge for itself alone seeks answers to such questions as 'How high is the sky?' or 'Why does a stone fall?' This is sheer curiosity – curiosity at its idlest and therefore perhaps at its most peremptory. After all, it serves no apparent purpose to know how high the sky is or why the stone falls. The lofty sky does not interfere with the ordinary business of life, and, as for the stone, knowing why it falls does not help us to dodge it more skilfully or soften the blow if it happens to hit us. Yet there have always been people who ask such apparently useless questions and try to answer them out of the sheer desire to know – out of the absolute necessity of keeping the brain working.

The obvious method of dealing with such questions is to make up an aesthetically satisfying answer: one that has sufficient analogies to what is already known to be comprehensible and plausible. The expression 'to make up' is rather bald and unromantic. The ancients liked to think of the process of discovery as the inspiration of the muses or a revelation from heaven. In any case, whether it was inspiration, revelation or the kind of creative thinking that goes into storytelling, their explanations depended heavily on analogy. The lightning bolt is destructive and terrifying, but it appears, after all, to be hurled like a weapon and does the damage of a hurled weapon – a fantastically violent one. Such a weapon must have a wielder similarly enlarged in scale, and so the thunderbolt becomes the hammer of Thor or the flashing spear of Zeus. The more-than-normal weapon is wielded by a more-than-normal man.

Thus a myth is born. The forces of nature are personified and become gods. The myths react on one another, are built up and improved by generations of myth-tellers until the original point may be obscured. Some may degenerate into pretty stories (or ribald ones), whereas others may gain an ethical content important enough to make them meaningful within the framework of a major religion.

Just as art may be fine or applied, so may mythology. Myths may be maintained for their aesthetic charm, or they may be bent to the physical uses of mankind. For instance, the earliest farmers would be intensely concerned with the phenomenon of rain and why it fell so capriciously. The fertilizing rain falling from the heavens on the earth presented an obvious analogy to the sex act, and, by personifying both heaven and earth, man found an easy explanation of the release or withholding of the rains. The earth-goddess, or the sky-god, was either pleased or offended, as the case might be. Once this myth was accepted, farmers had a plausible basis for bringing rain; namely, appeasing the god by appropriate rites. These rites might well be orgiastic in nature – an attempt to influence heaven and earth by example.

The Greek myths are among the prettiest and most sophisticated in our literary and cultural heritage. But it was the Greeks also who, in due course, introduced the opposite way of looking at the universe – that is, as something impersonal and inanimate. To the myth-makers, every aspect of nature was essentially human in its unpredictability. However mighty and majestic the personification, however superhuman Zeus or Marduk or Odin might be in powers, they were also – like mere men – frivolous, whimsical, emotional, capable of outrageous behaviour for petty reasons, susceptible to childish bribes. As long as the universe was in the control of such arbitrary and unpredictable deities, there was no hope of understanding it, only the shallow hope of appeasing it. But in the new view of the later Greek thinkers, the universe was a machine governed by inflexible laws. The Greek philosophers now devoted themselves to the exciting intellectual exercise of trying to discover just what the laws of nature might be.

The first to do so, according to Greek tradition, was Thales of Miletus, about 600 B.C. He was saddled with an almost impossible number of discoveries by later Greek writers, and it may be that he first brought the gathered Babylonian knowledge to the Greek world. His most spectacular achievement was that of predicting an eclipse for 585 B.C. – and having it take place.

In engaging in this intellectual exercise, the Greeks assumed, of course, that nature would play fair; that, if attacked in the proper manner, it would yield its secrets and would not change position or attitude in mid-play. (Thousands of years later Albert Einstein expressed this feeling when he said, 'God may be subtle, but he is not malicious.') There was also the feeling that the natural laws, when found, would be comprehensible. This Greek optimism has never entirely left the human race.

With confidence in the fair play of nature, man needed to work out an orderly system for learning how to determine the underlying laws from the observed data. To progress from one point to another by established rules of argument is to use 'reason'. A reasoner may use 'intuition' to guide his search for answers, but he must rely on sound logic to test his theories. To take a simple example: if brandy and water, whisky and water, vodka and water, and rum and water are all intoxicating beverages, one may jump to the conclusion that the intoxicating factor must be the ingredient these drinks hold in common – namely, water. There is something wrong with this reasoning, but the fault in the logic is not immediately obvious, and in more subtle cases the error may be hard indeed to discover.

The tracking-down of errors or fallacies in reasoning has amused thinkers from Greek times to the present. And of course we owe the earliest foundations of systematic logic to Aristotle of Stagira, who in the fourth century B.C. first summarized the rules of rigorous reasoning.

The essentials of the intellectual game of man-against-nature are three. First, you must collect observations about some facet of nature. Second, you must organize these observations into an orderly array. (The organization does not alter them but merely makes them easier to handle. This is plain in the game of bridge, for instance, where arranging the hand in suits and order of value does not change the cards or show the best course of play, but makes it easier to arrive at the logical plays.) Third, you must derive from your orderly array of observations some principle that summarizes the observations.

For instance, we may observe that marble sinks in water, wood floats, iron sinks, a feather floats, mercury sinks, olive oil floats, and so on. If we put all the sinkable objects in one list and all the floatable ones in another and look for a characteristic that differentiates all the objects in one group from all in the other, we will conclude: Heavy objects sink in water and light objects float.

The Greeks named their new manner of studying the universe *philosophia* ('philosophy'), meaning 'love of knowledge' or, in free translation, 'the desire to know'.

The Greeks achieved their most brilliant successes in geometry. These successes can be attributed mainly to their development of two techniques: abstraction and generalization.

Here is an example. Egyptian land surveyors had found a practical way to form a right angle: they divided a rope into twelve equal parts and made a triangle in which three parts formed one side, four parts another, and five parts the third side – the right angle lay where the three-unit side joined the four-unit side. There is no record of how the Egyptians discovered this method, and apparently their interest went no further than to make use of it. But the curious Greeks went on to investigate why such a triangle should contain a right angle. In the course of their analysis, they grasped the point that the physical construction itself was only incidental; it did not matter whether the triangle was made of rope or linen or wooden slats. It was simply a property of 'straight lines' meeting at angles. In conceiving of ideal straight lines, which were independent of any physical visualization and could exist only in imagination, they originated the method called abstraction – stripping away non-essentials and considering only those properties necessary to the solution of the problem.

The Greek geometers made another advance by seeking general solutions for classes of problems, instead of treating individual problems separately. For instance, one might discover by trial that a right angle appeared in triangles, not only with sides 3, 4, and 5 feet long, but also in those of 5, 12, and 13 feet and of 7, 24, and 25 feet. But these were merely numbers without meaning. Could some common property be found that would describe all right triangles? By careful reasoning the Greeks showed that a triangle was a right triangle if, and only if, the lengths of the sides had the relation $x^2+y^2 = z^2$, z being the length of the longest side. The right angle lay where the sides of length x and y met. Thus for the triangle with sides of 3, 4, and 5 feet, squaring the sides gives $9+16 = 25$; similarly, squaring the sides of 5, 12, and 13 gives $25+144 = 169$; and squaring 7, 24, and 25 gives $49+576 = 625$. These are only three cases out of an infinity of possible ones and, as such, trivial. What intrigued the Greeks was the discovery of a proof that the relation must hold in all cases. And they pursued geometry as an elegant means of discovering and formulating generalizations.

Various Greek mathematicians contributed proofs of relationships among the lines and points of geometric figures. The one involving the right triangle was reputedly worked out by Pythagoras of Samos about 525 B.C. and is still called the Pythagorean theorem in his honour.

About 300 B.C., Euclid gathered the mathematical theorems known in his time and arranged them in a reasonable order, such that each theorem could be proved through the use of theorems proved previously. Naturally, this system eventually worked back to something unprovable: If each theorem had to be proved with the help of one already proved, how could one prove theorem No. 1? The solution was to begin with a statement of truths so obvious and acceptable to all as to need no proof. Such a statement is called an 'axiom'. Euclid managed to reduce the accepted axioms of the day to a few simple statements. From these axioms alone, he built an intricate and majestic system of 'Euclidean geometry'. Never was so much constructed so well from so little, and Euclid's reward is that his textbook has remained in use, with but minor modification, for more than 2,000 years.

Working out a body of knowledge as the inevitable consequence of a set of axioms ('deduction') is an attractive game. The Greeks fell in love with it, thanks to the success of their geometry – sufficiently in love with it to commit two serious errors.

First, they came to consider deduction as the only respectable means of attaining knowledge. They were well aware that for some kinds of knowledge deduction was inadequate; for instance, the distance from Corinth to Athens could not be deduced from abstract principles but had to be measured. The Greeks were willing to look at nature when necessary; however, they were always ashamed of the necessity and considered that the highest type of knowledge was that arrived at by cerebration. They tended to undervalue knowledge which was too directly involved with everyday life. There is a story that a student of Plato, receiving mathematical instruction from the master, finally asked impatiently: 'But what is the use of all this?' Plato, deeply offended, called a slave and ordered him to give the student a coin. 'Now,' he said, 'you need not feel your instruction has been entirely to no purpose.' With that, the student was expelled.

There is a well-worn belief that this lofty view arose from the Greeks' slave-based culture, in which all practical matters were relegated to the slaves. Perhaps so, but I incline to the view that the

Greeks felt that philosophy was a sport, an intellectual game. We regard the amateur in sports as a gentleman socially superior to the professional who makes his living at it. In line with this concept of purity, we take almost ridiculous precautions to make sure that the contestants in the Olympic games are free of any taint of professionalism. The Greek rationalization for the 'cult of uselessness' may similarly have been based on a feeling that to allow mundane knowledge (such as the distance from Athens to Corinth) to intrude on abstract thought was to allow imperfection to enter the Eden of true philosophy. Whatever the rationalization, the Greek thinkers were severely limited by their attitude. Greece was not barren of practical contributions to civilization, but even its great engineer, Archimedes of Syracuse, refused to write about his practical inventions and discoveries; to maintain his amateur status, he broadcast only his achievements in pure mathematics. And lack of interest in earthly things – in invention, in experiment, in the study of nature – was but one of the factors that put bounds on Greek thought. The Greeks' emphasis on purely abstract and formal study – indeed, their very success in geometry – led them into a second great error and, eventually, to a dead end.

Seduced by the success of the axioms in developing a system of geometry, the Greeks came to think of the axioms as 'absolute truths' and to suppose that other branches of knowledge could be developed from similar 'absolute truths'. Thus in astronomy they eventually took as self-evident axioms the notions that (1) the earth was motionless and the centre of the universe, and (2) whereas the earth was corrupt and imperfect, the heavens were eternal, changeless, and perfect. Since the Greeks considered the circle the perfect curve and since the heavens were perfect, it followed that all the heavenly bodies must move in circles around the earth. In time their observations (arising from navigation and calendar making) showed that the planets did not move in perfectly simple circles, and so they were forced to allow planets to move in ever more complicated combinations of circles, which, about A.D. 150, were formulated as an uncomfortably complex system by Claudius Ptolemaeus (Ptolemy) at Alexandria. Similarly, Aristotle worked up fanciful theories of motion from 'self-evident' axioms, such as the proposition that the speed of an object's fall was proportional to its weight. (Anyone could see that a stone fell faster than a feather.)

Now this worship of deduction from self-evident axioms was bound to wind up at the edge of a precipice, with no place to go. After the Greeks had worked out all the implications of the axioms, further important discoveries in mathematics or astronomy seemed out of the question. Philosophic knowledge appeared complete and perfect, and, for nearly 2,000 years after the Golden Age of Greece, when questions involving the material universe arose, there was a tendency to settle matters to the satisfaction of all by saying, 'Aristotle says . . .' or 'Euclid says . . .'

Having solved the problems of mathematics and astronomy, the Greeks turned to more subtle and challenging fields of knowledge. One was the field of the human soul.

Plato was far more interested in such questions as 'What is justice?' or 'What is virtue?' than in why rain fell or how the planets moved. As the supreme moral philosopher of Greece, he superseded Aristotle, the supreme natural philosopher. The Greek thinkers of the Roman period found themselves drawn more and more to the subtle delights of moral philosophy and away from the apparent sterility of natural philosophy. The last development in ancient philosophy was an exceedingly mystical 'neo-Platonism' formulated by Plotinus about A.D. 250.

Christianity, with its emphasis on the nature of God and His relation to man, introduced an entirely new dimension into the subject matter of moral philosophy and increased its superiority as an intellectual pursuit over natural philosophy. From A.D. 200 to A.D. 1200, Europeans concerned themselves almost exclusively with moral philosophy, in particular with theology. Natural philosophy was nearly forgotten.

The Arabs, however, managed to preserve Aristotle and Ptolemy through the Middle Ages, and, from them, Greek natural philosophy eventually filtered back to Western Europe. By 1200, Aristotle had been rediscovered. Further infusions came from the dying Byzantine Empire, which was the last area in Europe that maintained a continuous cultural tradition from the great days of Greece.

The first and most natural consequence of the rediscovery of Aristotle was the application of his system of logic and reason to theology. About 1250, the Italian theologian Thomas Aquinas established the system called 'Thomism', based on Aristotelian principles, which still represents the basic theology of the Roman Catholic Church. But men soon began to apply the revival of Greek thought to secular fields as well.

Because the leaders of the Renaissance shifted emphasis from matters concerning God to the works of humanity, they were called 'humanists', and the study of literature, art, and history is still referred to as 'the humanities'.

To the Greek natural philosophy, the Renaissance thinkers brought a fresh outlook, for the old views no longer entirely satisfied. In 1543 the Polish astronomer Nicolaus Copernicus published a book that went so far as to reject a basic axiom of astronomy: he proposed that the sun, not the earth, be considered the centre of the universe. (He retained the notion of circular orbits for the earth and other planets, however.) This new axiom allowed a much simpler explanation of the observed motions of heavenly bodies. Yet the Copernican axiom of a moving earth was far less 'self-evident' than the Greek axiom of a motionless earth, and so it is not surprising that it took nearly a century for the Copernican theory to be accepted.

In a sense, the Copernican system itself was not a crucial change. Copernicus had merely switched axioms; and Aristarchus of Samos had already anticipated this switch to the sun as the centre 2,000 years earlier. This is not to say that the changing of an axiom is a minor matter. When mathematicians of the nineteenth century challenged Euclid's axioms and developed 'non-Euclidean geometries' based on other assumptions, they influenced thought on many matters in a most profound way: today the very history and form of the universe are thought to conform to a non-Euclidean (Riemannian) geometry rather than the 'commonsense' geometry of Euclid. But the revolution initiated by Copernicus entailed not just a shift in axioms but eventually involved a whole new approach to nature. This revolution was carried through in the person of the Italian Galileo Galilei.

The Greeks, by and large, had been satisfied to accept the 'obvious' facts of nature as starting points for their reasoning. It is not on record that Aristotle ever dropped two stones of different weight to test his assumption that the speed of fall was proportional to an object's weight. To the Greeks, experimentation seemed irrelevant. It interfered with and detracted from the beauty of pure deduction. Besides, if an experiment disagreed with a deduction, could one be certain that the experiment was correct? Was it likely that the imperfect world of reality would agree completely with the perfect world of abstract ideas, and, if it did not, ought one to adjust the perfect to the demands of the imperfect? To test a perfect theory with imperfect instruments did not impress the Greek philosophers as a valid way to gain knowledge.

Experimentation began to become philosophically respectable in Europe with the support of such philosophers as Roger Bacon (a contemporary of Thomas Aquinas) and his later namesake Francis Bacon. But it was Galileo who overthrew the Greek view and effected the revolution. He was a convincing logician and a genius as a publicist. He described his experiments and his point of view so clearly and so dramatically that he won over the European learned community. And they accepted his methods along with his results.

According to the best-known story about him, Galileo tested Aristotle's theories of falling bodies by asking the question of nature in such a way that all Europe could hear the answer. He is supposed to have climbed to the top of the Leaning Tower of Pisa and dropped a ten-pound sphere and a one-pound sphere simultaneously; the thump of the two balls hitting the ground in the same split second killed Aristotelian physics.

Actually Galileo probably did not perform this particular experiment, but the story is so typical of his dramatic methods that it is no wonder it has been widely believed through the centuries.

Galileo undeniably did roll balls down inclined planes and measured the distance that they travelled in given times. He was the first to conduct time experiments, the first to use measurement in a systematic way.

His revolution consisted in elevating 'induction' above deduction as the logical method of science. Instead of building conclusions on an assumed set of generalizations, the inductive method starts with observations and derives generalizations (axioms, if you will) from them. Of course, even the Greeks obtained their axioms from observation; Euclid's axiom that a straight line is the shortest distance between two points was an intuitive judgement based on experience. But whereas the Greek philosopher minimized the role played by induction, the modern scientist looks on induction as the essential process of gaining knowledge, the only way of justifying generalizations. Moreover, he realizes that no generalization can be allowed to stand unless it is repeatedly tested by newer and still newer experiments – unless it withstands the continuing test of further induction.

The present general viewpoint is just the reverse of the Greeks'. Far from considering the real world an imperfect representation of ideal truth, we consider generalizations to be only imperfect representatives of the real world. No amount of inductive testing can render a generalization completely and absolutely valid. Even though billions of observers tend to bear out a generalization, a single observation that contradicts or is inconsistent with it must force its modification. And no matter how many times a theory meets its tests successfully, there can be no certainty that it will not be overthrown by the next observation.

This, then, is a cornerstone of modern natural philosophy. It makes no claim of attaining ultimate truth. In fact, the phrase 'ultimate truth' becomes meaningless, because there is no way in which enough observations can be made to make truth certain, and therefore 'ultimate'. The Greek philosophers recognized no such limitation. Moreover, they saw no difficulty in applying exactly the same method of reasoning to the question 'What is justice?' as to the question 'What is matter?' Modern science, on the other hand, makes a sharp distinction between the two types of question. The inductive method cannot make generalizations about what it cannot observe, and, since the nature of the human soul, for example, is not observable by any direct means yet known, this subject lies outside the realm of the inductive method.

The victory of modern science did not become complete until it established one more essential principle – namely, free and co-operative communication among all scientists. Although this necessity seems obvious to us now, it was not obvious to the philosophers of ancient and medieval times. The Pythagoreans of ancient Greece were a secret society who kept their mathematical discoveries to themselves. The alchemists of the Middle Ages deliberately obscured their writings to keep their so-called findings within as small an inner circle as possible. In the sixteenth century, the Italian mathematician Niccolo Tartaglia, who discovered a method of solving cubic equations, saw nothing wrong in attempting to keep it a secret. When Geronimo Cardano, a fellow mathematician, wormed the secret out of Tartaglia and published it as his own, Tartaglia naturally was outraged, but aside from Cardano's trickery in claiming the credit, he was certainly correct in his reply that such a discovery had to be published.

Nowadays no scientific discovery is reckoned a discovery if it is kept secret. The English chemist Robert Boyle, a century after Tartaglia and Cardano, stressed the importance of publishing all scientific observations in full detail. A new observation or discovery, moreover, is no longer considered valid, even after publication, until at least one other investigator has repeated the observation and 'confirmed' it. Science is the product not of individuals but of a 'scientific community'.

One of the first groups (and certainly the most famous) to represent such a scientific community was the Royal Society of London for Improving Natural Knowledge, usually called simply the 'Royal Society'. It grew out of informal meetings, beginning about 1645, of a group of gentlemen interested in the new scientific methods originated by Galileo. In 1660, the Society was formally chartered by King Charles II.

The members of the Royal Society met and discussed their findings openly, wrote letters describing them in English rather than Latin, and pursued their experiments with vigour and vivacity. Nevertheless, through most of the seventeenth century they remained in a defensive position. The attitude of many of their learned contemporaries might be expressed by a cartoon, after the modern fashion, showing the lofty shades of Pythagoras, Euclid, and Aristotle staring down haughtily at children playing with marbles, labelled 'Royal Society'.

All this was changed by the work of Isaac Newton, who became a member of the society. From the observations and conclusions of Galileo, of the Danish astronomer Tycho Brahe, and of the German astronomer Johannes Kepler, who figured out the elliptical nature of the orbits of the planets, Newton arrived by induction at his three simple laws of motion and his great fundamental generalization – the law of universal gravitation. The educated world was so impressed with this discovery that Newton was idolized, almost deified, in his own lifetime. This majestic new universe, built upon a few simple assumptions, now made the Greek philosophers look like boys playing with marbles. The revolution that Galileo had initiated at the beginning of the seventeenth century was triumphantly completed by Newton at the century's end.

It would be pleasant to be able to say that science and man have lived happily ever since. But the truth is that the real difficulties of both were only beginning. As long as science had remained deductive, natural philosophy could be part of the general culture of all educated men. But inductive science became an immense labour – of observation, learning, and analysis. It was no longer a game for amateurs. And the complexity of science grew each decade. During the century after Newton, it was still possible for a man of unusual attainments to master all fields of scientific knowledge. But, by 1800, this had become entirely impracticable. As time went on, it was increasingly necessary for a scientist to limit himself to a portion of the field if he intended an intensive concern with it. Specialization was forced on science by its own inexorable growth. And with each generation of scientists, specialization has grown more and more intense.

The publications of scientists concerning their individual work have never been so copious – and so unreadable for anyone but their fellow specialists. This has been a great handicap to science itself, for basic advances in scientific knowledge often spring from the cross-fertilization of knowledge from different specialities. What is even more ominous is that science has increasingly lost touch with non-scientists. Under such circumstances scientists come to be regarded almost as magicians – feared rather than admired. And the impression that science is incomprehensible magic, to be understood only by a chosen few who are suspiciously different from ordinary mankind, is bound to turn many youngsters away from science.

In the 1960s, indeed, strong feelings of outright hostility towards science were to be found among the young – even among the educated young in the colleges. Our industrialized society is based on the scientific discoveries of the last two centuries, and our society finds it is plagued by undesirable side-effects of its very success.

Improved medical techniques have brought about a runaway increase in population; chemical industries and the internal-combustion engine are fouling our water and our air; the demand for materials and for energy is depleting and destroying the earth's crust. And this is all too easily blamed on 'science' and 'scientists' by those who do not quite understand that if knowledge can create problems, it is not through ignorance that we can solve them.

Yet modern science need not be so complete a mystery to non-scientists. Much could be accomplished towards bridging the gap if scientists accepted the responsibility of communication – explaining their own fields of work as simply and to as many as possible – and if non-scientists, for their part, accepted the responsibility of listening. To gain a satisfactory appreciation of the developments in a field of science, it is not essential to have a total understanding of the science. After all, no one feels that he must be capable of writing a great work of literature in order to appreciate Shakespeare. To listen to a Beethoven symphony with pleasure does not require the listener to be capable of composing an equivalent symphony of his own. By the same token, one can appreciate and take pleasure in the achievements of science even though he does not himself have a bent for creative work in science.

But what, you may ask, would be accomplished? The first

answer is that no one can really feel at home in the modern world and judge the nature of its problems – and the possible solutions to those problems – unless he has some intelligent notion of what science is up to. But beyond this, initiation into the magnificent world of science brings great aesthetic satisfaction, inspiration to youth, fulfilment of the desire to know, and a deeper appreciation of the wonderful potentialities and achievements of the human mind.

It is with this in mind that I have undertaken to write this book.

2 The Universe

The Size of the Universe

There is nothing about the sky that makes it look particularly distant to a casual observer. Young children have no great trouble in accepting the fantasy that 'the cow jumped over the moon' or 'he jumped so high, he touched the sky'. The ancient Greeks, in their myth-telling stage, saw nothing ludicrous in allowing the sky to rest on the shoulders of Atlas. Of course, Atlas might have been astronomically tall, but another myth suggests otherwise. Atlas was enlisted by Hercules to help him with the eleventh of his famous twelve labours – fetching the golden apples (oranges?) of the Hesperides ('the far west' – Spain?). While Atlas went off to fetch the apples, Hercules stood on a mountain and held up the sky. Granted that Hercules was a large specimen, he was nevertheless not a giant. It follows then that the early Greeks took quite calmly to the notion that the sky cleared the mountain tops by only a few feet.

It is natural to suppose, to begin with, that the sky is simply a hard canopy in which the shining heavenly bodies are set like diamonds. (Thus the Bible refers to the sky as the 'firmament', from the same Latin root as the word 'firm'.) As early as the sixth to fourth centuries B.C., Greek astronomers realized that there must be more than one canopy. For while the 'fixed' stars moved around the earth in a body, apparently without changing their relative positions, this was not true of the sun, moon, and five bright starlike objects (Mercury, Venus, Mars, Jupiter, and Saturn) – in fact, each moved in a separate path. These seven bodies were called planets (from a Greek word meaning 'wanderer'), and it seemed obvious that they could not be attached to the vault of the stars.

The Greeks assumed that each planet was set in an invisible vault of its own and that the vaults were nested one above the other, the nearest belonging to the planet that moved fastest. The quickest motion belonged to the moon, which circled the sky in about twenty-nine and a half days. Beyond it lay in order (so thought the Greeks) Mercury, Venus, the sun, Mars, Jupiter, and Saturn.

The first scientific measurement of any cosmic distance came about 240 B.C. Eratosthenes of Cyrene, the head of the Library at Alexandria, then the most advanced scientific institution in the world, pondered the fact that on 21 June, when the noonday sun was exactly overhead at the city of Syene in Egypt, it was not quite at the zenith at noon in Alexandria, 500 miles north of Syene. Eratosthenes decided that the explanation must be that the surface of the earth curved away from the sun. From the length of the shadow in Alexandria at noon on the solstice, straightforward geometry could yield an answer as to the amount by which the earth's surface curved in the 500-mile distance from Syene to Alexandria. From that one could calculate the circumference and the diameter of the earth, assuming that the earth was spherical in shape – a fact which Greek astronomers of the day were ready to accept.

Eratosthenes worked out the answer (in Greek units), and, as nearly as we can judge, his figures in our units came out at about 8,000 miles for the diameter and 25,000 miles for the circumference of the earth. This, as it happens, is just about right. Unfortunately, this accurate value for the size of the earth did not prevail. About 100 B.C. another Greek astronomer, Posidonius of Apamea, repeated Eratosthenes' work, but reached the conclusion that the earth was but 18,000 miles in circumference.

It was the smaller figure that was accepted by Ptolemy and, therefore, throughout medieval times. Columbus accepted the smaller figure and thought that a 3,000-mile westward voyage would take him to Asia. Had he known the earth's true size, he might not have ventured. It was not until 1521–3, when Magellan's fleet (or rather the one remaining ship of the fleet) finally circumnavigated the earth, that Eratosthenes' correct value was finally established.

In terms of the earth's diameter, Hipparchus of Nicaea, about

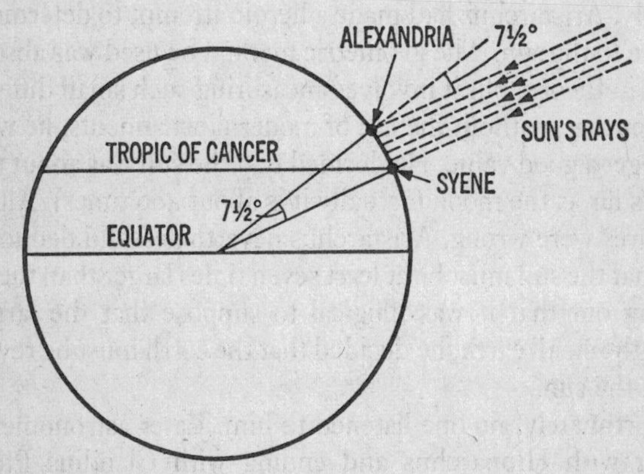

Eratosthenes measured the size of the earth from its curvature. At noon, 21 June, the sun is directly overhead at Syene, which lies on the Tropic of Cancer. But at this same time, the sun's rays, seen from farther north in Alexandria, fall at an angle of 7·5° to the vertical and therefore cast a shadow. Knowing the distance between the two cities and the length of the shadow in Alexandria, Eratosthenes made his calculations.

150 B.C., worked out the distance to the moon. He used a method that had been suggested a century earlier by Aristarchus of Samos, the most daring of all Greek astronomers. The Greeks had already surmised that eclipses of the moon were caused by the earth coming between the sun and the moon. Aristarchus saw that the curve of the earth's shadow as it crossed the moon should indicate the relative sizes of the earth and the moon. From this, geometric methods offered a way to calculate how far distant the moon was in

terms of the diameter of the earth. Hipparchus, repeating this work, calculated that the moon's distance from the earth was thirty times the earth's diameter. Taking Eratosthenes' figure of 8,000 miles for the earth's diameter, that meant the moon must be about 240,000 miles from the earth. This again happens to be about correct.

But finding the moon's distance was as far as Greek astronomy managed to carry the problem of the size of the universe – at least correctly. Aristarchus had made a heroic attempt to determine the distance to the sun. The geometric method he used was absolutely correct in theory, but it involved measuring such small differences in angles that, without the use of modern instruments, he was unable to get a good value. He decided that the sun was about twenty times as far as the moon (actually it is about 400 times). Although his figures were wrong, Aristarchus nevertheless did deduce from them that the sun must be at least seven times larger than the earth. Pointing out that it was illogical to suppose that the large sun circled the small earth, he decided that the earth must be revolving around the sun.

Unfortunately, no one listened to him. Later astronomers, beginning with Hipparchus and ending with Claudius Ptolemy, worked out all the heavenly movements on the basis of a motionless earth at the centre of the universe, with the moon 240,000 miles away and other objects an undetermined distance farther. This scheme held sway until 1543, when Nicolaus Copernicus published his book, which returned to the viewpoint of Aristarchus and forever dethroned earth's position as the centre of the universe.

The mere fact that the sun was placed at the centre of the solar system did not in itself help determine the distance of the planets. Copernicus adopted the Greek value for the distance of the moon, but he had no notion of the distance of the sun. It was not until 1650 that a Belgian astronomer, Godefroy Wendelin, repeated Aristarchus's observations with improved instruments and decided that the sun was not twenty times the moon's distance (5 million miles) but 240 times (60 million miles). The value was still too small, but it was much better than before.

In 1609, meanwhile, the German astronomer Johannes Kepler opened the way to accurate distance determinations with his discovery that the orbits of the planets were ellipses, not circles. For the first time, it became possible to calculate planetary orbits accurately and, furthermore, to plot a scale map of the solar system. That is, the relative distances and orbit-shapes of all the known bodies in the system could be plotted. This meant that if the distance between any two bodies in the system could be determined in miles, all the other distances could be calculated at once. The distance to the sun, therefore, need not be calculated directly, as Aristarchus and Wendelin had attempted to do. The determination of the distance of the nearer body, such as Mars or Venus, outside the earth–moon system would do.

One method by which cosmic distances can be calculated involves the use of parallax. It is easy to illustrate what this term means. Hold your finger about three inches before your eyes and look at it first with just the left eye and then with just the right. Your finger will shift position against the background, because you have changed your point of view. Now if you repeat this procedure with your finger farther away, say at arm's length, the finger again will shift against the background, but this time not so much. Thus the amount of shift can be used to determine the distance of the finger from your eye.

Of course, for an object fifty feet away the shift in position from one eye to the other begins to be too small to measure; we need a wider 'baseline' than just the distance between our two eyes. But all we have to do to widen the change in point of view is to look at the object from one spot, then move twenty feet to the right and look at it again. Now the parallax is large enough to be measured easily and the distance can be determined. Surveyors make use of just this method for determining the distance across a stream or ravine.

The same method, precisely, can be used to measure the distance to the moon, with the stars playing the role of background. Viewed from an observatory in California, for instance, the moon will be in one position against the stars. Viewed at the same instant from an observatory in England, it will be in a slightly different position.

From this change in position, and the known distance between the two observatories (in a straight line through the earth), the distance of the moon can be calculated. Of course, we can, in theory, enlarge the baseline by making observations from observatories at directly opposite sides of the earth; the length of the baseline is then 8,000 miles. The resulting angle of parallax, divided by two, is called the 'geocentric parallax'.

The shift in position of a heavenly body is measured in degrees or sub-units of a degree – minutes and seconds. One degree is 1/360 of the circuit around the sky; each degree is split into sixty minutes of arc, and each minute into sixty seconds of arc. A minute of arc is therefore 1/(360×60) or 1/21,600 of the circuit of the sky, while a second of arc is 1/(21,600×60) or 1/1,296,000 of the circuit of the sky.

By trigonometry, Claudius Ptolemy was able to measure the distance of the moon from its parallax, and his result agreed with the earlier figure of Hipparchus. It turned out that the geocentric parallax of the moon is fifty-seven minutes of arc (nearly a full degree). The shift is about equal to the width of a 2p piece as seen at a distance of five feet. This is easy enough to measure even with the naked eye. But when it came to measuring the parallax of the sun or a planet, the angles involved were too small. The only conclusion that could be reached was that the other bodies were much farther than the moon. How much farther, no one could tell.

Trigonometry alone, in spite of its refinement by the Arabs during the Middle Ages and by European mathematicians of the sixteenth century, could not give the answer. But measurement of small angles of parallax became possible with the invention of the telescope (which Galileo first built and turned to the sky in 1609, after hearing of a magnifying tube that had been made some months earlier by a Dutch spectacle-maker).

The method of parallax passed beyond the moon in 1673, when the Italian-born French astronomer Giovanni Domenico Cassini determined the parallax of Mars. He determined the position of Mars against the stars while, on the same evenings, the French astronomer Jean Richer, in French Guiana, was making the same observation. Combining the two, Cassini obtained his parallax and calculated the scale of the solar system. He arrived at a figure of 86 million miles for the distance of the sun from the earth, a figure that was only 7 per cent less than the actual value.

Since then, various parallaxes in the solar system have been measured with increasing accuracy. In 1931, a vast international project was made out of the determination of the parallax of a small planetoid named Eros, which happened at that time to approach the earth more closely than any heavenly body except the moon. Eros on this occasion showed a large parallax that could be measured with considerable precision, and the scale of the solar system was determined more accurately than ever before. From these calculations and by the use of methods still more accurate than those involving parallax, the distance of the sun from the earth is now known to average approximately 92,965,000 miles, give or take a thousand miles or so. (Because the earth's orbit is elliptical, the actual distance varies from 91·4 million to 94·6 million miles.)

This average distance is called an 'astronomical unit' (A. U.), and other distances in the solar system are given in this unit. Saturn, for instance, turned out to be, on the average, 887 million miles from the sun, or 9·54 A. U. As the outer planets – Uranus, Neptune, and Pluto – were discovered, the boundaries of the solar system were successively enlarged. The extreme diameter of Pluto's orbit is 7,300 million miles, or 79 A. U. And some comets are known to recede to even greater distances from the sun.

By 1830, the solar system was known to stretch across thousands of millions of miles of space, but obviously this was by no means the full size of the universe. There were still the stars.

Astronomers felt certain that the stars were spread throughout space and that some were closer than others, if only because some were so much brighter than others. This should mean that the nearer stars would show a parallax when compared with the more distant ones. However, no such parallax could be detected. Even when the astronomers used as their baseline the full diameter of the earth's orbit around the sun (186 million miles), looking at the stars from the opposite ends of the orbit at half-year intervals, they still could observe no parallax. This meant, of course, that even the nearest stars must be extremely distant. As better and better telescopes failed to show a stellar parallax, the estimated distance of the stars had to be increased more and more. That they were visible at all at the vast distances to which they had to be pushed made it quite plain that they must be tremendous balls of flame like our own sun.

But telescopes and other instruments continued to improve. In the 1830s, the German astronomer Friedrich Wilhelm Bessel made use of a newly invented device, called the 'heliometer' ('sun measure') because it was originally intended to measure the diameter of the sun with great precision. It could, of course, be used equally well to measure other distances in the heavens, and Bessel used it to measure the distance between two stars. By noticing the change in this distance from month to month, he finally succeeded in measuring the parallax of a star. He chose a small star in the constellation Cygnus, called 61 Cygni. His reason for choosing it was that it showed an unusually large shift in position from year to year against the background of the other stars, which could only mean that it was nearer than the others. (This steady, but very

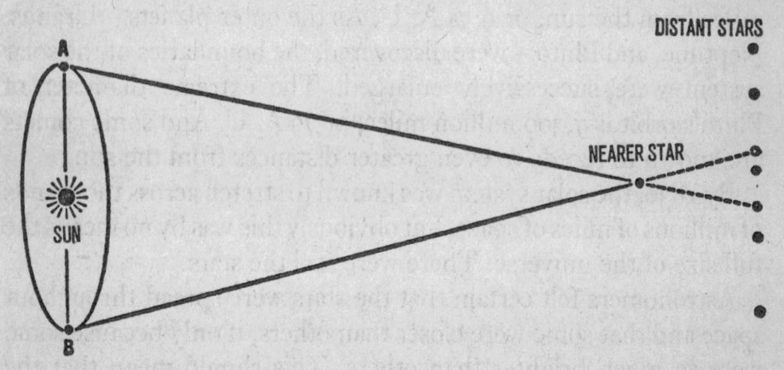

Parallax of a star measured from opposite points on the earth's orbit around the sun.

slow, motion across the sky, called 'proper motion', should not be confused with the back-and-forth shift against the background that indicates parallax.) Bessel pinpointed the successive positions of 61 Cygni against the 'fixed' neighbouring stars (presumably much more distant) and continued his observations for more than a year. Then, in 1838, he reported that 61 Cygni had a parallax of 0·31 second of arc – the width of a 2p piece as seen from a distance of 10 miles! This parallax, observed with the diameter of the earth's orbit as the baseline, meant that 61 Cygni was about 64 billion (64,000,000,000,000) miles away. That is 9,000 times the width of our solar system. Thus, compared to the distance of even the nearest stars, the solar system shrinks to an insignificant dot in space.

Because distances in billions of miles are inconvenient to handle, astronomers shrink the numbers by giving the distances in terms of the speed of light – 186,282 miles per second. In a year, light travels 5,880,000,000,000 (nearly 6 billion) miles. That distance is therefore called a 'light-year'. In terms of this unit, 61 Cygni is about 11 light-years away.

Two months after Bessel's success (so narrow a margin by which to lose the honour of being the first!), the British astronomer Thomas Henderson reported the distance of the star Alpha Centauri. This star, located low in the southern skies and not visible from the United States or Europe, is the third brightest in the heavens. It turned out that Alpha Centauri has a parallax of 0·75 second of arc, more than twice that of 61 Cygni. Alpha Centauri was therefore correspondingly closer. In fact, it is only 4·3 light-years from the solar system and is our nearest stellar neighbour. Actually it is not a single star, but a cluster of three.

In 1840, the German-born Russian astronomer, Friedrich Wilhelm von Struve announced the parallax of Vega, the fourth brightest star in the sky. He was a little off in his determination, as it turned out, but this was understandable, because Vega's parallax was very small and it was much farther away – 27 light-years.

By 1900, the distances of about seventy stars had been determined by the parallax method (and by 1950, nearly 6,000). One hundred light-years is about the limit of the distance that can be measured with any accuracy, even with the best instruments. And beyond this are countless stars at much greater distances.

With the naked eye we can see about 6,000 stars. The invention of the telescope at once made plain that this was only a fragment of the universe. When Galileo raised his telescope to the heavens in 1609, he not only found new stars previously invisible, but, on turning to the Milky Way, received an even more profound shock. To the naked eye, the Milky Way is merely a luminous band of foggy light. Galileo's telescope broke down this foggy light into myriads of stars, as numerous as the grains in talcum powder.

The first man to try to make sense out of this was the German-born English astronomer William Herschel. In 1785, Herschel suggested that the stars of the heavens were arranged in a lens-shape. If we look towards the Milky Way, we see a vast number of stars, but when we look out to the sky at right angles to this wheel, we see relatively few stars. Herschel deduced that the heavenly bodies formed a flattened system, with the long axis in the direction of the Milky Way. We now know that, within limits, this picture is correct, and we call our star system the Galaxy, which is actually another term for Milky Way, because 'galaxy' comes from the Greek word for 'milk'.

Herschel tried to estimate the size of the Galaxy. He assumed that all the stars had about the same intrinsic brightness, so that one could tell the relative distance of a star by its brightness. (By a well-known law, brightness decreases as the square of the distance, so if star A is one-ninth the brightness of star B, it should be three times as far as star B.)

By counting samples of stars in various spots of the Milky Way, Herschel estimated that there were about 100 million stars in the Galaxy altogether. From the levels of their brightness, he decided that the diameter of the Galaxy was 850 times the distance to the bright star Sirius and that its thickness was 155 times that distance.

We now know that the distance to Sirius is 8·8 light-years, so Herschel's estimate was equivalent to a Galaxy about 7,500 light-years in diameter and 1,300 light-years thick. This turned out to be far too conservative. But like Aristarchus's over-conservative measure of the distance to the sun, it was a step in the right direction. (Furthermore, Herschel used his statistics to show that the sun was moving at 12 miles per second in the direction of the constellation Hercules. The sun did move after all, but not in the fashion the Greeks had thought.)

Beginning in 1906, the Dutch astronomer Jacobus Cornelis Kapteyn conducted another survey of the Milky Way. He had photography at his disposal and knew the true distance of the nearer stars, so he was able to make a better estimate than Herschel had. Kapteyn decided that the dimensions of the Galaxy were 23,000 light-years by 6,000. Thus Kapteyn's model of the Galaxy was four times as wide and five times as thick as Herschel's; but it was still over-conservative.

To sum up, by 1900 the situation with respect to stellar distances was the same as that with respect to planetary distances in 1700. In 1700, the moon's distance was known but the distance of the farther planets could only be guessed at. In 1900, the distance of the nearer stars was known, but that of the more distant stars could only be guessed at.

The next major step forward was the discovery of a new measuring rod – certain variable stars which fluctuated in brightness. This part of the story begins with a fairly bright star called Delta Cephei, in the constellation Cepheus. On close study, the star was found to have a cycle of varying brightness: from its dimmest stage it rather quickly doubled in brightness, then slowly faded to its dim point again. It did this over and over with great regularity. Astronomers found a number of other stars that varied in the same regular way, and in honour of Delta Cephei all were named 'Cepheid variables', or simply 'Cepheids'.

The Cepheids' periods (the time from dim point to dim point) vary from less than a day to as long as nearly two months. Those nearest our sun seem to have a period in the neighbourhood of a week. The period of Delta Cephei itself is 5·3 days, while the nearest Cepheid of all (the Pole Star, no less) has a period of 4 days. (The Pole Star, however, varies only slightly in luminosity; not enough to be noticeable to the unaided eye.)

The importance of the Cepheids to astronomers involves their brightness, which is a subject that requires a small digression.

Ever since Hipparchus, the brightness of stars has been measured by the term 'magnitude'. The brighter the star, the lower the magnitude. The twenty brightest stars he called 'first magnitude'. Somewhat dimmer stars are 'second magnitude'. Then, third, fourth, and fifth, until the dimmest, those just barely visible, are of the 'sixth magnitude'.

In modern times – 1856, to be exact – Hipparchus's notion was made quantitative by the English astronomer Norman Robert Pogson. He showed that the average first-magnitude star was about 100 times brighter than the average sixth-magnitude star. Allowing this interval of five magnitudes to represent a ratio of one hundred in brightness, the ratio for one magnitude must be 2·512. A star of magnitude 4 is 2·512 times as bright as a star of magnitude 5, and 2·512×2·512, or about 6·3 times as bright as a star of magnitude 6.

Among the stars, 61 Cygni is a dim star with a magnitude of 5·0 (modern astronomical methods allow magnitudes to be fixed to the nearest tenth and even to the nearest hundredth in some cases). Capella is a bright star, with a magnitude of 0·9; Alpha Centauri still brighter, with a magnitude of 0·1. And the measure goes on to still greater brightnesses which are designated by magnitude 0 and beyond this by negative numbers. Sirius, the brightest star in the sky, has a magnitude of −1·6. The planet Venus attains a magnitude of −4; the full moon, −12; the sun, −26.

These are the 'apparent magnitudes' of the stars as we see them – not their absolute luminosities independent of distance. But if we know the distance of a star and its apparent magnitude, we can calculate its actual luminosity. Astronomers base the scale of 'absolute magnitudes' on the brightness at a standard distance, which has been established at ten 'parsecs', or 32·6 light-years. (The 'parsec' is the distance at which a star would show a parallax of one second of arc; it is equal to a little more than 19 billion miles, or 3·26 light-years.)

Although Capella looks dimmer than Alpha Centauri and Sirius, actually it is a far more powerful emitter of light than either of them. It merely happens to be a great deal farther away. If all were at the standard distance, Capella would be much the brightest of the three. Capella has an absolute magnitude of −0·1, Sirius 1·3, and Alpha Centauri 4·8. Our own sun is just about as bright as Alpha Centauri, with an absolute magnitude of 4·86. It is an ordinary, medium-sized star.

Now to get back to the Cepheids. In 1912, Miss Henrietta Leavitt, an astronomer at the Harvard Observatory, was studying the smaller of the Magellanic Clouds – two huge star systems in the Southern Hemisphere named after Ferdinand Magellan, because they were first observed during his voyage around the globe. Among the stars of the Small Magellanic Cloud, Miss Leavitt detected twenty-five Cepheids. She recorded the period of variation of each and to her surprise found that the longer the period, the brighter the star.

This was not true of the Cepheid variables in our own neighbourhood; why should it be true of the Small Magellanic Cloud? In our own neighbourhood, we know only the apparent magnitudes of the Cepheids; not knowing their distances or absolute brightnesses, we have no scale for relating the period of a star to its brightness. But in the Small Magellanic Cloud, all the stars are effectively at about the same distance from us, because the cloud itself is so far away. It is as though a person in New York were trying to calculate his distance from each person in Chicago. He would conclude that all the Chicagoans were about equally distant from himself – what is a difference of a few miles in a total distance of a thousand? Similarly, a star at the far end of the Cloud is not significantly farther away than one at the near end.

With the stars in the Small Magellanic Cloud at about the same distance from us, their apparent magnitude could be taken as a measure of their comparative absolute magnitude. So Miss Leavitt could consider the relationship she saw a true one: that is, the period of the Cepheid variables increased smoothly with the absolute magnitude. She was thus able to establish a 'period-luminosity curve' – a graph which showed what period a Cepheid of any absolute magnitude must have, and conversely what absolute magnitude a Cepheid of a given period must have.

If Cepheids everywhere in the universe behaved as they did in the Small Magellanic Cloud (a reasonable assumption), then astronomers had a *relative* scale for measuring distances, as far out as Cepheids could be detected in the best telescopes. If they spotted two Cepheids with equal periods, they could assume that both were equal in absolute magnitude. If Cepheid A seemed four times as bright as Cepheid B, Cepheid B must be twice as distant from us. In this way, the relative distances of all the observable Cepheids could be plotted on a scale map. Now if the actual distance of just one of the Cepheids could be determined, so could the distances of all the rest.

Unfortunately, even the nearest Cepheid, the Pole Star, is hundreds of light-years away, much too far to measure its distance by parallax. Astronomers had to use less direct methods. One usable clue was proper motion: on the average, the more distant a star is, the smaller its proper motion. (Recall that Bessel decided 61 Cygni was relatively close because it had a large proper motion.) A number of devices were used to determine the proper motions of groups of stars, and statistical methods were brought to bear. The procedure was complicated, but the results gave the approximate distances of various groups of stars which contained Cepheids. From the distances and the apparent magnitudes of those Cepheids, their absolute magnitudes could be determined, and these could be compared with the periods.

In 1913, the Danish astronomer Ejnar Hertzsprung found that a Cepheid of absolute magnitude −2·3 had a period of 6·6 days. From that, and using Miss Leavitt's period–luminosity curve, he could determine the absolute magnitude of any Cepheid. (It turned out, incidentally, that Cepheids generally were large, bright stars, much more luminous than our sun. Their variations in brightness are probably the result of pulsations. The stars seem to expand and contract steadily, as though they are ponderously breathing in and out.)

A few years later, the American astronomer Harlow Shapley repeated the work and decided that a Cepheid of absolute magnitude −2·3 had a period of 5·96 days. The agreement was close enough to allow astronomers to go ahead. They had their yardstick.

In 1918, Shapley began observing the Cepheids of our own Galaxy in an attempt to determine the Galaxy's size by this new method. He concentrated on the Cepheids found in groups of stars called 'globular clusters' – densely packed spherical aggregates of tens of thousands to tens of millions of stars, with diameters of the order of 100 light-years.

These clusters (the nature of which was first observed by Herschel a century earlier) present an astronomical environment quite different from that prevailing in our own neighbourhood in space. At the centre of the larger clusters, stars are packed together with a density of 500 per ten cubic parsecs, as compared with one star per ten cubic parsecs in our own neighbourhood. Starlight under such conditions would be far brighter than moonlight on earth, and a planet situated at the centre of such a cluster would know no night.

There are about one hundred known globular clusters in our Galaxy and probably as many again that have not yet been detected. Shapley calculated the distance of the various globular clusters at from 20,000 to 200,000 light-years from us. (The nearest cluster, like the nearest star, is in the constellation Centaurus. It is visible to the naked eye as a star-like object, 'Omega Centauri'. The most distant, NGC 2419, is so far off as scarcely to be considered a member of the Galaxy.)

Shapley found the clusters were distributed in a large sphere that the plane of the Milky Way cut in half; they surrounded a portion of the main body of the Galaxy like a halo. Shapley made the natural assumption that they encircled the centre of the Galaxy. His calculations placed the central point of this halo of globular clusters within the Milky Way in the direction of the constellation Sagittarius and about 50,000 light-years from us. This meant that our solar system, far from being at the centre of the Galaxy, as Herschel and Kapteyn had thought, was far out towards one edge.

Shapley's model pictured the Galaxy as a giant lens about 300,000 light-years in diameter. This time its size was overestimated, as another method of measurement soon showed.

From the fact that the Galaxy had a disk shape, astronomers from William Herschel onwards assumed that it was rotating in space. In 1926, the Dutch astronomer Jan Oort set out to measure this rotation. Since the Galaxy is not a solid object, but is composed of numerous individual stars, it is not to be expected that it rotates in one piece, as a wheel does. Instead, stars close to the gravitational centre of the disk must revolve round it faster than those farther away (just as the planets closest to the sun travel fastest in their orbits). This means that the stars towards the centre of the Galaxy (i.e., in the direction of Sagittarius) should tend to drift ahead of our sun, whereas those farther from the centre (in the direction of the constellation Gemini) should tend to lag behind us in their revolution. And the farther a star is from us, the greater this difference in speed should be.

On these assumptions, it became possible to calculate the rate of rotation around the galactic centre from the relative motions of the stars. The sun and nearby stars, it turned out, travel at about 150 miles a second relative to the galactic centre and make a complete revolution around the centre in approximately 200 million years. (The sun travels in a nearly circular orbit, but the orbit of some stars, such as Arcturus, are quite elliptical. The fact that the various stars do not rotate in perfectly parallel orbits accounts for the sun's relative motion towards the constellation Hercules.)

Having estimated a value for the rate of rotation, astronomers were then able to calculate the strength of the gravitational field of the galactic centre and, therefore, its mass. The galactic centre (which contains most of the mass of the Galaxy) turns out to be well over 100,000 million times as massive as our sun. Since our sun is a star of average mass, our Galaxy therefore contains perhaps 100,000 to 200,000 million stars – up to 2,000 times the number estimated by Herschel.

From the curve of the orbits of the revolving stars, it is also possible to locate the centre around which they are revolving. The centre of the Galaxy in this way has been confirmed to be in the direction of Sagittarius, as Shapley found, but only 27,000 light-years from us, and the total diameter of the Galaxy comes to 100,000 light-years instead of 300,000. In this new model, now believed to be correct, the thickness of the disk is some 20,000 light-years at the centre and falls off towards the edge: at the location of our sun, which is two thirds of the way out towards the extreme edge, the disk is perhaps 3,000 light-years thick. But these are only rough figures, because the Galaxy has no sharply definite boundaries.

If the sun is so close to the edge of the Galaxy, why is not the Milky Way much brighter in the direction towards the centre than in the opposite direction, where we look towards the edge? Looking towards Sagittarius, we face the main body of the Galaxy with nearly 100,000 million stars, whereas out towards the edge there is only a scattering of some millions. Yet in each direction the band of the Milky Way seems of about the same brightness. The answer appears to be that huge clouds of obscuring dust hide much of the centre of the Galaxy from us. As much as half the mass of the galactic outskirts may be composed of such clouds of dust and gas. Probably we see no more than 1/10,000 (at most) of the light of the galactic centre.

This explains why Herschel and other early students of the Galaxy thought our solar system was at the centre, and also, it seems, why Shapley originally overestimated the size of the Galaxy. Some of the clusters he studied were dimmed by the intervening dust, so that the Cepheids in them seemed dimmer and therefore more distant than they really were.

Even before the size and mass of the Galaxy itself had been determined, the Cepheid variables of the Magellanic Clouds (where Miss Leavitt had made the crucial discovery of the period-luminosity curve) were used to determine the distance of the clouds. They proved to be more than 100,000 light-years away. The best modern figures place the Large Magellanic Cloud at about 150,000 light-years from us and the Small Magellanic Cloud at 170,000 light-years. The Large Cloud is no more than half the size of our Galaxy in diameter; the Small Cloud, no more than a fifth. Besides, they seem to be less densely packed with stars. The Large Magellanic Cloud contains 5,000 million stars (only 1/20 or less the number in our Galaxy), while the Small Magellanic Cloud has only 1,500 million.

That was the situation as it stood in the early 1920s. The known universe was less than 200,000 light-years in diameter and consisted of our Galaxy and its two neighbours. The question then arose as to whether anything existed outside that.

Suspicion rested upon small patches of luminous fog, called nebulae (from the Greek word for 'cloud'), which astronomers had long noted. The French astronomer Charles Messier had catalogued 103 of them about 1800. (Many are still known by the numbers he gave them, preceded by the letter 'M' for Messier.)

Were these nebulosities merely clouds as they seemed to be? Some, such as the Orion Nebula (first discovered in 1656 by the Dutch astronomer Christian Huygens), seemed to be just that. It was a cloud of gas and dust, equal in mass to about 500 suns like ours, and illuminated by hot stars contained within them. Others, on the other hand, turned out to be globular clusters – huge assemblages of stars.

But there remained patches of luminous cloud that seemed to contain no stars at all. Why, then, were they luminous? In 1845, the British astronomer William Parsons (third Earl of Rosse) using a 72-inch telescope he had spent his life building, had ascertained that some of these patches had a spiral structure; thus they came to be called 'spiral nebulae'; but that did not help explain the source of the luminosity.

The most spectacular of these patches, known as M-31, or the Andromeda Nebula (because it is in the constellation Andromeda) was first studied in 1612 by the German astronomer Simon Marius. It is an elongated oval of dim light about half the size of the full moon. Could it be composed of stars so distant that they could not be made out separately even in large telescopes? If so, the Andromeda Nebula must be incredibly far away and incredibly large to be visible at all at such a distance. (As long ago as 1755, the German philosopher Immanuel Kant had speculated on the existence of such far-distant star collections. 'Island universes', he called them.)

In 1924, the American astronomer Edwin Powell Hubble turned the new 100-inch telescope at Mount Wilson in California on the Andromeda Nebula. The powerful new instrument resolved portions of the Nebula's outer edge into individual stars. This showed at once that the Andromeda Nebula, or at least parts of it, resembled the Milky Way and that there might be something to this 'island universe' notion.

Among the stars at the edge of the Andromeda Nebula were Cepheid variables. With these measuring rods it was found that the Nebula was nearly a million light-years away! So the Andromeda Nebula was far, far outside our Galaxy. Allowing for its distance, its apparent size showed that it must be a huge conglomeration of stars, almost rivalling our own Galaxy.

Other nebulosities, too, turned out to be conglomerations of stars, even farther away than the Andromeda Nebula. These 'extra-galactic nebulae' all had to be recognized as galaxies – new universes which reduced our own to just one of many in space. Once again the universe had expanded. It was larger than ever – not merely hundreds of thousands of light-years across, but perhaps hundreds of millions.

Through the 1930s, astronomers wrestled with several nagging puzzles about these galaxies. For one thing, on the basis of their assumed distances, all of them turned out to be much smaller than our own. It seemed an odd coincidence that we should be inhabiting by far the largest galaxy in existence. For another thing, globular clusters surrounding the Andromeda galaxy seemed to be only one half or one third as luminous as those of our own Galaxy. (Andromeda is about as rich in globular clusters as our own

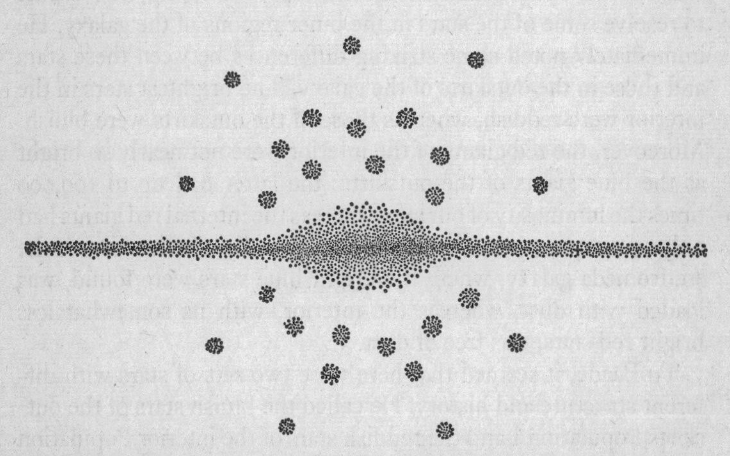

A model of our Galaxy seen edgewise. Globular clusters are arrayed around the central portion of the Galaxy. The position of our sun is indicated by +.

Galaxy, and its clusters are spherically arranged about Andromeda's centre. This seems to show that Shapley's assumption that our own clusters were so arranged was a reasonable one. Some galaxies are amazingly rich in globular clusters. The galaxy M-87, in Virgo, possesses at least a thousand.)

The most serious puzzle is that the distances of the galaxies seemed to imply that the universe was only about 2,000 million years old (for reasons I shall discuss later in this chapter). This was puzzling, for the earth itself was considered by geologists to be older than that, on what was thought to be the very best kind of evidence.

The beginning of an answer came during the Second World War, when the German-born American astronomer Walter Baade discovered that the yardstick by which the galaxies' distances had been measured was wrong.

In 1942 Baade took advantage of the wartime blackout of Los Angeles, which cleared the night sky at Mount Wilson, to make a detailed study of the Andromeda galaxy with the 100-inch Hooker telescope (named after John B. Hooker, who had provided the funds for its construction). With the improved seeing, he was able to resolve some of the stars in the inner regions of the galaxy. He immediately noted some striking differences between these stars and those in the outskirts of the galaxy. The brightest stars in the interior were reddish, whereas those of the outskirts were bluish. Moreover, the red giants of the interior were not nearly so bright as the blue giants of the outskirts: the latter had up to 100,000 times the luminosity of our sun, whereas the internal red giants had only up to 1,000 times that luminosity. Finally, the outskirts of the Andromeda galaxy, where the bright blue stars were found, was loaded with dust, whereas the interior, with its somewhat less bright red stars, was free of dust.

To Baade, it seemed that here were two sets of stars with different structure and history. He called the bluish stars of the outskirts Population I and the reddish stars of the interior Population II. Population I stars, it turns out, are relatively young, with high metal content, and follow nearly circular orbits about the galactic centre in the median plane of the galaxy. Population II stars are relatively old, with low metal content, with orbits that are markedly elliptical, and with considerable inclination to the median plane of the galaxy. Both populations have been broken down into finer sub-groups since Baade's discovery.

When the new 200-inch Hale telescope (named after the American astronomer, George Ellery Hale, who supervised its construction) was set up on Palomar Mountain after the war, Baade continued his investigations. He found certain regularities in the distribution of the two populations and these depended on the nature of the galaxies involved. Galaxies of the class called 'elliptical' (systems with the shape of an ellipse and rather uniform internal structure) apparently were made up mainly of Population II stars, as were globular clusters in any galaxy. On the other hand, in 'spiral galaxies' (that is, with arms which make them look like a pinwheel) the spiral arms were composed of Population I, set against a Population II background.

It is estimated that only about 2 per cent of the stars in the universe are of the Population I type. But our own sun and the familiar stars in our neighbourhood fall into this class. From this fact alone, we can deduce that ours is a spiral galaxy and that we lie on one of the spiral arms. (This explains why there are so many dust clouds, both light and dark, in our neighbourhood: the spiral arms of a galaxy are clogged with dust.) Photographs show that the Andromeda galaxy also is of the spiral type.

Now to get back to the yardstick. Baade began to compare the Cepheid stars found in globular clusters (Population II) with those found in our spiral arm (Population I). It turned out that the Cepheids in the two populations were really of two different types, as far as the relation between period and luminosity was concerned. Cepheids of Population II followed the period–luminosity curve set up by Leavitt and Shapley. With this yardstick, Shapley had accurately measured the distances to the globular clusters and the size of our Galaxy. But the Cepheids of Population I, it now developed, were a different yardstick altogether! A Population I Cepheid was four or five times as luminous as a Population II Cepheid of the same period. This meant that use of the Leavitt scale would result in miscalculation of the absolute magnitude of a

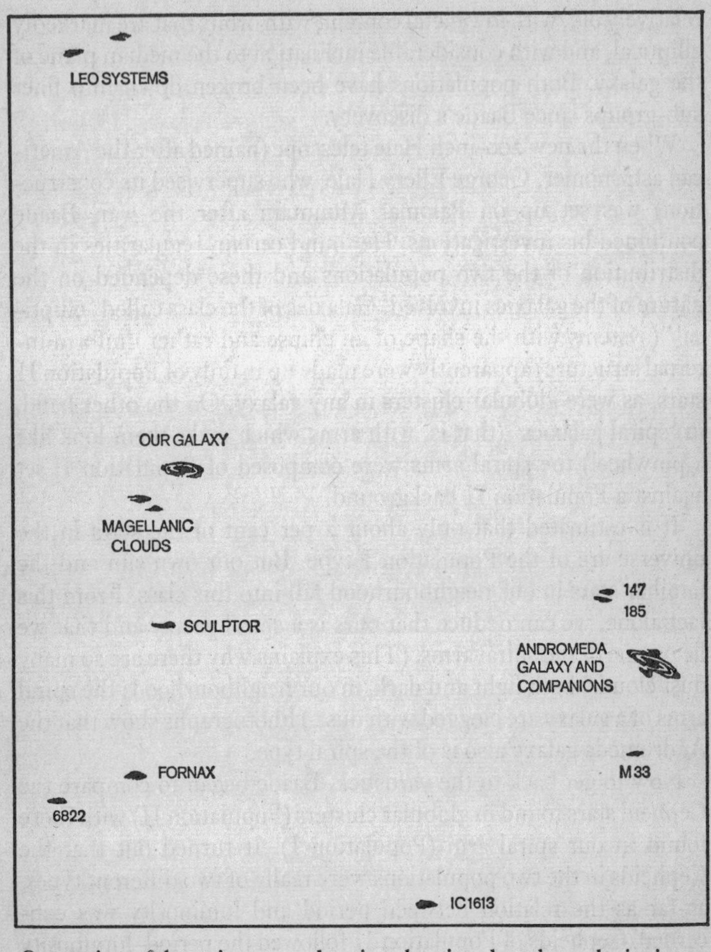

Our region of the universe – a drawing showing the other galaxies in our neighbourhood.

Population I Cepheid from its period. And if the absolute magnitude was wrong, the calculation of distance must be wrong; the star would actually be much farther away than the calculation indicated.

Hubble had gauged the distance of the Andromeda galaxy from the Cepheids (of Population I) in its outskirts – the only ones that could be resolved at the time. Now, with the revised yardstick, the galaxy proved to be about 2·5 million light-years away, instead of less than a million. And other galaxies had to be moved out in proportion. (The Andromeda galaxy is still a close neighbour, however. The average distance between galaxies is estimated to be something like 20 million light-years.)

At one stroke, the size of the known universe was more than doubled. This instantly solved the problems that had plagued the 1930s. Our Galaxy was no longer larger than all the others; the Andromeda galaxy, for instance, was definitely more massive than ours. Second, it now appeared that the Andromeda galaxy's globular clusters were as luminous as ours; they had seemed less bright only because of the misjudgement of their distance. Finally, for reasons I will explain later, the new scale of distances allowed the universe to be considered much older – at least 5,000 million years old and very likely considerably more than that – which brought it into line with the geologists' estimates of the age of the earth.

Doubling the distance of the galaxies does not end the problem of size. We must now consider the possibility of still larger systems – of clusters of galaxies and super-galaxies.

Actually, modern telescopes have shown that clusters of galaxies do exist. For instance, in the constellation of Coma Berenices there is a large, ellipsoidal cluster of galaxies about 8 million light-years in diameter. The 'Coma Cluster' contains about 11,000 galaxies, separated by an average distance of only 300,000 light-years (as compared with an average of something like 3 million light-years between galaxies in our own vicinity).

Our own Galaxy seems to be part of a 'local cluster' that includes the Magellanic Clouds, the Andromeda galaxy, and three small 'satellite galaxies' near it, plus some other galaxies for a total of nineteen members altogether. Two of these, called 'Maffei One' and 'Maffei Two' (after Paolo Maffei, the Italian astronomer, who first reported them), were discovered only in 1971. The lateness of the discovery was owing to the fact that they can only be detected through dust clouds that lie between them and ourselves.

Of the local cluster, only our own Galaxy, Andromeda, and the two Maffeis are giants, whereas the rest are dwarfs. One of the dwarfs, IC 1613, may contain only 60 million stars; hence it is scarcely more than a large globular cluster. Among galaxies, as among stars, dwarfs far outnumber giants.

If galaxies do form clusters and clusters of clusters, does that mean that the universe goes on for ever and that space is infinite? Or is there some end, both to the universe and to space? Well, astronomers can make out objects up to an estimated 9,000 million light-years away, and there is no sign of an end of the universe – yet. At the theoretical level, there are arguments both for an end of space and for no end, for a beginning in time and for no beginning. Having considered space, let us consider time next.

The Birth of the Universe

Myth-makers have invented many fanciful creations of the universe (usually concentrating on the earth itself, with all the rest dismissed quickly as 'the sky' or 'the heavens'). Generally, the time of creation is set not very far in the past (although we should remember that to people in the pre-literate stage, a time of a thousand years was even more impressive than a billion years is today).

The creation story with which we are most familiar is, of course, that given in the first chapters of Genesis, which, some people hold, is an adaptation of Babylonian myths, intensified in poetic beauty and elevated in moral grandeur.

Various attempts have been made to work out the date of the Creation on the basis of the data given in the Bible (the reigns of the various kings, the time from the Exodus to the dedication of Solomon's temple, the ages of the patriarchs both before and after the flood). Medieval Jewish scholars put the date of the Creation at 3760 B.C. and the Jewish calendar still counts its years from that date. In A.D. 1656, Archbishop James Ussher of the Anglican Church calculated the date of the Creation to be 4004 B.C., at 8 p.m. of 22 October of that year, to be exact. Some theologians of the Greek Orthodox Church put Creation as far back as 5508 B.C.

Even as late as the eighteenth century, the Biblical version was accepted by the learned world, and the age of the universe was considered to be only six or seven thousand years at most. This view received its first major blow in 1785 in the form of a book entitled *Theory of the Earth*, by a Scottish naturalist named James Hutton. Hutton started with the proposition that the slow natural processes working on the surface of the earth (mountain building and erosion, the cutting of river channels, and so on) had been working at about the same rate throughout the earth's history. This 'uniformitarian principle' implied that the processes must have been working for a stupendously long time to produce the observed phenomena. Therefore the earth must be not thousands but many millions of years old.

Hutton's views were immediately derided. But the ferment worked. In the early 1830s, the British geologist Charles Lyell reaffirmed Hutton's views, and in a three-volume work entitled *Principles of Geology* presented the evidence with such clarity and force that the world of science was won over. The modern science of geology can be dated from that work.

Attempts were made to calculate the age of the earth on the basis of the uniformitarian principle. For instance, if one knew the amount of sediment laid down by the action of water each year (a modern estimate is one foot in 880 years), one could calculate the age of a layer of sedimentary rock from its thickness. It soon became obvious that this approach could not determine the earth's age accurately, because the record of the rocks was obscured by processes of erosion, crumbling, upheavals, and other forces. Nevertheless, even the incomplete evidence indicated that the earth must be at least 500 million years old.

Another way of measuring the age of the earth was to estimate the rate of accumulation of salt by the oceans, a suggestion first advanced by the English astronomer Edmund Halley as long ago as 1715. Rivers steadily washed salt into the sea; since only fresh water left it by evaporation, the salt concentration rose. Assuming that the ocean had started as fresh water, the time necessary for the rivers to have endowed the oceans with their salt content of over 3 per cent could have been as long as a thousand million years.

This great age was very agreeable to the biologists, who during the latter half of the nineteenth century were trying to trace the slow development of living organisms from primitive one-celled creatures to the complex higher animals. They needed long aeons for the development to take place, and a thousand million years gave them sufficient time.

However, by the mid nineteenth century astronomical considerations brought in sudden complications. For instance, the principle of the 'conservation of energy' raised an interesting problem with respect to the sun. The sun was pouring out energy in colossal quantities and had been doing so throughout recorded history. If the earth had existed for countless aeons, where had all this energy come from? It could not have come from the usual sources familiar to mankind. If the sun had started as solid coal burning in an atmosphere of oxygen, it would have been reduced to a cinder (at the rate it was delivering energy) in the space of about 2,500 years.

The German physicist Hermann Ludwig Ferdinand von Helmholtz, one of the first to enunciate the law of conservation of energy, was particularly interested in the problem of the sun. In 1854, he pointed out that if the sun were contracting, its mass would gain energy as it fell towards its centre of gravity, just as a rock gains energy when it falls. This energy could be converted into radiation. Helmholtz calculated that a contraction of the sun by a mere ten thousandth of its radius could provide it with a 2,000-year supply of energy.

The British physicist William Thomson (later Lord Kelvin) did more work on the subject and decided that on this basis the earth could not be more than 50 million years old, for at the rate the sun had spent energy, it must have contracted from a gigantic size, originally as large as the earth's orbit round the sun. (This meant, of course, that Venus must be younger than the earth and Mercury still younger.) Lord Kelvin went on to estimate that if the earth itself had started as a molten mass, the time needed to cool to its present temperature, and therefore its age, would be about 20 million years.

By the 1890s the battlelines seemed drawn between two invincible armies. The physicists seemed to have shown conclusively that the earth could not have been solid for more than a few million years, while geologists and biologists seemed to have proved just as conclusively that the earth must have been solid for not less than a thousand million years.

And then something new and completely unexpected turned up, and the physicists found themselves with their case crumbling.

In 1896, the discovery of radioactivity made it clear that the earth's uranium and other radioactive substances were liberating large quantities of energy and had been doing so for a very long time. This finding made Kelvin's calculations meaningless, as was pointed out first, in 1904, by the New Zealand-born British physicist, Ernest Rutherford, in a lecture – with the aged (and disapproving) Kelvin, himself, in the audience.

There is no point in trying to decide how long it would take the earth to cool if you don't take into account the fact that heat is being constantly supplied by radioactive substances. With this new factor, it might take the earth thousands of millions of years, rather than millions, to cool from a molten mass to its present temperature. The earth might even be warming with time.

Actually, radioactivity itself eventually gave the most conclusive evidence of the earth's age, for it allowed geologists and geochemists to calculate the age of rocks directly from the quantity of uranium and lead they contain. By the clock of radioactivity, some of the earth's rocks are now known to be nearly 4,000 million years old, and there is every reason to think that the earth is somewhat older than that. An age of 4,700 million years for the earth in its present solid form is now accepted as likely. And, indeed, some of the rocks brought back from our neighbouring world, the moon, have proved to be just about that old.

And what of the sun? Radioactivity, together with discoveries concerning the atomic nucleus, introduced a new source of energy, much larger than any previously known. In 1930, the British physicist Sir Arthur Eddington set a train of thought working when he suggested that the temperature and pressure at the centre of the sun must be outrageously high: the temperature might be as high as 15 million degrees. At such temperatures and pressures, the nuclei of atoms could undergo reactions which could not take place in the bland mildness of the earth's environment. The sun is known to consist largely of hydrogen. If four hydrogen nuclei combined (forming a helium atom), they would liberate large amounts of energy.

Then, in 1938, the German-born American physicist Hans Albrecht Bethe worked out the possible ways in which this combination of hydrogen to helium could take place. There were two processes by which this could occur under the conditions at the centre of stars like the sun. One involved the direct conversion of hydrogen to helium; the other involved a carbon atom as an intermediate in the process. Either set of reactions can occur in stars; in our own sun, the direct hydrogen conversion seems to be the dominant mechanism. Either brought about the conversion of mass to energy. (Einstein, in his Special Theory of Relativity, had shown that mass and energy were different aspects of the same thing and could be interconverted; furthermore, that a great deal of energy could be liberated by the conversion of a small amount of mass.)

The rate of radiation of energy by the sun requires the disappearance of solar mass at the rate of 4·2 million tons per second. At first blush this seems a frightening loss, but the total mass of the sun is 2,200,000,000,000,000,000,000,000,000 tons, so the sun loses only 0·00000000000000000002 per cent of its mass each second. Assuming the sun to have been in existence for 6,000 million years, as astronomers now believe, and if it has been radiating at its present rate all that time, it would have expended only 1/40,000 of its mass. It is easy to see, then, that the sun can continue to radiate energy at its present rate for thousands of millions of years to come.

By 1940, then, an age of 6,000 million years for the solar system as a whole seemed reasonable. The whole matter of the age of the universe might have been settled, but astronomers had thrown another monkey wrench into the machinery. Now the universe as a whole seemed too youthful to account for the age of the solar system. The trouble arose from an examination of the distant galaxies by the astronomers and from a phenomenon first discovered in 1842 by an Austrian physicist named Christian Johann Doppler.

The 'Doppler effect' is familiar enough; it is most commonly illustrated by the whistle of a passing locomotive, which rises in pitch as it approaches and drops in pitch as it recedes. The change in pitch is due simply to the fact that the number of sound waves striking the eardrum per second changes because of the source's motion.

As Doppler suggested, the Doppler effect applies to light waves as well as to sound. When light from a moving source reaches the eye, there is a shift in frequency – that is, colour – when the source is moving fast enough. For instance, if the source is travelling towards us, more light waves are crowded into each second and the light perceived shifts towards the high-frequency violet end of the visible spectrum. On the other hand, if the source is moving away, fewer waves arrive per second and the light shifts towards the low-frequency red end of the spectrum.

Astronomers have been studying the spectra of stars for a long time, and they are well acquainted with the normal picture – a pattern of bright lines against a dark background or dark lines against a bright background showing the emission or absorption of light by atoms at certain wavelengths, or colours. They have been able to calculate the velocity of stars moving towards or away from us (i.e., radial velocity) by measuring the displacement of the usual spectral lines towards the violet or red end of the spectrum.

It was the French physicist Armand Hippolyte Louis Fizeau who, in 1848, pointed out that the Doppler effect in light could best be observed by noting the position of the spectral lines. For that reason, the Doppler effect is called the 'Doppler–Fizeau effect' where light is concerned.

The Doppler–Fizeau effect has been used in a variety of ways. Within our solar system, it could be used to demonstrate the rota-

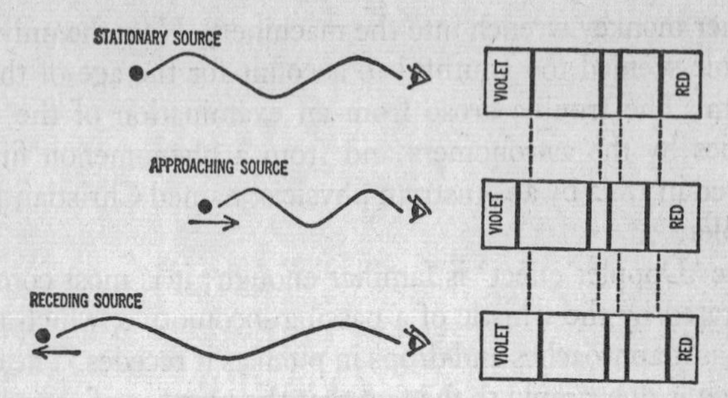

The Doppler–Fizeau effect. The lines in the spectrum shift towards the violet end (*left*) when the light source is approaching. When the source recedes, the spectral lines shift towards the red end (*right*).

tion of the sun in a new way. The spectral lines originating from that limb of the sun being carried towards us in the course of its vibration would be shifted towards the violet (a 'violet shift'). The lines from the other limb would show a 'red shift' since that limb was receding.

To be sure, the motion of sunspots is a better and more obvious way of detecting and measuring solar rotation (which turns out to have a period of about 25 days, relative to the stars). However, the effect can also be used to determine the rotation of featureless objects, such as the rings of Saturn.

The Doppler–Fizeau effect can be used for objects at any distance, so long as those objects can be made to produce a spectrum for study. Its most dramatic victories, therefore, were in connection with the stars.

In 1868, the British astronomer Sir William Huggins measured the radial velocity of Sirius and announced that it was moving away from us at 29 miles per second. (We have better figures now, but he came reasonably close for a first try.) By 1890, the American astronomer James Edward Keeler, using better instruments, was producing reliable results in quantity; he showed, for instance, that Arcturus was approaching us at a rate of $3\frac{3}{4}$ miles per second.

The effect could even be used to determine the existence of star systems, the details of which could not be made out by telescope.

In 1782, for instance, an English astronomer, John Goodricke (a deaf-mute who died at twenty-two; a first-rate brain in a tragically defective body), studied the star Algol, which increased and decreased regularly in brightness. Goodricke explained it by supposing that a dark companion circled Algol. Periodically the dark companion passed in front of Algol, eclipsing it and dimming its light.

A century passed before this plausible hypothesis was supported by additional evidence. In 1889, the German astronomer Hermann Carl Vogel showed that the lines of Algol's spectrum underwent alternate red and violet shifts that matched its brightening and dimming. First it receded while the dark companion approached and then approached while the dark companion receded. Algol was an 'eclipsing binary star'.

In 1890, Vogel made a similar and more general discovery. He found that some stars were both advancing and receding. That is, the spectral lines showed both a red shift and a violet shift, appearing to have doubled. Vogel interpreted this as indicating that the star was an eclipsing binary with the two stars (both bright) so close together that they appeared as a single star even in the best telescopes. Such stars are 'spectroscopic binaries'.

But there was no need to restrict the Doppler–Fizeau effect to the stars of our Galaxy. Objects beyond the Milky Way could be studied in this way, too. In 1912, the American astronomer Vesto Melvin Slipher found, on measuring the radial velocity of the Andromeda galaxy, that it was moving towards us at approximately 125 miles per second. But when he went on to examine other galaxies, he discovered that most of them were moving away from us. By 1914, Slipher had figures on 15 galaxies; of these, 13 were receding, all at the healthy clip of several hundred miles per second.

As research along these lines continued, the situation grew more remarkable. Except for a few of the nearest galaxies, all were fleeing from us. Furthermore, as techniques improved so that fainter, more distant galaxies could be tested, the red shift increased.

In 1929, Hubble at Mount Wilson suggested that there was a regular increase in these velocities of recession in proportion to the distance of the galaxy involved. If galaxy A was twice as far from us as galaxy B, then galaxy A receded at twice the velocity of galaxy B. This is sometimes known as 'Hubble's law'.

Hubble's law certainly continued to be borne out by observations. Beginning in 1929, Milton La Salle Humason at Mount Wilson used the 100-inch telescope to obtain spectra of dimmer and dimmer galaxies. The most distant galaxies he could test were receding at 25,000 miles per second. When the 200-inch telescope came into use, still more distant galaxies could be studied, and by the 1960s, objects were detected so distant that their recession velocities were as high as 150,000 miles per second.

Why should this be? Well, imagine a balloon with small dots painted on it. When the balloon is inflated, the dots move apart. To a manikin standing on any one of the dots, all the other dots would seem to be receding, and the farther away from him a particular dot was, the faster it would be receding. It would not matter on which particular dot he was standing; the effect would be the same.

The galaxies behave as though the universe were expanding like a balloon. Astronomers have now generally accepted the fact of this expansion, and Einstein's 'field equations' in his General Theory of Relativity can be construed to fit an expanding universe.

But this raises profound questions. Does the visible universe have a limit? The farthest objects we can now see (about 9,000 million light-years away) are receding from us at four-fifths the speed of light. If Hubble's law of the increase in recession velocity holds, at about 11,000 million light-years from us the galaxies are receding with the speed of light. But the speed of light is the maximum possible velocity, according to Einstein's theory. Does that mean there can be no visible galaxies more distant?

There is also the age question. If the universe has been expanding constantly, it is logical to suppose that it was smaller in the past than it is now, and that at some time in the distant past it began as a dense core of matter. And that is where the conflict over the age of the universe lay in the 1940s. From its rate of expansion and the present distance of the galaxies, it appeared that the universe could not be more than 2,000 million years old. But the geologists, thanks to radioactivity, were now certain that the earth must be nearly 4,000 million years old, at least.

Fortunately, the revision of the Cepheid yardstick in 1952 saved the situation. By doubling, possibly tripling, the size of the universe, it doubled or tripled its age, and so the rocks and the red shift now agreed that both the solar system and the galaxies were 5,000 or 6,000 million years old.

By the 1960s, the situation was thrown into some confusion again. The British astronomer Fred Hoyle, after analysing the probable composition of Population I and Population II stars, decided that, of the two processes by which stars burn hydrogen to form helium, the slower one was predominant. On that basis, he estimated that some stars must be at least 10,000, perhaps 15,000 million years old. Then the American astronomer Allen Sandage found that stars in a cluster called NGC 188 appeared to be at least 24,000 million years old, while the Swiss-American astronomer Fritz Zwicky speculated on ages as great as a thousand billion years. Such ages would not conflict with the rocks' evidence on the age of the earth, for the earth could certainly be younger than the universe, but if the universe has been expanding at the present rate for 24,000 million years or more, it would seem that it should be more spread out than it is. So the astronomers have a new problem to resolve.

Assuming that the universe expands and that Einstein's field equations agree with that interpretation, the question still arises inexorably: Why? The easiest, and almost inevitable, explanation is that the expansion is the result of an explosion at the beginning. In 1927, the Belgian mathematician Abbé Georges Édouard Lemaître suggested that all matter came originally from a tremendously dense 'cosmic egg', which exploded and gave birth to the universe as we know it. Fragments of the original sphere of matter formed galaxies, which are still rushing outwards in all directions as a result of that unimaginably powerful multi-million-year-old explosion.

The Russian-American physicist George Gamow has elaborated

this notion. His calculations led him to believe that the various elements as we know them were formed in the first half-hour after the explosion. For 250 million years after the explosion, radiation predominated over matter, and the universe's matter, as a consequence, remained dispersed as a thin gas. After a critical point in the expansion, however, matter came into predominance and began to condense into the beginnings of galaxies. Gamow believes that the expansion will probably continue until all the galaxies, except for those in our own cluster, have receded beyond the reach of our most powerful instruments. We will then be alone in the universe.

Where did the matter in the 'cosmic egg' come from? Some astronomers suggest that the universe started as an extremely thin gas, gradually contracted under the force of gravitation to a superdense mass, and then exploded. In other words, it began an eternity ago in the form of almost complete emptiness, went through a contracting stage to the 'cosmic egg', exploded, and is going through an expanding stage back to an eternity of almost complete emptiness. We just happen to be living during the very temporary period (an instant in eternity) of the fullness of the universe.

Other astronomers, notably W. B. Bonnor of England, argue that the universe has gone through an unending series of such cycles, each lasting perhaps tens of thousands of millions of years – in other words, an 'oscillating universe'. In 1956, Sandage suggested a period of 82,000 million years for each oscillation.

Whether the universe is simply expanding, or contracting and then expanding, or oscillating, all these theories have in mind an 'evolutionary universe'.

In 1948, the British astronomers Hermann Bondi and Thomas Gold put forward a theory, since extended and popularized by another British astronomer, Fred Hoyle, that forbade evolution. Their universe is called the 'steady-state universe' or 'continuous-creation universe'. They agree that the galaxies are receding and that the universe is expanding. When the farthest galaxies reach the speed of light, so that no light from them can reach us, they may be said to have left our universe. However, while the galaxies and clusters of galaxies of our universe move apart, new galaxies are continually forming among the old ones. For every galaxy that disappears over the speed-of-light edge of the universe, another joins our midst. Therefore the universe remains in a steady state, with galaxies always at the same density in space.

Of course, new matter has to be created continually to replace the galaxies that leave, and no such creation of matter has been detected. This is not surprising, however. In order to supply new matter to form galaxies at the necessary rate, only one atom of hydrogen need be formed per year in a thousand million litres of space. This is creation at far too slow a rate to be detected by the instruments now at our disposal.

If we are to suppose matter to be created continuously, at however slow a rate, we must ask, 'Where does this new matter come from?' What happens to the law of conservation of mass-energy? Surely matter cannot be made out of nothing. Hoyle replies that the energy for the creation of new matter may be siphoned from the energy of the expansion. In other words, the universe may be expanding a bit more slowly than it would be if matter were not being formed, and the matter being formed could be manufactured at the expense of the energy being pumped into expansion.

The argument between the proponents of the evolutionary and steady-state views has been hot. The best way of deciding between the two theories would be to study the far-distant reaches of the universe, thousands of millions of light-years away.

If the steady-state theory is correct, then the universe is much the same everywhere, and its appearance thousands of millions of light-years away ought to be equivalent to its appearance in our own neighbourhood. By the evolutionary theory, however, the universe thousands of millions of light-years away would be seen by light that had been formed thousands of millions of years ago. That light had been formed when the universe was young, and not long after the 'big bang'. What we see in the young universe ought to be quite different from what we see in our own neighbourhood where the universe is old.

Unfortunately, it is hard to make out clearly what we see by telescope in the very distant galaxies and, until the 1960s, the information gathered was insufficient. When the evidence finally began to come in, it involved (as we shall see) radiation other than that of ordinary light.

The Death of the Sun

Whether the universe is evolutionary or steady-state is a point that does not affect individual galaxies or clusters of galaxies directly. Even if all the distant galaxies recede and recede until they are out of range of the best possible instruments, our own Galaxy will remain intact, its component stars held firmly within its gravitational field. Nor will the other galaxies of the local cluster leave us. But changes within our Galaxy, possibly disastrous to our planet and its life, are by no means excluded.

The whole conception of changes in heavenly bodies is a modern one. The ancient Greek philosophers, Aristotle in particular, believed that the heavens were perfect and unchangeable. All change, corruption, and decay were confined to the imperfect regions that lay below the nethermost sphere – the moon. This seemed only common sense, for certainly from generation to generation and century to century there was no important change in the heavens. To be sure, there were the mysterious comets that occasionally materialized out of nowhere – erratic in their comings and goings, ghost-like as they shrouded stars with a thin veil, baleful in appearance, for the filmy tail looked like the streaming hair of a distraught creature prophesying evil (in fact, the word 'comet' comes from the Latin word for 'hair'). About twenty-five of these objects are visible to the naked eye each century.

Aristotle tried to reconcile these apparitions with the perfection of the heavens by insisting that they belonged to the atmosphere of the corrupt and changing earth. This view prevailed until late in the sixteenth century. But, in 1577, the Danish astronomer Tycho Brahe attempted to measure the parallax of a bright comet and discovered that it could not be measured (this was before the days of the telescope). Since the moon's parallax *was* measurable, Tycho

Brahe was forced to conclude that the comet lay far beyond the moon and that there was change and imperfection in the heavens. (The Roman philosopher Seneca had suspected this in the first century A.D.)

Actually, changes even in the stars had been noticed much earlier, but apparently they had aroused no great curiosity. For instance, there are the variable stars that change noticeably in brightness from night to night, even to the naked eye. No Greek astronomer made any reference to variations in the brightness of any star. It may be that we have lost the records of such references; on the other hand, perhaps the Greek astronomers simply chose not to see these phenomena. One interesting case in point is Algol, the second brightest star in the constellation Perseus, which suddenly loses two thirds of its brightness then regains it, and does this regularly every 69 hours. (We know now, thanks to Goodricke and Vogel, that Algol has a dim companion star that eclipses it and diminishes its light at 69-hour intervals.) The Greek astronomers made no mention of the dimming of Algol, nor did the Arab astronomers of the Middle Ages. Nevertheless, the Greeks placed the star in the head of Medusa, the demon who turned men to stone, and the very name 'Algol', which is Arabic, means 'the ghoul'. Clearly, the ancients felt uneasy about this strange star.

A star in the constellation Cetus, called Omicron Ceti, varies irregularly. Sometimes it is as bright as the Pole Star; sometimes it vanishes from sight. Neither the Greeks nor the Arabs said a word about it, and the first man to report it was a Dutch astronomer, David Fabricius, in 1596. It was later named Mira (Latin for 'wonderful'), astronomers having grown less frightened of heavenly change by then.

Even more remarkable was the sudden appearance of 'new stars' in the heavens. This the Greeks could not altogether ignore. Hipparchus is said to have been so impressed by the sighting of such a new star in the constellation Scorpio in 134 B.C. that he designed the first star map, in order that new stars in the future might be more easily detected.

In A.D. 1054, in the constellation Taurus, another new star was sighted – a phenomenally bright one, in fact. It surpassed Venus in brightness, and for weeks it was visible in broad daylight. Chinese and Japanese astronomers recorded its position accurately, and their records have come down to us. In the Western world, however, the state of astronomy was so low at the time that no European record of this remarkable occurrence has survived, probably because none was kept.

It was different in 1572, when a new star as bright as that of 1054 appeared in the constellation Cassiopeia. European astronomy was reviving from its long sleep. The young Tycho Brahe carefully observed the new star and wrote a book entitled *De Nova Stella*. It is from the title of that book that the word 'nova' was adopted for any new star.

In 1604, still another remarkable nova appeared, in the constellation Serpens. It was not quite as bright as that of 1572, but it was bright enough to outshine Mars. Johannes Kepler observed this one, and he too wrote a book about the subject.

After the invention of the telescope, novae became less mysterious. They were not new stars at all, of course; merely faint stars that had suddenly brightened to visibility.

Increasing numbers of novae were discovered with time. They would brighten many thousandfold, sometimes within the space of a few days, and then dim slowly over a period of months to their previous obscurity. Novae showed up at the average rate of twenty per year per galaxy (including our own).

From an investigation of the Doppler–Fizeau shifts that took place during nova formation and from certain other fine details of their spectra, it became plain that the novae were exploding stars. In some cases the star material blown into space could be seen as a shell of expanding gas, illuminated by the remains of the star. Such stars are called 'planetary nebulae'.

This sort of nova formation does not necessarily signify the death of a star. It is a tremendous catastrophe, of course, for the luminosity of such a star may increase a millionfold in less than a day. (If our sun were to become a nova, it would destroy all life on earth and possibly vaporize the planet.) But the explosion apparently ejects only 1 or 2 per cent of the star's mass, and afterward the star settles back to a reasonably normal life. In fact, some stars seem to undergo such explosions periodically and still survive.

The most remarkable nova that appeared after the invention of the telescope was one that was discovered by the German astronomer Ernst Harwig in the Andromeda galaxy in 1885 and was given the name 'S Andromedae'. It was just below visibility to the naked eye; in the telescope it looked one tenth as bright as the entire Andromeda galaxy. At the time, no one realized how distant the Andromeda galaxy was, or how large, and so the brightness of its nova occasioned no particular excitement. But after Hubble worked out the distance of the Andromeda galaxy, the brilliance of that nova of 1885 suddenly staggered astronomers. Hubble eventually discovered a number of novae in the Andromeda galaxy, but none even approached the 1885 nova in brightness. The nova of 1885 must have been 10,000 times as bright as ordinary novae. It was a 'supernova'. Looking back now, we realize that the novae of 1054, 1572, and 1604 also were supernovae. What is more, they must have been in our own Galaxy, which would account for their extreme brightness. In 1965, Bernard Goldstein of Yale presented evidence to the effect that a fourth supernova in our Galaxy may have flared in 1006, if an obscure note by an Egyptian astrologer of the period is to be accepted.

Supernovae apparently are quite different in physical behaviour from ordinary novae, and astronomers are eager to study their spectra in detail. The main difficulty is that they are so rare. About three per thousand years is the average for any one galaxy, according to Zwicky. Although astronomers have managed to spot about fifty so far, all these are in distant galaxies and cannot be studied in detail. The 1885 supernova of Andromeda, the closest to us in the last 350 years, appeared a couple of decades before photography in astronomy had been fully developed; consequently, no permanent record of its spectrum exists. (However, the distribution of supernovae in time is random. In one galaxy recently, three supernovae were detected in just seventeen years. Astronomers on earth may yet prove lucky.)

The brightness of a supernova (absolute magnitudes range from -14 to an occasional -17) could only come about as a result of a complete explosion – a star literally tearing itself to pieces. What would happen to such a star? Well, let us go back a little....

As early as 1834, Bessel (the astronomer who was later to be the first to measure the parallax of a star) noticed that Sirius and Procyon shifted position very slightly from year to year in a manner which did not seem related to the motion of the earth. Their motions were not in a straight line but wavy, and Bessel decided that each must actually be moving in an orbit round something.

From the manner in which Sirius and Procyon were moving in these orbits, the 'something' in each case had to have a powerful gravitational attraction that could belong to nothing less than a star. Sirius's companion, in particular, had to be as massive as our own sun to account for the bright star's motions. So the companions were judged to be stars, but since they were invisible in telescopes of the time, they were referred to as 'dark companions'. They were believed to be old stars growing dim with time.

Then in 1862 the American instrument maker, Alvan Clark, testing a new telescope, sighted a dim star near Sirius and, sure enough, on further observation this turned out to be the companion. Sirius and the dim star circled about a mutual centre of gravity in a period of about fifty years. The companion of Sirius ('Sirius B' it is now called, with Sirius itself as 'Sirius A') has an absolute magnitude of only 11·2, and so it is only about 1/400 as bright as our sun, though it is just as massive.

This seemed to check with the notion of a dying star. But in 1914 the American astronomer Walter Sydney Adams, after studying the spectrum of Sirius B, decided that the star had to be as hot as Sirius A itself and hotter than our sun. The atomic vibrations that gave rise to the particular absorption lines found in its spectrum could only be taking place at very high temperatures. But if Sirius B was so hot, why was its light so faint? The only possible answer was that it was considerably smaller than our sun. Being hotter, it radiated more light per unit of surface, but to account for the small total amount of light, its total surface had to be small. In fact, the star could not be more than 16,000 miles in diameter – only twice the earth's diameter. Yet Sirius B had a mass equal to that of our own sun! Adams found himself trying to imagine this mass mashed down into a volume as small as that of Sirius B. The star's density would have to be nearly 3,000 times that of platinum.

This represented nothing less than a completely new state of matter. Fortunately, by this time physicists had no trouble in suggesting the answer. They knew that in ordinary matter the atoms are composed of very tiny particles, so tiny that most of the volume of an atom is 'empty' space. Under extreme pressure the sub-atomic particles could be forced together into a super-dense mass. Yet even in super-dense Sirius B, the sub-atomic particles are far enough apart to move about freely so that the far-denser-than-platinum substance still acts as a gas. The English physicist Ralph Howard Fowler suggested in 1925 that this be called a 'degenerate gas', and the Soviet physicist Lev Davidovich Landau pointed out in the 1930s that even ordinary stars such as our own sun ought to consist of degenerate gas at the centre.

The companion of Procyon ('Procyon B'), first detected in 1896 by J. M. Schaberle at Lick Observatory, was also found to be a super-dense star although only five eighths as massive as Sirius B and, as the years passed, more examples were found. These stars are called 'white dwarfs', because they combine small size with high temperature and white light. White dwarfs are probably quite numerous and may make up as much as 3 per cent of all stars. However, because of their small size, only those in our own neighbourhood are likely to be discovered in the forseeable future. (There are also 'red dwarfs', considerably smaller than our sun, but not as small as white dwarfs. Red dwarfs are cool and of ordinary density. They may be the most common of all stars but, because of their dimness, are as difficult to detect as are white dwarfs. A pair of red dwarfs, a mere 6 light-years distant from us, was only discovered in 1948. Of the 36 stars known to be within 14 light-years of the sun, 21 are red dwarfs, and 3 are white dwarfs. There are no giants among them, and only two, Sirius and Procyon, are distinctly brighter than our sun.)

The year after Sirius B was found to have its astonishing properties, Albert Einstein presented his General Theory of Relativity, which was mainly concerned with new ways of looking at gravity.

Einstein's views of gravity led to the prediction that light emitted by a source possessing a very strong gravitational field should be displaced towards the red (the 'Einstein shift'). Adams, fascinated by the white dwarfs he had discovered, carried out careful studies of the spectrum of Sirius B and found that there was indeed the red shift predicted by Einstein. This was not only a point in favour of Einstein's theory but also a point in favour of the super-density of Sirius B, for in an ordinary star such as our sun the red-shift effect would be only one thirtieth as great. Nevertheless, in the early 1960s this very small Einstein shift produced by our sun was detected, and the General Theory of Relativity was further confirmed.

But what have white dwarfs to do with supernovae, the subject that prompted this discussion? To answer that, let us go back to the supernova of 1054. In 1844 the Earl of Rosse, investigating the location in Taurus where the original astronomers had reported finding the 1054 supernova, studied a small cloudy object. Because of its irregularity and its clawlike projections, he named the object the 'Crab Nebula'. Continued observation over decades showed that the patch of gas was slowly expanding. The actual rate of expansion could be calculated from the Doppler–Fizeau effect, and this, combined with the apparent rate of expansion, made it possible to compute the distance of the Crab Nebula as 3,500 light-years from us. From the expansion rate it was also determined that the gas had started its expansion from a central explosion point nearly 900 years ago, which agrees well with the date 1054. So there can be little doubt that the Crab Nebula, which now spreads over a volume of space some 5 light-years in diameter, represents the remnants of the 1054 supernova.

No similar region of turbulent gas has been observed at the reported sites of the supernovae of Tycho and Kepler, although small spots of nebulosity have been observed close to each site. There are some 150 planetary nebulae, however, in which doughnut-shaped rings of gas may represent large stellar explosions. A particularly extended and thin gas cloud, the Veil Nebula in Cygnus, may be what is left of a supernova explosion 30,000 years ago. When it took place it must have been even closer and brighter than the supernova of 1054 – but no civilization existed on earth to record the spectacle.

There are even suggestions that a very faint nebulosity enveloping the constellation Orion may be what is left of a still older supernova.

In all these cases, though, what happened to the star that exploded? The difficulty, or impossibility, of locating it points to exceeding dimness and that, in turn, suggests a white dwarf. If so, are all white dwarfs the remnants of stars that have exploded? In that case, why do some white dwarfs, such as Sirius B, lack an enclosing envelope of gas? Will our own sun some day explode and become a white dwarf? These queries lead us into the problem of the evolution of stars.

Of the stars near us, the bright ones seem to be hot and the dim ones cooler, according to a fairly regular brightness–temperature scale. If the surface temperatures of various stars are plotted against their absolute magnitudes, most of the familiar stars fall within a narrow band, increasing steadily from dim coolness to bright hotness. This band is called the 'main sequence'. It was first plotted in 1913 by the American astronomer Henry Norris Russell, following work along similar lines by Hertzsprung (the astronomer who first determined the absolute magnitudes of the Cepheids). A graph showing the main sequence is therefore called a Hertzsprung–Russell diagram, or H–R diagram.

Not all stars belong in the main sequence. There are some red stars that, despite their rather low temperature, have large absolute magnitudes, because their substance is spread out in rarefied fashion into tremendous size. Among these 'red giants', the best known are Betelgeuse and Antares. They are so cool (it was discovered in 1964) that many have atmospheres rich in water vapour, which would decompose to hydrogen and oxygen at the higher temperatures of our own sun. The high-temperature white dwarfs also fall outside the main sequence.

In 1924, Eddington pointed out that the interior of any star must be very hot. Because of a star's great mass, its gravitational force is immense. If the star is not to collapse, this huge force must be

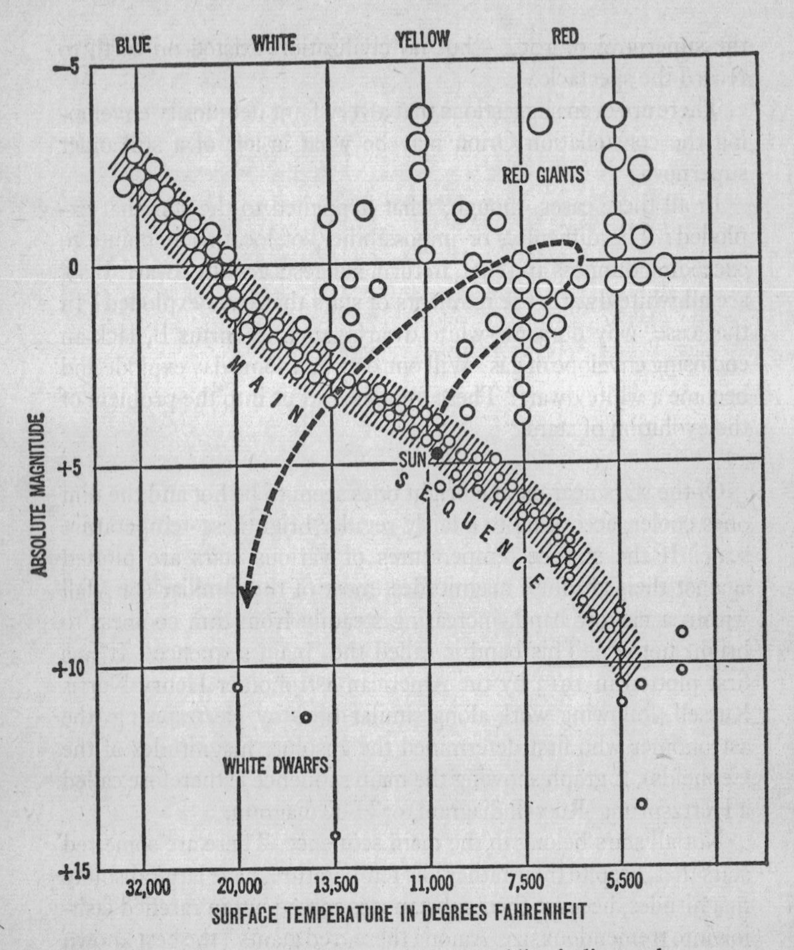

The Hertzsprung–Russell diagram. The dotted line indicates the evolution of a star. The relative size of the stars are given only schematically, not according to scale.

balanced by an equal internal pressure – from radiation energy. The more massive the star, the higher the central temperature required to balance the gravitational force. To maintain this high temperature and radiation pressure, the more massive stars must be burning energy faster, and they must be brighter than less massive ones. This is the 'mass–luminosity law'. The relationship is a drastic one, for luminosity varies as the sixth or seventh power of the mass. If the mass is increased by 3 times, then the luminosity increases by a factor of six or seven 3s multiplied together, say 750-fold.

It follows that the massive stars are spendthrift with their hydrogen fuel and have a shorter life. Our sun has enough hydrogen to last it at its present radiation rate for many thousands of millions of years. A bright star such as Capella must burn out in about 20 million years, and some of the brightest stars – for example, Rigel – cannot possibly last more than one or two million years. This means that the very brightest stars must be very youthful. New stars are perhaps even now being formed in regions where space is dusty enough to supply the raw material.

Indeed, the American astronomer George Herbig detected two stars in the dust of the Orion Nebula, in 1955, that were not visible in photographs of the region taken some years back. They may represent stars that were actually born as we watched.

By 1965, hundreds of stars were located that were so cool, they didn't quite shine. They were detected by their infra-red radiation and are therefore called 'infra-red giants' because they are made up of large quantities of rarefied matter. Presumably, these are quantities of dust and gas, gathering together and gradually growing hotter. Eventually, they will become hot enough to shine, and whether they will join the main sequence at some point then depends on the total mass of the gathered-together matter.

The next advance in the study of the evolution of stars came from analysis of the stars in globular clusters. The stars in a cluster are all about the same distance from us, so their apparent magnitude is proportional to their absolute magnitude (as in the case of the Cepheids in the Magellanic Clouds). Therefore, with their magnitude known, an H–R diagram of these stars can be prepared. It is found that the cooler stars (burning their hydrogen slowly) are on the main sequence, but the hotter ones tend to depart from it. In accordance with their high rate of burning, and with their rapid ageing, they follow a definite line showing various stages of evolution, first towards the red giants and then back, across the main sequence again, and down towards the white dwarfs.

From this and from certain theoretical considerations as to the manner in which sub-atomic particles can combine at certain high temperatures and pressures, Fred Hoyle has drawn a detailed picture of the course of a star's evolution. According to Hoyle, in its early stages a star changes very little in size or temperature. (This is the position our sun is in now and will continue to be in for a long time.) As it converts its hydrogen in the extremely hot interior into helium, the helium accumulates at the centre of the star. When this helium core reaches a certain size, the star starts to change its size and temperature dramatically. It becomes cooler and expands enormously. In other words, it leaves the main sequence and moves in the red-giant direction. The more massive the star, the more quickly it reaches this point. In the globular clusters, the more massive ones have already progressed varying lengths along the road.

The expanded giant releases more heat, despite its lower temperature, because of its larger surface area. In the far distant future, when the sun leaves the main sequence, or even somewhat before, it will have heated to the point where life will be impossible on the earth. That point, however, is still thousands of millions of years in the future.

Until recently the hydrogen-to-helium conversion was the only source of energy recognized in stars, and that raised a problem with respect to the red giants. By the time a star has reached the red-giant stage, most of its hydrogen is gone. How, then, can it go on radiating energy in such large quantities? Hoyle suggested that the helium core itself contracts, and as a result it rises to a temperature at which the helium nuclei can fuse to form carbon, with the liberation of additional energy. In 1959, the American physicist David Elmer Alburger showed in the laboratory that this reaction actually can take place. It is a very rare and unlikely sort of reaction, but there are so many helium atoms in a red giant that enough such fusions can occur to supply the necessary quantities of energy.

Hoyle goes further. The new carbon core heats up still more, and still more complicated atoms, such as those of oxygen and neon, begin to form. While this is happening, the star is contracting and getting hotter again; it moves back towards the main sequence. By now the star has begun to acquire a series of layers, like an onion. It has an oxygen–neon core, then a layer of carbon, then one of helium, the whole is enveloped in a skin of still-unconverted hydrogen.

As the temperature at the centre continues to increase, more and more complex types of reaction can go on. The neon in the new core can combine further to magnesium, which can combine in turn to form silicon, and then, in turn, iron. At a late stage in its life, the star may be built up of more than half a dozen concentric shells, in each of which a different fuel is being consumed. The central temperature may have reached 3,000 to 4,000 million degrees by then.

However, in comparison with its long life as a hydrogen consumer, the star is on a quick toboggan slide through the remaining fuels. Its life off the main sequence is a merry one, but short. Once the star begins to form iron, it has reached a dead end, for iron atoms represent the point of maximum stability and minimum energy content. To alter iron atoms in the direction of more complex atoms, or of less complex atoms, requires an input of energy.

Furthermore, as central temperatures rise with age, radiation pressure rises, too, and in proportion to the fourth power of the temperature. When the temperature doubles, the radiation pressure increases sixteen-fold, and the balance between it and gravitation becomes even more delicate. A temporary imbalance will have more and more drastic results, and, if radiation pressure shoots up a little too quickly, the explosion of a nova can result. The loss of some mass probably relieves the situation, at least temporarily, and the star may then continue to age without further catastrophe for another million years or so.

It may be, though, that the balance is maintained and the star does not relieve the situation by a minor explosion. In that case, the central temperatures may rise so high, according to Hoyle's suggestion, that the iron atoms are driven apart into helium. But for this to happen, as I have just said, energy must be poured into the atoms. The only place the star can get this energy from is its gravitational field. When the star shrinks, the energy it gains can be used to convert iron to helium. The amount of energy needed is so great, however, that the star must shrink drastically to a tiny fraction of its former volume, and this must happen, according to Hoyle, 'in about a second'.

In the blink of an eye, then, the ordinary star is gone, and a white dwarf takes its place. That is the fate the far, far future holds in store for our sun, and stars currently brighter than the sun will reach that stage sooner, perhaps within 5,000 million years.

All this purports to explain the formation of a white dwarf without an explosion. It may be the story of dwarfs such as Sirius B and Procyon B. But where do supernovae come in?

The Indian astronomer Subrahmanyan Chandrasekhar, working at Yerkes Observatory, calculated that no star more than 1·4 times the mass of our sun (now called 'Chandrasekhar's limit') could become a white dwarf by the 'normal' process Hoyle described. And in fact all the white dwarfs so far observed turn out to be below Chandrasekhar's limit in mass. It also turns out, though, that the Crab Nebula, which is accepted as the remnant of a supernova explosion, and which, it seems certain, has a white dwarf at its centre, possesses more than 1·4 times the mass of our sun, if we count the mass of the ejected gas.

So we now have to explain how the original over-the-limit star could have become a white dwarf. The reason for Chandrasekhar's limit is that, the more massive the star, the more it has to shrink (i.e., the denser it has to become) to provide the energy necessary to reconvert its iron to helium, and there is a limit to the possible shrinkage, so to speak. However, a very massive star can get around that limit. When such a star starts to collapse, its iron core is still surrounded with a voluminous outer mantle of atoms not yet built up to a maximum stability. As the outer regions collapse and their temperature rises, these still combinable substances 'take fire' all at once. The result is an explosion which blasts the outer material away from the body of the star. The white dwarf left at the conclusion of such an explosion is then below Chandrasekhar's limit, although the original star was above it.

This may be the explanation not only of the Crab Nebula but also of all supernovae. Our sun, by the way, being below Chandrasekhar's limit, may become a white dwarf some day but apparently will never become a supernova.

Hoyle suggests that the matter blasted into space by a supernova may spread through the galaxies and serve as raw material for the formation of new, 'second-generation' stars, rich in iron and other metallic elements. Our own sun is probably a second-generation star, much younger than the old stars of some of the dust-free globular clusters. Those 'first-generation' stars are low in metals and rich in hydrogen. The earth, formed out of the same debris of which the sun was born, is extraordinarily rich in iron – iron which once may have existed at the centre of a star that exploded many thousands of millions of years ago.

As for white dwarfs, though they are dying, it seems that their death will be indefinitely prolonged. Their only source of energy is their gravitational contraction, but this force is so immense that it can supply the charily radiating white dwarfs with enough energy to last tens of thousands of millions of years before they dim out altogether and become 'black dwarfs'.

Or, perhaps, as we shall see later in the chapter, it may be that even the white dwarf is not the most extreme case of stellar evolution, but that stars may exist even more shrunken – where the subatomic particles making them up approach until they are in virtual contact and the mass of an entire star is compressed into a globe perhaps no more than ten miles across.

The detection of such extremes had to await new methods of probing the universe, taking advantage of radiations other than those of visible light.

The Windows to the Universe

Man's greatest weapons in his conquest of knowledge are his understanding mind and the inexorable curiosity that drives it on. And his resourceful mind has continually invented new instruments which have opened up horizons beyond the reach of his unaided sense organs.

The best-known example is the vast surge of new knowledge that followed the invention of the telescope in 1609. The telescope, essentially, is simply an oversized eye. In contrast to the quarter-inch pupil of the human eye, the 200-inch telescope on Palomar Mountain has more than 31,000 square inches of light-gathering area. Its light-collecting power intensifies the brightness of a star

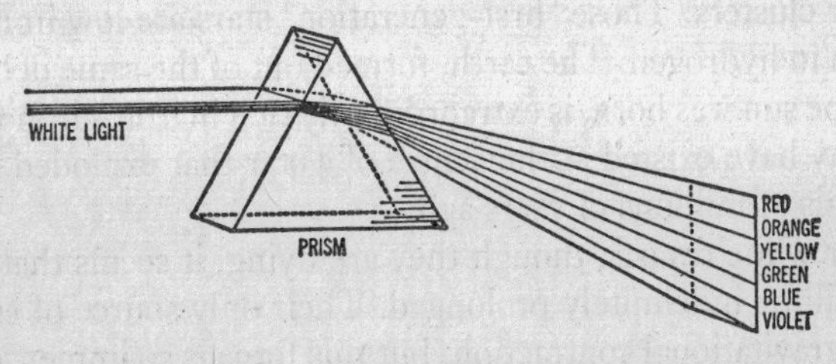

Newton's experiment splitting the spectrum of white light.

about a million times, as compared with what the naked eye can see. This telescope, first put into use in 1948, is the largest in use today, but the Soviet Union, whose largest telescope now is a 102-incher, is constructing two more of that size and a 236-inch telescope which will be the largest in the world when it is done. Meanwhile, during the 1950s Merle A. Ture developed an image tube which electronically magnified the faint light gathered by a telescope, tripling its power. Nevertheless, the law of diminishing returns is setting in. To make still bigger telescopes will be useless, for the light absorption and temperature variations of the earth's atmosphere are what now limits the ability to see fine detail. If bigger telescopes are to be built, it will have to be for use in an airless observatory, perhaps an observatory on the moon.

But mere magnification and light-intensification are not the full measure of the telescope's gift to man. The first step towards making it something more than a mere light collector came in 1666 when Isaac Newton discovered that light could be separated into what he called a 'spectrum' of colours. He passed a beam of sunlight through a triangular-shaped prism of glass and found that the beam spread out into a band made up of red, orange, yellow, green, blue, and violet light, each colour fading gently into the next. (The phenomenon itself, of course, has always been familiar in the form of the rainbow, the result of sunlight passing through water droplets, which act like tiny prisms.)

What Newton showed was that sunlight, or 'white light', is a mixture of many specific radiations (that we now recognize as wave forms of varying wavelengths) which impress the eye as so many different colours. A prism separates the colours because, on passing from air into glass, and from glass into air, light is bent, or 'refracted', and each wavelength undergoes a different amount of refraction – the shorter the wavelength, the greater the refraction. The short wavelengths of violet light are refracted most; the long wavelengths of red, least.

Among other things, this explains an important flaw in the very earliest telescopes, which was that objects viewed through them were surrounded by obscuring rings of colour, because the lenses through which light passed dispersed that light into spectra.

Newton despaired of correcting this as long as lenses of any sort were used. He therefore designed and built a 'reflecting telescope' in which a parabolic mirror, rather than a lens, was used to magnify an image. Light of all wavelengths was reflected alike so that no spectra were formed on reflection and rings of colour ('chromatic aberration') were not to be found.

In 1757, the English optician John Dollond prepared lenses of two different kinds of glass; one kind cancelling out the spectrum-forming tendency of the other. In this way, 'achromatic' ('no colour') lenses could be built. Using such lenses, 'refracting telescopes' became popular again. The largest such telescope, with a 40-inch lens, is at Yerkes Observatory near Williams Bay, Wisconsin, and was built in 1897. No larger refracting telescopes have been built since or are likely to be built, for still larger lenses would absorb so much light as to cancel their superior magnifying powers. The giant telescopes of today are all of the reflecting variety, in consequence, since the reflecting surface of a mirror absorbs very little light.

In 1814, a German optician, Joseph von Fraunhofer, went beyond Newton. He passed a beam of sunlight through a narrow slit before allowing it to be refracted by a prism. The spectrum that resulted was actually a series of images of the slit in light of every possible wavelength. There were so many slit images that they melted together to form the spectrum. Fraunhofer's prisms were so excellently made and produced such sharp slit images that it was possible to see that some of the slit images were missing. If particular wavelengths of light were missing in sunlight, no slit image would be formed at that wavelength and the sun's spectrum would be crossed by dark lines.

Fraunhofer mapped the location of the dark lines he detected, recording over 700. They have been known as 'Fraunhofer lines' ever since. In 1842, the lines of the solar spectrum were first photographed by the French physicist Alexandre Edmond Becquerel. Such photography greatly facilitated spectral studies, and, with the use of modern instruments, more than 30,000 dark lines have been detected in the solar spectrum and their wavelengths measured.

In the 1850s, a number of scientists toyed with the notion that the lines were characteristic of the various elements present in the sun. The dark lines would represent absorption of light at the wavelengths in question by certain elements; bright lines would represent characteristic emissions of light by elements. About 1859, the German chemists Robert Wilhelm Bunsen and Gustav Robert Kirchhoff worked out a system for identifying elements in this way. They heated various substances to incandescence, spread out their glow into spectra, measured the location of the lines (in this case, bright lines of emission, against a dark background) on a background scale, and matched up each line with a particular element. Their 'spectroscope' was quickly applied to discovering new elements by means of new spectral lines not identifiable with known elements. Within a couple of years Bunsen and Kirchhoff discovered cesium and rubidium in this manner.

The spectroscope was also applied to the light of the sun and the stars and soon turned up an amazing quantity of new information, chemical and otherwise. In 1862, the Swedish atronomer Anders

Jonas Ångström identified hydrogen in the sun by the presence of spectral lines characteristic of that element.

Hydrogen could also be detected in the stars, although, by and large, the spectra of the stars varied among themselves because of differences in their chemical constitution (and other properties, too). In fact, stars could be classified according to the general nature of their spectral line pattern. Such a classification was first worked out by the Italian astronomer Pietro Angelo Secchi in the mid nineteenth century, on the basis of a few scattered spectra. By the 1890s, the American astronomer Edward Charles Pickering was studying stellar spectra by the tens of thousands, and the spectral classification could be made finer.

Originally, the classification was by capital letters in alphabetical order, but as more and more was learned about the stars, it became necessary to alter that order to put the spectral classes into a logical arrangement. If the letters are arranged in order of stars of decreasing temperature, we have O, B, A, F, G, K, M, R, N, and S. Each classification can be further subdivided by numbers from 1 to 10. The sun is a star of intermediate temperature with a spectral class of G-0, while Alpha Centauri is G-2. The somewhat hotter Procyon is F-5, while the considerably hotter Sirius is A-0.

Just as the spectroscope could locate new elements on earth, so it could locate them in the heavens. In 1868, the French astronomer Pierre Jules César Janssen was observing a total eclipse of the sun in India, and reported sighting a spectral line he could not identify with any produced by any known element. The English astronomer Sir Norman Lockyer, sure that the line represented a new element, named it 'helium', from the Greek word for 'sun'. Not until nearly thirty years later was helium found on the earth.

The spectroscope eventually became a tool for measuring the radial velocity of stars, as we saw earlier in this chapter, and for exploring many other matters – the magnetic characteristics of a star, its temperature, whether the star is single or double, and so on.

Moreover, the spectral lines were a veritable encyclopedia of information about atomic structure, which, however, could not properly be utilized until after the 1890s, when the sub-atomic particles within the atom were first discovered. For instance, in

1885, the German physicist Johann Jakob Balmer showed that hydrogen produced a whole series of lines that were regularly spaced according to a rather simple formula. This was used, a generation later, to deduce an important picture of the structure of the hydrogen atom (see p. 402).

Lockyer himself showed that the spectral lines produced by a given element altered at high temperatures. This indicated some change in the atoms. Again, this was not appreciated until it was later found that an atom consisted of smaller particles, some of which were driven off at high temperatures, altering the atomic structure and the nature of the lines the atom produced. (Such altered lines were sometimes mistaken for indications of new elements but, alas, helium remained the only new element ever discovered in the heavens.)

When, in 1830, the French artist, Louis Jacques Mandé Daguerre produced the first 'daguerreotypes' and thus introduced photography, this, too, soon became an invaluable instrument for astronomy. Through the 1840s, various American astronomers photographed the moon, and one picture, by the American astronomer George Phillips Bond, was a sensation at the Great Exhibition of 1851 in London. They also photographed the sun. In 1860, Secchi made the first photograph of a total eclipse of the sun. By 1870, photographs of such eclipses had proved that the corona and prominences were part of the sun and not of the moon.

Meanwhile, beginning in the 1850s, astronomers were also making pictures of the distant stars. By 1887, the Scottish astronomer David Gill was making stellar photography routine. Photography was well on its way to becoming more important than the human eye in observing the universe.

The technique of photography with telescopes steadily improved. A major stumbling block was the fact that a large telescope can cover only a very small field. If an attempt is made to enlarge the field, distortion creeps in at the edges. In 1930, the Russian-German optician Bernard Schmidt designed a method for introducing a correcting lens that would prevent such distortion. With such a lens, a wide swatch of sky can be photographed at one swoop and studied for interesting objects that can then be studied intensely by an ordinary telescope. Since such telescopes are almost invariably used for photographic work, they are called 'Schmidt cameras'.

The largest Schmidt cameras now in use are a 53-inch instrument, first put to use in 1960 in Tautenberg, East Germany, and a 48-inch instrument used in conjunction with the 200-inch Hale telescope on Mount Palomar. The third largest is a 39-inch instrument put into use at an observatory in Soviet Armenia in 1961.

About 1800, William Herschel (the astronomer who first guessed the shape of our Galaxy) performed a very simple but interesting experiment. In a beam of sunlight transmitted through a prism, he held a thermometer beyond the red end of the spectrum. The mercury climbed! Plainly some form of invisible radiation existed at wavelengths below the visible spectrum. The radiation Herschel had discovered became known as 'infra-red' – below the red – and, as we now know, fully 60 per cent of the sun's radiation is in the infra-red.

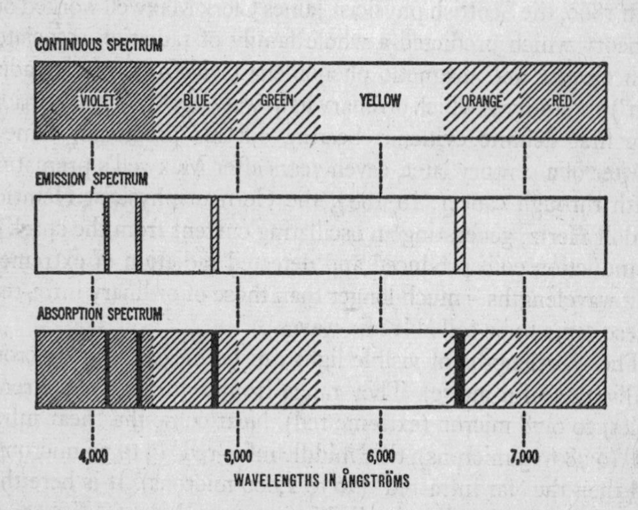

The visible spectrum, indicating emission and absorption lines.

At about the same time the German physicist Johann Wilhelm Ritter was exploring the other end of the spectrum. He found that silver nitrate, which breaks down to metallic silver and darkens when it is exposed to blue or violet light, would break down even more rapidly if it were placed beyond the point in the spectrum where violet faded out. Thus Ritter discovered the 'light' now called 'ultra-violet' (beyond the violet). Between them, Herschel and Ritter had widened the time-honoured spectrum and crossed into new realms of radiation.

These new realms bear promise of yielding much information. The ultra-violet portion of the solar spectrum, invisible to the eye, shows up in nice detail by way of photography. In fact, if a quartz prism is used (quartz transmits ultra-violet light, whereas ordinary glass absorbs most of it), quite a complicated ultra-violet spectrum can be recorded, as was first demonstrated in 1852 by the British physicist George Gabriel Stokes. Unfortunately, the atmosphere transmits only the 'near ultra-violet' – that part with wavelength almost as long as violet light. The 'far ultra-violet', with its particularly short wavelengths, is absorbed in the upper atmosphere.

In 1860, the Scottish physicist James Clerk Maxwell worked out a theory which predicted a whole family of radiation associated with electric and magnetic phenomena ('electromagnetic radiation') – a family of which ordinary light was only one small portion. The first definite evidence bearing out his prediction came a quarter of a century later, seven years after Maxwell's premature death through cancer. In 1887, the German physicist Heinrich Rudolf Hertz, generating an oscillating current from the spark of an induction coil, produced and detected radiation of extremely long wavelengths – much longer than those of ordinary infra-red. These came to be called 'radio waves'.

The wavelengths of visible light can be measured in microns (millionths of a metre). They range from 0·39 micron (extreme violet) to 0·78 micron (extreme red). Next come the 'near infra-red' (0·78 to 3 microns), the 'middle infra-red' (3 to 30 microns), and then the 'far infra-red' (30 to 1,000 microns). It is here that radio waves begin: the so-called 'microwaves' run from 1,000 to

160,000 microns and long-wave radio goes as high as many thousand million microns.

Radiation can be characterized not only by wavelength, but also by 'frequency', the number of waves of radiation produced in each second. This value is so high for visible light and the infra-red that it is not commonly used in these cases. For the radio waves, however, frequency reaches down into lower figures and comes into its own. One thousand waves per second is a 'kilocycle', while a million waves per second is a 'megacycle'. The microwave region runs from 300,000 megacycles down to 1,000 megacycles. The much longer radio waves used in ordinary radio stations are down in the kilocycle range.

Within a decade after Hertz's discovery, the other end of the spectrum opened up similarly. In 1895, the German physicist Wilhelm Konrad Roentgen accidentally discovered a mysterious radiation which he called 'X-rays'. Their wavelengths turned out to be shorter than ultra-violet. Later, 'gamma rays', associated with radioactivity, were shown by Rutherford to have wavelengths even smaller than those of X-rays.

The short-wave half of the spectrum is now divided roughly as follows: The wavelengths from 0·39 down to 0·17 micron belong to the 'near ultra-violet', from 0·17 down to 0·01 micron to the 'far ultra-violet', from 0·01 to 0·00001 micron to X-rays, and gamma rays range from this down to less than a thousand millionth of a micron.

Newton's original spectrum was thus expanded enormously. If we consider each doubling of wavelength as equivalent to one octave (as is the case in sound), the electromagnetic spectrum over the full range studied amounts to almost sixty octaves. Visible light occupies just one octave near the centre of the spectrum.

With a wider spectrum, of course, we can get a fuller view of the stars. We know, for instance, that sunshine is rich in ultra-violet and in infra-red. Our atmosphere cuts off most of these radiations; but in 1931, quite by accident, a radio window to the universe was discovered.

Karl Jansky a young radio engineer at the Bell Telephone Laboratories, was studying the static that always accompanies

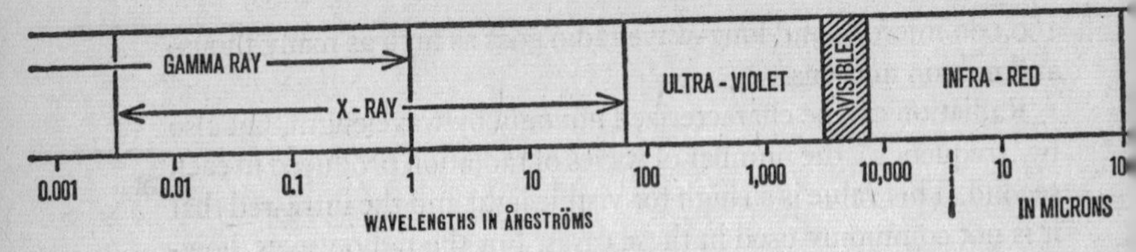

The spectrum of electromagnetic radiation.

radio reception. He came across a very faint, very steady noise which could not be coming from any of the usual sources. He finally decided that the static was caused by radio waves from outer space.

At first the radio signals from space seemed strongest in the direction of the sun, but day by day the direction of strongest reception slowly drifted away from the sun and made a circuit of the sky. By 1933, Jansky decided the radio waves were coming from the Milky Way and, in particular, from the direction of Sagittarius, towards the centre of the Galaxy.

Thus was born 'radio astronomy'. Astronomers did not take to it immediately, for it had serious drawbacks. It gave no neat pictures – only wiggles on a chart which were not easy to interpret. More important, radio waves are much too long to resolve a source as small as a star. The radio signals from space had wavelengths hundreds of thousands and even millions of times the wavelength of light, and no ordinary radio receiver could give anything more than a general idea of the direction they were coming from.

These difficulties obscured the importance of the new discovery, but a young radio ham named Grote Reber carried on, for no reason other than personal curiosity. Through 1937 he spent time and money building in his backyard a small 'radio telescope' with a parabolic 'dish' about 30 feet in diameter to receive and concentrate the radio waves. Beginning in 1938, he found a number of sources of radio waves other than the one in Sagittarius – one in the constellation Cygnus, for instance, and another in Cassiopeia. (Such sources of radiation were at first called 'radio stars', whether the sources were actually stars or not, but are now usually called 'radio sources'.)

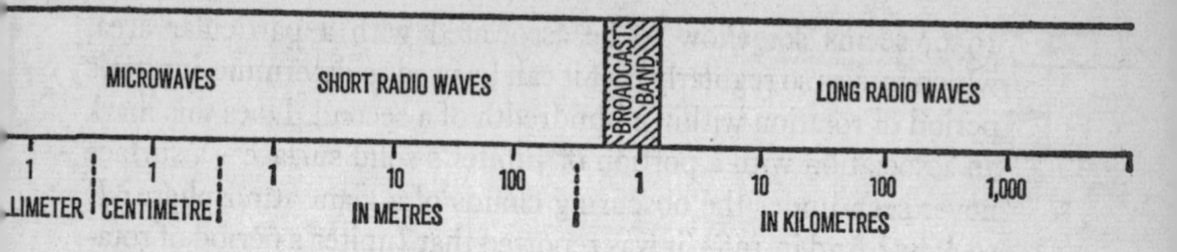

During the Second World War, while British scientists were developing radar, they discovered that the sun was interfering by sending out signals in the microwave region. This aroused their interest in radio astronomy, and after the war the British pursued their tuning-in on the sun. In 1950, they found that much of the sun's radio signals were associated with sunspots. (Jansky had conducted his experiments during a period of minimal sunspot activity, which is why he detected the galactic radiation rather than that of the sun.)

The British pioneered in building large antennae and arrays of widely separated receivers (a technique first used in Australia) to sharpen reception and pinpoint radio stars. Their 250-foot dish at Jodrell Bank in England, built under the supervision of Sir Bernard Lovell, was the first really large radio telescope.

In 1947 the Australian astronomer John C. Bolton narrowed down the third strongest radio source in the sky, and it proved to be none other than the Crab Nebula. Of the 2,000 or so radio sources detected here and there in the sky, this was the first to be pinned down to an actual visible object. It seemed unlikely that a white dwarf was giving rise to the radiation, since other white dwarfs did not. The source was much more likely to be the cloud of expanding gas in the nebula.

This strengthened other evidence that cosmic radio signals arise primarily from turbulent gas. The turbulent gas of the outer atmosphere of the sun gives rise to radio waves, so that what is called the 'radio sun' is much larger than the visible sun. Then, too, Jupiter, Saturn, and Venus, each with a turbulent atmosphere, have been found to be radio emitters. In the case of Jupiter, however, the radiation first detected in 1955, in records going back to

1950, seems somehow to be associated with a particular area, which moves so regularly that it can be used to determine Jupiter's period of rotation within a hundredth of a second. Does this mark an association with a portion of Jupiter's solid surface – a surface never seen under the obscuring clouds of a giant atmosphere? If so, why? And in 1964, it was reported that Jupiter's period of rotation had altered quite suddenly, though, to be sure, only very slightly. Again, why? It was also shown, in 1965, that Jupiter is most likely to emit a strong burst of radio waves when its satellite Io is in the first or third quarter (to the left or right of Jupiter as seen from Earth). Still again, why? So far, radio studies have raised more questions about the planets than they have answered, but there is nothing so stimulating to science and scientists as a good, unanswered question.

Jansky, who started it all, was largely unappreciated in his lifetime and died in 1950 at the age of forty-four, just as radio astronomy was hitting its stride. He received posthumous recognition in that the strength of radio emission is now measured in 'janskies'.

Radio astronomy probed far out into space. Within our Galaxy, there is a strong radio source (the strongest outside the solar system) which is called 'Cass' because it is located in Cassiopeia. Walter Baade and Rudolph Minkowski at Palomar trained the 200-inch telescope on the spot where this source was pinpointed by British radio telescopes, and they found streaks of turbulent gas. It is possible that these may be remnants of the supernova of 1604, which Kepler observed in Cassiopeia.

A still more distant discovery was made in 1951. The second strongest radio source lies in the constellation Cygnus. Reber first reported it in 1944. As radio telescopes later narrowed down its location, it began to appear that this radio source was outside our Galaxy – the first to be pinpointed beyond the Milky Way. Then, in 1951, Baade, studying the indicated portion of the sky with the 200-inch telescope, found an odd galaxy in the centre of the field. It had a double centre and seemed to be distorted. Baade at once suspected that this odd, distorted, double-centred galaxy was not one galaxy but two, joined broadside-to like a pair of clashing cymbals. Baade thought they were two colliding galaxies – a possibility which he had already discussed with other astronomers.

It took another year to settle the matter. The spectroscope showed absorption lines which could only be explained by supposing the dust and gas of the two galaxies to be coming into collision. The collision is now accepted to be a fact. Moreover, it seems likely that galactic collisions are fairly common, especially in dense clusters, where galaxies may be separated by distances not much more than their own diameters.

When two galaxies collide, the stars themselves are not likely to encounter one another: they are so widely spaced that one galaxy could pass through the other without any stars coming even close. But dust and gas clouds are stirred into vast turbulence and thereby generate very powerful radio radiation. The colliding galaxies in Cygnus are 260 million light-years away, yet their radio signals reaching us are stronger than those of the Crab Nebula, only 3,500 light-years away. By this token, we should be able to detect colliding galaxies at far greater distances than we can see with the optical telescope. The 250-foot Jodrell Bank radio telescope, for instance, should outrange the 200-inch Hale telescope.

And yet as the number of radio sources found among the distant galaxies increased and passed the hundred mark, astronomers grew uneasy. Surely they could not all be brought about by colliding galaxies. That would be overdoing a good thing.

In fact, the whole notion of galactic collisions in the sky grew shaky. The Soviet astrophysicist Victor Amazaspovich Ambartsumian advanced theoretical reasons in 1955 for supposing that radio galaxies were exploding rather than colliding. By the early 1960s, Fred Hoyle was backing this view and suggesting that radio galaxies might be subjected to whole series of supernovae. In the crowded centre of a galactic nucleus, a supernova may explode and may heat a nearby star to just the point where it, too, lets go in a supernova explosion. The second explosion sets off a third and that sets off a fourth, and so on, domino fashion. In a sense, the whole centre of a galaxy is exploding.

The possibility that this may be so has been greatly strengthened by the discovery, in 1963, that the galaxy M-82, in the constellation of Ursa Major (a strong radio source about 10 million light-years away), is such an 'exploding galaxy'.

Investigation of M-82 with the 200-inch Hale telescope, making use of the light of a particular wavelength, showed great jets of matter up to a thousand light-years long emerging from the galactic centre. From the amount of matter exploding outwards, the distance it had travelled, and its rate of travel, it seems likely that 1·5 million years ago the light of some 5 million stars exploding in the nucleus, almost simultaneously, first reached us.

The New Objects

By the time astronomers had entered the 1960s, it might have been easy for them to suppose that there were few surprises left among the physical objects in the heavens. New theories, new insights, yes; but surely little in the way of startling new varieties of stars, galaxies, or anything else could remain after three centuries of observation with steadily more sophisticated instruments.

If any astronomers thought this, they were due for an enormous shock – the first coming as a result of the investigation of certain radio sources that looked unusual but not surprising.

The radio sources first studied in deep space seemed to exist in connection with extended bodies of turbulent gas: the Crab Nebula, distant galaxies, and so on. There did exist a few radio sources, however, that seemed unusually small. As radio telescopes grew more refined and as the view of the radio sources was sharpened, it began to seem possible that radio waves were being emitted by individual stars.

Among these compact radio sources were several known as 3C48, 3C147, 3C196, 3C273, and 3C286. The '3C' is short for *Third Cambridge Catalogue of Radio Stars*, a listing compiled by the English astronomer Martin Ryle and his co-workers, while the remaining numbers represent the placing of the source on that list.

In 1960, the areas containing these compact radio sources were combed by Sandage with the 200-inch telescope, and in each case a star did indeed seem to be the source. The first star to be detected was that associated with 3C48. In the case of 3C273, the brightest of the objects, the precise position was obtained by Cyril Hazard, in Australia, who recorded the moment of radio blackout as the moon passed before it.

The stars involved had been recorded on previous photographic sweeps of the sky and had always been taken to be nothing more than faint members of our own galaxy. Painstaking photography, spurred by their unusual radio-emission, now showed, however, that that was *not* all there was to it. Faint nebulosities proved to be associated with some of the objects, and 3C273 showed signs of a tiny jet of matter emerging from it. In fact, there were two radio sources in connection with 3C273: one from the star and one from the jet. Another point of interest that arose after close inspection was that these stars were unusually rich in ultra-violet light.

It would seem then that the compact radio sources, although they looked like stars, might not be ordinary stars after all. They eventually came to be called 'quasi-stellar sources' ('quasi-stellar' means 'star-resembling'). As the term became more and more important to astronomers, quasi-stellar radio sources became too inconvenient a mouthful and in 1964 it was shortened by the Chinese-American physicist Hong Yee Chiu to 'quasar' ('*quas*i-stell*ar*'), an uneuphonious word that is now firmly embedded in astronomic terminology.

Clearly, the quasars were interesting enough to warrant investigation with the full battery of astronomic techniques, and that meant spectroscopy. Such astronomers as Allen Sandage, Jesse L. Greenstein, and Maarten Schmidt laboured to obtain the spectra. When they accomplished the task in 1960, they found themselves with strange lines they could not identify. Furthermore, the lines in the spectra of one quasar did not match those in any other.

In 1963, Schmidt returned to the spectrum of 3C273, which, as the brightest of these puzzling objects, showed the clearest spectrum. Six lines were present, of which four were spaced in such a way as to seem to resemble a series of hydrogen lines – except that no such series ought to exist in the place in which they were found. What, though, if those lines were located elsewhere but were found where they were because they had been displaced towards the red end of the spectrum? If so, it was a large displacement, one that indicated a recession at the velocity of over 25,000 miles per second. This seemed unbelievable, and yet, if such a displacement existed, the other two lines could also be identified: one represented oxygen minus two electrons, the other magnesium minus two electrons.

Schmidt and Greenstein turned to the other quasar spectra and found that the lines there could also be identified, provided huge red shifts were assumed.

Such enormous red shifts could be brought about by the general expansion of the universe; but if the red shift were equated with distance in accordance with Hubble's law, it turned out that the quasars could not be ordinary stars of our own galaxy at all. They had to be among the most distant objects known – thousands of millions of light-years away.

By the end of the 1960s, a concentrated search had uncovered 150 quasars. The spectra of about 110 of them were studied. Every single one of these showed a large red shift – larger ones, indeed, than that of 3C273. The distance of a couple of them is estimated to be about 9 thousand million light-years.

If the quasars are indeed as far away as the red shift makes them seem, astronomers are faced with some puzzling and difficult points. For one thing, they must be extraordinarily luminous to appear as bright as they do at such a distance; they must be anywhere from thirty to a hundred times as luminous as an entire ordinary galaxy.

Yet if this is so, and if the quasars had the form and appearance of a galaxy, they ought to contain up to a hundred times as many stars as an ordinary galaxy and be up to five or six times as large in each dimension. Even at their enormous distance they ought to show up as distinct oval blotches of light in large telescopes. Yet they don't. They remain star-like points in even the largest telescope, which seems to indicate that, despite their unusual luminosity, they were far smaller in size than ordinary galaxies.

The smallness in size was accentuated by another phenomenon, for as early as 1963 the quasars were found to be variable in the energy they emitted, both in the visible-light region and in the radio-wave region. Increases and decreases of as much as three magnitudes were recorded over the space of a few years.

For radiation to vary so markedly in so short a time, a body must be small. Small variations might result from brightenings and dimmings in restricted regions of a body, but large variations must involve the body as a whole. If the body is involved as a whole, some effect must make itself felt across the full width of the body within the time of variation. But no effect can travel faster than light, so that if a quasar varies markedly over a period of a few years, it cannot be more than a light-year or so in diameter. Actually, some calculations indicate quasars may be as little as a light-week (500,000 million miles) in diameter.

Bodies which are at once so small and so luminous must be expending energy at a rate so great that the reserves cannot last long (unless there is some energy source as yet undreamed-of, which is not impossible, of course). Some calculations indicate that a quasar can only deliver energy at this enormous rate for a million years or so. In that case, the quasars we see only became quasars a short time ago, cosmically speaking, and there must be a number of objects that were once quasars but are quasars no longer.

Sandage, in 1965, announced the discovery of objects that may indeed be aged quasars. They seemed like ordinary bluish stars, but they possessed huge red shifts as quasars do. They were as distant, as luminous, as small quasars, but they lacked the radio-wave emission. Sandage called them 'blue stellar objects', which can be abbreviated to BSOs.

The BSOs seem to be more numerous than quasars: a 1967 estimate places the total number of BSOs within reach of our telescopes at 100,000. There are so many more BSOs than quasars because the bodies last so much longer in BSO form than in quasar form.

The chief interest in the quasars (aside from the knotty puzzle of what they actually are) lies in the fact that they are at once so unusual and so distant. Perhaps they represent a kind of body that existed only in the youth of the universe. (After all, a body that is 9,000 million light-years away is seen only by light that left it 9,000 million years ago, and that is a time that may have been only shortly after the explosion of the 'cosmic egg'. If so, then it is clear that the appearance of the universe was radically different thousands of millions of years ago. The universe, in that case, evolved, as the proponents of the 'big bang' theory maintain, and is not eternally changeless-on-the-average, as the proponents of 'continuous creation' insist.

The use of the quasars as evidence in favour of the 'big bang' did not go unchallenged. A number of astronomers advance evidence to the effect that the quasars are not really very distant and therefore cannot be taken to represent objects characteristic of the youth of the universe. Astronomers holding this view must then explain the enormous red shifts in the quasar spectra by some effect other than vast distance, and this is not easy to do. On the whole, though the question is far from settled and there are enormous difficulties involved in both views, the weight of opinion seems to be on the side of quasars as very distant objects.

Even if they are, there seems to be reason to question whether they are entirely characteristic only of the youth of the universe. Are they distributed fairly evenly, so that they are to be found in the universe at all ages?

Thus, back in 1943, the American astronomer Carl Seyfert observed an odd galaxy, one with a very bright and very small nucleus. Others of the sort have since been observed, and the entire group is now referred to as 'Seyfert galaxies'. Though only a dozen were known by the end of the 1960s, there is reason to suspect that as many as 1 per cent of all galaxies may be of the Seyfert type.

Can it be that Seyfert galaxies are objects intermediate between ordinary galaxies and quasars? Their bright centres show light variations that would make those centres almost as small as quasars. If the centres were further intensified and the rest of the galaxy further dimmed, they would become indistinguishable from a quasar and one Seyfert galaxy, 3C120, is almost quasar-like in appearance.

The Seyfert galaxies have only moderate red shifts and are not enormously distant. Can it be that the quasars are very distant Seyfert galaxies, so distant that we can only see the luminous and small centres; and so distant that we can only see the largest, so that we get the impression that quasars are extraordinarily luminous, whereas we should rightly suspect that only a few very large Seyfert galaxies make up the quasars we can see despite their distance?

But if we consider the Seyfert galaxies near us to be either small quasars or, perhaps, large quasars in the course of development, then it may be that the quasar distribution is not characteristic only of the youth of the universe and that their existence is not strong evidence, after all, of the 'big bang' theory.

As it happened, however, the 'big bang' received strong support in another direction. In 1949, Gamow had calculated that the radiation associated with the 'big bang' should have died down with the expansion of the universe to the point where it would now consist of radio-wave radiation coming equally from all parts of the sky as a kind of radio-background. He suggested that the radiation would be that to be expected of objects at a temperature of 5°K (that is, 5° above absolute zero or −268°C).

In 1965, just such a background radio-wave radiation was detected and reported by A. A. Penzias and R. W. Wilson of Bell Telephone Laboratories in New Jersey. The temperature associated with the radiation was 3°K, which was in not-too-bad agreement with Gamow's prediction. No explanation, other than that of the 'big bang', has as yet been offered for the existence of this background radiation, so that for the moment the 'big bang' theory of the evolutionary universe beginning with the explosion of a large volume of condensed matter seems to hold the field.

If radio-wave radiation had given rise to that peculiar and puzzling astronomical body, the quasar, research at the other end of the spectrum suggested another body – just as peculiar, if not quite as puzzling.

In 1958, the American astrophysicist Herbert Friedman discovered that the sun produced a considerable quantity of X-rays. These could not be detected from the earth's surface for the atmosphere absorbed them; but rockets, shooting beyond the atmosphere and carrying appropriate instruments, could detect the radiation with ease.

For a while, the source of solar X-rays was a puzzle. The temperature of the sun's surface is only 6,000°C – high enough to vaporize any form of matter, but not high enough to produce X-rays. The source had to lie in the sun's corona, a tenuous halo of gases stretching outwards from the sun in all directions for many millions of miles. Although the corona delivers fully half as much light as the full moon, it is completely masked by the light of the sun itself and is visible only during eclipses, at least under ordinary circumstances. In 1930, the French astronomer Bernard Ferdinand Lyot invented a telescope which, at high altitudes and on clear days, could observe the inner corona even in the absence of an eclipse.

The corona was felt to be the X-ray source because, even before the rocket studies of X-rays, it had been suspected of possessing unusually high temperatures. Studies of the spectrum of the corona (during eclipses) have revealed lines that could not be associated with any known element. A new element was suspected and named 'coronium'. In 1941, however, it was found that the lines of coronium could be produced by iron atoms that had had many sub-atomic particles broken away from them. To break off all those particles, however, required a temperature of something like a million degrees, and such a temperature would certainly be enough to produce X-rays.

X-ray radiation increases sharply when a solar flare erupts into the corona. The X-ray intensity at that time implies a temperature as high as 100 million degrees in the corona above the flare. The reason for such enormous temperatures in the thin gas of the corona is still a matter of controversy. (Temperature here has to be distinguished from heat. The temperature is a measure of the kinetic energy of the atoms or particles in the gas, but since the particles are few, the actual heat content per unit of volume is low. The X-rays are produced by collisions between the extremely energetic particles.)

X-rays came from beyond the solar system, too. In 1963, rocket-borne instruments were launched by Bruno Rossi and others to see if solar X-rays were reflected from the moon's surface. They detected, instead, two particularly concentrated X-ray sources elsewhere in the sky. The weaker ('Tau X-1' because it was in the constellation Taurus) was quickly associated with the Crab Nebula. In 1966, the stronger, in the constellation Scorpio ('Sco X-1'), was found to be associated with an optical object which seemed the remnant (like the Crab Nebula) of an old nova. Since then several dozen other, and weaker, X-ray sources have been detected in the sky.

To be giving off energetic X-rays with an intensity sufficient to be detected across an interstellar gap required a source of extremely high temperature and large mass. The concentration of X-rays emitted by the sun's corona would not do at all.

To be at once massive and have a temperature of a million degrees suggested a kind of 'super-white dwarf'. As long ago as 1934, Zwicky had suggested that the sub-atomic particles of a white dwarf might, under certain conditions, combine into unchanged particles called 'neutrons'. These could then be forced together until actual contact was made. The result would be a sphere no more than ten miles across which would yet retain the mass of a full-sized star. In 1939, the properties of such a 'neutron star' were worked out in some detail by the American physicist J. Robert Oppenheimer. Such an object would attain so high a surface temperature, at least in the initial stages after its formation, as to emit X-rays in profusion.

The search by Friedman for actual evidence of the existence of such 'neutron stars' centred on the Crab Nebula, where it was felt that the terrific explosion that had formed it might have left behind, not a condensed white dwarf but a super-condensed neutron star. In July 1964, the moon passed across the Crab Nebula and a rocket was sent beyond the atmosphere to record the X-ray emission. If it were coming from a neutron star, then the X-ray emission would be cut off entirely and at once as the moon passed before the tiny object. If the X-ray emission were from the Crab Nebula generally, then it would drop off gradually as the moon eclipsed the nebula bit by bit. The latter proved to be the case, and the Crab Nebula seemed to be but a larger and much more intense corona, about a light-year in diameter.

For a moment, the possibility that neutron stars might actually exist and be detectable dwindled, but in the same year that the Crab Nebula failed its test, a new discovery was made in another direction. The radio waves from certain sources seemed to indicate a very rapid fluctuation in intensity. It was as though there were 'radio twinkles' here and there.

Astronomers quickly designed instruments capable of catching very short bursts of radio-wave radiation. They felt this would make it possible to study these fast changes in greater detail. One astronomer making use of such a radio telescope was Anthony Hewish at Cambridge University Observatory.

He had hardly begun operating the telescope with its new detector when he detected bursts of radio-wave energy from a place midway between Vega and Altair. It was not difficult to detect and would have been found years earlier if astronomers had expected to find quite such short bursts and had developed the equipment to detect them. The bursts were, as it happened, astonishingly brief, lasting only 1/30 of a second. What was even more astonishing, the bursts followed one another with remarkable regularity at intervals of 1 1/3 seconds. The intervals were so regular, in fact, that the period could be worked out to a hundred-millionth of a second: it was 1·33730109 seconds.

Naturally, there was no way of telling what these pulses represented, at least not at first. Hewish could only think of it as a 'pulsating star', each pulsation giving out a burst of energy. This was shortened almost at once to 'pulsar', and it is by that name that the new object came to be known.

One should speak of the new objects in the plural, for once Hewish found the first, he searched for others. By February 1968, when he announced the discovery, he had located four. Other astronomers avidly began searching, and more were quickly discovered. In two more years, nearly forty more pulsars were located.

Two thirds of them are located very close to the galactic equator, which is a good sign that pulsars generally are part of our own

Galaxy. Some may be as close as a hundred light-years or so. (There is no reason to suppose they don't exist in other galaxies, too, but at the distance of other galaxies they are probably too faint to detect.)

All the pulsars are characterized by extreme regularity of pulsation, but the exact period varies from pulsar to pulsar. One had a period as long as 3·7 seconds. In November 1968, astronomers at Green Bank, West Virginia, detected a pulsar in the Crab Nebula that had a period of only 0·033089 seconds. It was pulsing thirty times a second.

Naturally, the question is, What can produce such short flashes with such fantastic regularity? Some astronomical body must be undergoing some very regular change at intervals rapid enough to produce the pulses. Could it be a planet that was circling a star in such a way that once each revolution it moved beyond the star (as seen from the direction of earth) and, as it emerged, emitted a powerful flash of radio waves? Or else could a planet be rotating and, each time it did so, would some particular spot on its surface, which leaked radio waves in vast quantity, sweep past our direction?

To do this, however, a planet must revolve about a star or rotate about its axis in a period of seconds or fractions of a second, and this was unthinkable. For pulses as rapid as those of pulsars, some object must be rotating or revolving at enormous velocities. That requires very small size combined with huge temperatures, or huge gravitational fields, or both.

This instantly brought white dwarfs to mind, but even white dwarfs could not revolve about each other, or rotate on their axes, or pulsate, with a period short enough to account for pulsars. White dwarfs were still too large, and their gravitational fields were still too weak.

Thomas Gold at once suggested that a neutron star was involved. He pointed out that a neutron star was small enough and dense enough to be able to rotate about its axis in four seconds or less. What's more, it had already been theorized that a neutron star would have an enormously intense magnetic field, with magnetic poles that need not be at the pole of rotation. Electrons would be held so tightly by the neutron star's gravity that they could emerge only at the magnetic poles. As they were thrown off, they would lose energy, in the form of radio waves. This would mean that there would be a steady sheaf of radio waves emerging from two opposite points on the neutron star's surface.

If, as the neutron star rotates, one or both of those sheafs of radio waves sweeps past our direction, then we will detect a short burst of radio-wave energy once or twice each revolution. If this is so, we would detect only pulsars which happen to rotate in such a way as to sweep at least one of the magnetic poles in our direction. Some astronomers estimate that only one neutron star out of a hundred would do so. They guess that there might be as many as 100,000 neutron stars in the galaxy, but that only 1,000 would be detectable from earth.

Gold went on to point out that if his theory were correct, the neutron star would be leaking energy at the magnetic poles and its rate of rotation would be slowing down. This means that the shorter the period of a pulsar, the younger it is and the more rapidly it would be losing energy and slowing down.

The most rapid pulsar known is in the Crab Nebula. It might well be the youngest, since the supernova explosion that would have left the neutron star behind took place only a thousand years ago.

The period of the Crab Nebula pulsar was studied carefully, and it was indeed found to be slowing, just as Gold had predicted. The period was increasing by 36·48 thousand millionths of a second each day. The same phenomenon was discovered in other pulsars as well, and as the 1970s opened, the neutron-star hypothesis was widely accepted.

Sometimes a pulsar will suddenly speed up its period very slightly, then resume the slowing trend. Some astronomers suspect this may be the result of a 'starquake', a shifting of mass distribution within the neutron star. Or perhaps it might be the result of some sizeable body plunging into the neutron star and adding its own momentum to that of the star.

There was, of course, no reason why the electrons emerging from the neutron star should lose energy only as microwaves. This phenomenon should produce waves all along the spectrum. It should produce visible light, too.

Keen attention was focused on the sections of the Crab Nebula where visible remnants of the old explosion might exist. Sure enough, in January 1969, it was noted that the light of a dim star within the Nebula *did* flash on and off in precise time with the microwave pulses. It would have been detected earlier if astronomers had had the slightest idea that they ought to search for such rapid alternations of light and darkness. The Crab Nebula pulsar was the first optical pulsar discovered – the first visible neutron star.

The Crab Nebula pulsar released X-rays, too. About 5 per cent of all the X-rays from the Crab Nebula emerged from that tiny flickering light. The connection between X-rays and neutron stars, which seemed extinguished in 1964, thus came triumphantly back to life.

Nor is even the neutron star the limit. When Oppenheimer worked out the properties of the neutron star in 1939, he predicted also that it was possible for a star that was massive enough and cool enough to collapse altogether to nothingness. When such collapse proceeded past the neutron-star stage, the gravitational field would become so intense that no matter, no light could escape from it. Nothing could be seen of it; it would simply be a 'black hole' in space.

Will it be possible at some time in the future to detect such black holes – surely the ultimate in strange new objects in the universe? That remains to be seen.

Are quasars large clusters of neutron stars? Are they single neutron stars of galactic mass? Are they phenomena associated with black-hole formation? That, too, remains to be seen.

But if there are objects in the universe that surprise us, there are also surprises in the vast not-so-empty spaces between the stars. The non-emptiness of 'empty space' has proved to be a matter of difficulty for astronomers in observations relatively close to home.

In a sense, the galaxy hardest for us to see is our own. For one thing, we are imprisoned within it, while the other galaxies can be viewed as a whole from outside. It is like the difference between trying to view a city from the roof of a low building and seeing it from an aeroplane. Furthermore, we are far out from the centre and, to make matters worse, we lie in a spiral arm clogged with dust. In other words, we are on a low roof on the outskirts of the city on a foggy day.

The space between stars, generally speaking, is not a perfect vacuum under the best of conditions. There is a thin gas spread generally through interstellar space within galaxies. Spectral absorption lines due to such 'interstellar gas' were first detected in 1904 by the German astronomer Johannes Franz Hartmann. That would be supportable. The trouble is, however, that in the outskirts of a galaxy the concentration of gas and dust becomes much thicker. We can see such dark fogs of dust rimming the nearer galaxies.

We can actually 'see' the dust clouds, in a negative way, within our own Galaxy as dark areas in the Milky Way. Examples are the dark Horsehead Nebula, outlined starkly against the surrounding brilliance of millions of stars, and the even more dramatically named Coal Sack in the Southern Cross, a region of scattered dust particles 30 light-years in diameter and about 400 light-years away from us.

Although the gas and dust clouds hide the spiral arms of the Galaxy from direct vision, they do not hide the structure of the arms from the spectroscope. Hydrogen atoms in the clouds are ionized (broken up into electrically charged sub-atomic particles) by the energetic radiation from the bright Population I stars in the arms. Beginning in 1951, streaks of ionized hydrogen were found by the American astronomer William Wilson Morgan, marking out the lines of the blue giants, i.e., the spiral arms. Their spectra were similar to the spectra shown by the spiral arms of the Andromeda galaxy.

The nearest such streak of ionized hydrogen includes the blue giants in the constellation of Orion, and this streak is therefore called the 'Orion Arm'. Our solar system is in that arm. Two other arms were located in the same way. One lies farther out from the galactic centre than our own and includes giant stars in the constellation Perseus (the 'Perseus Arm'). The other lies closer to the galactic centre and contains bright clouds in the constellation Sagittarius (the 'Sagittarius Arm'). Each arm seems to be about 10,000 light-years long.

Then radio came along as a still more powerful tool. Not only could it pierce through the obscuring clouds, but it made the clouds themselves tell their story – through their own voice. This came about as a result of the work of the Dutch astronomer H. C. Van de Hulst. In 1944, the Netherlands was ground under the heavy boot of the Nazi army, and astronomic observation was nearly impossible. Van de Hulst, confining himself to pen and paper work, studied the characteristics of ordinary non-ionized hydrogen atoms, of which most of the interstellar gas is composed.

He suggested that every once in a while such atoms, on colliding, might change their energy state and, in so doing, emit a weak radiation in the radio part of the spectrum. A particular hydrogen atom might do so only once in 11 million years, but among the vast numbers present in intergalactic space, enough would be radiating each moment to produce a continuously detectable emission.

Van de Hulst calculated that the wavelength of the radiation should be 21 centimetres. Sure enough, with the development of new radio techniques after the war, this 'song of hydrogen' was detected in 1951 by Edward Mills Purcell and Harold Irving Ewen at Harvard University.

By turning in on the 21-centimetre radiation of collections of hydrogen, astronomers were able to trace out the spiral arms and follow them for long distances – in most cases nearly all the way around the Galaxy. More arms were found, and maps of the concentration of hydrogen show half a dozen or more streaks.

What is more, the song of hydrogen told something about its movements. Like all waves, this radiation is subject to the Doppler–Fizeau effect. It allows astronomers to measure the velocity of the moving hydrogen clouds, and thereby to explore, among other things, the rotation of our Galaxy. This new technique confirmed that the Galaxy rotates in a period (at our distance from the centre) of 200 million years.

In science, each new discovery unlocks doors leading to new mysteries. And the greatest progress comes from the unexpected – the discovery that overthrows previous notions. An interesting example at the moment is a puzzling phenomenon brought to light by radio study of a concentration of hydrogen at the centre of our Galaxy. The hydrogen seems to be expanding, yet is confined to the equatorial plane of the Galaxy. The expansion itself is surprising, because there is no theory to account for it. And if the hydrogen is expanding, why has it not all dissipated away during the long lifetime of the Galaxy? Is it a sign perhaps that, some 10 million years ago, as Oort suspects, its centre exploded, as that of M-82 did much more recently? Then too, the plane of hydrogen is not perfectly flat. It bends downwards on one end of the Galaxy and upwards on the other. Why? No good explanation has yet been offered.

Hydrogen is not, or should not be, unique as far as radio waves are concerned. Every different atom, or combination of atoms, is capable of emitting characteristic radio-wave radiation, or of absorbing characteristic radio-wave radiation from a general background. Naturally, then, astronomers sought to find the telltale fingerprints of atoms other than the supremely common hydrogen.

Almost all the hydrogen that occurs in nature is of a particularly simple variety called 'hydrogen 1'. There is a more complex form, which is 'deuterium' or 'hydrogen 2'. The radio-wave radiations from various spots in the sky were combed for the wavelengths that theory predicted. In 1966, it was detected, and the indications are that the quantity of hydrogen 2 in the universe is about 5 per cent that of hydrogen 1.

Next to the varieties of hydrogen as common components of the universe are helium and oxygen. An oxygen atom can combine with a hydrogen atom to form a 'hydroxyl group'. This combination would not be stable on earth, for the hydroxyl group is very active and would combine with almost any other atom or molecule it encountered. It would, notably, combine with a second hydrogen atom to form a molecule of water. In interstellar space, however, where the atoms are spread so thin that collisions are few and far between, a hydroxyl group, once formed, would persist undisturbed for long periods of time. This was pointed out in 1953 by the Soviet astronomer I. S. Shklovskii.

Such a hydroxyl group would, calculations showed, emit or absorb four particular wavelengths of radio waves. In October 1963, two of them were detected by a team of radio engineers at Lincoln Laboratory of M.I.T.

Since the hydroxyl group is some seventeen times as massive as the hydrogen atom alone, it is more sluggish and moves at only one fourth the velocity of the hydrogen atom at any given temperature. In general, movement blurs the wavelengths so that the hydroxyl wavelengths are sharper than those of hydrogen. Its shifts are easier to determine, and it is easier to tell whether a gas cloud, containing hydroxyl, is approaching or receding.

Astronomers were pleased, but not entirely astonished, at finding evidence of a two-atom combination in the vast reaches between the stars. Automatically, they began to search for other combinations, but not with a great deal of hope. Atoms are spread out so thin in interstellar space that the chance of more than two atoms coming together long enough to form a combination seemed remote. The chance that atoms less common than oxygen (such as those of carbon and nitrogen, which are next most common of those that are able to form combinations) would be involved seemed out of the question.

But then, beginning in 1968, came the real surprise. In November of that year, they discovered the telltale radio-wave fingerprints of water molecules (H_2O). Those molecules were made up of two hydrogen atoms and an oxygen atom – three atoms altogether. In the same month, even more astonishingly, ammonia molecules (NH_3) were detected. These were composed of four-atom combinations: three atoms of hydrogen and one of nitrogen.

In 1969, another four-atom combination, including a carbon atom, was detected. This was formaldehyde (H_2CO).

In 1970, a number of new discoveries were made, including the presence of a five-atom molecule, cyanoacetylene, which contained a chain of three carbon atoms (HCCCN). And then, as a climax (at least for that year), came methyl alcohol, a molecule of six atoms (CH_3OH).

Astronomers found themselves with a totally new, and quite unexpected, subdivision of the science before them: 'astrochemistry'.

How those atoms came together to form molecules so complicated, and how such molecules manage to remain in being despite the flood of hard radiation from the stars, which ordinarily might be expected to smash them apart, astronomers can't say. Presumably, these molecules are formed under conditions that are not quite as empty as we assumed interstellar space to be – perhaps in regions where dust clouds are thickening towards star-formation.

If so, still more complicated molecules may be detected, and their presence may revolutionize our views on the formation of planets and on the development of life on those planets. Astronomers are combing the radio-wave radiation bands avidly for additional and different molecular traces.

3 The Earth

Birth of the Solar System

However glorious the unimaginable depths of the universe and however puny the earth in comparison, it is on the earth that we live and to the earth that we must return.

By the time of Newton, it had become possible to speculate intelligently about the creation of the earth and the solar system as a separate problem from the creation of the universe as a whole. The picture of the solar system showed it to be a structure with certain unifying characteristics.

1. All the major planets circle the sun in approximately the plane of the sun's equator. In other words, if you were to prepare a three-dimensional model of the sun and its planets, you would find it could be made to fit into a very shallow cakepan.

2. All the major planets circle the sun in the same direction – counter-clockwise if you were to look down on the solar system from the direction of the North Star.

3. Each major planet (with some exceptions) rotates around its axis in the same counter-clockwise sense as its revolution around the sun, and the sun itself also rotates counter-clockwise.

4. The planets are spaced at smoothly increasing distances from the sun and have nearly circular orbits.

5. All the satellites, with minor exceptions, revolve about their respective planets in nearly circular orbits in the plane of the planetary equator and in a counter-clockwise direction.

The general regularity of this picture naturally suggested that some single process had created the whole system.

What, then, is the process that produced the solar system? All the theories so far proposed fall into two classes: catastrophic and evolutionary. The catastrophic view is that the sun was created in single blessedness and gained a family as the result of some violent event. The evolutionary ideas hold that the whole system came into being in an orderly way.

In the eighteenth century, when scientists were still under the spell of the Biblical stories of such great events as the Flood, it was fashionable to assume that the history of the earth was full of violent catastrophes. Why not one super-catastrophe to start the whole thing going? One popular theory was the proposal of the French naturalist Georges Louis Leclerc de Buffon that the solar system had been created out of the debris resulting from a collision between the sun and a comet. Buffon's theory collapsed, however, when it was discovered that comets were only wisps of extremely thin dust.

In the nineteenth century, as such concepts of long-drawn-out natural processes as Hutton's uniformitarian principle (see page 45) won favour, catastrophes went out of fashion. Instead, scientists turned more and more to theories involving evolutionary processes, following Newton rather than the Bible.

Newton himself had suggested that the solar system might have been formed from a thin cloud of gas and dust that slowly condensed under gravitational attraction. As the particles came together, the gravitational field would become more intense, the condensation would be hastened, and finally the whole mass would collapse into a dense body (the sun), made incandescent by the energy of the contraction.

In essence, this is the basis of the most popular theories of the origin of the solar system today. But a great many thorny problems had to be solved to answer specific questions. How, for instance, could a highly dispersed gas be brought together by the extremely weak force of gravitation? In recent years, scientists have proposed another plausible mechanism – the pressure of light. Now particles in space are bombarded by radiation from all sides, but, if two

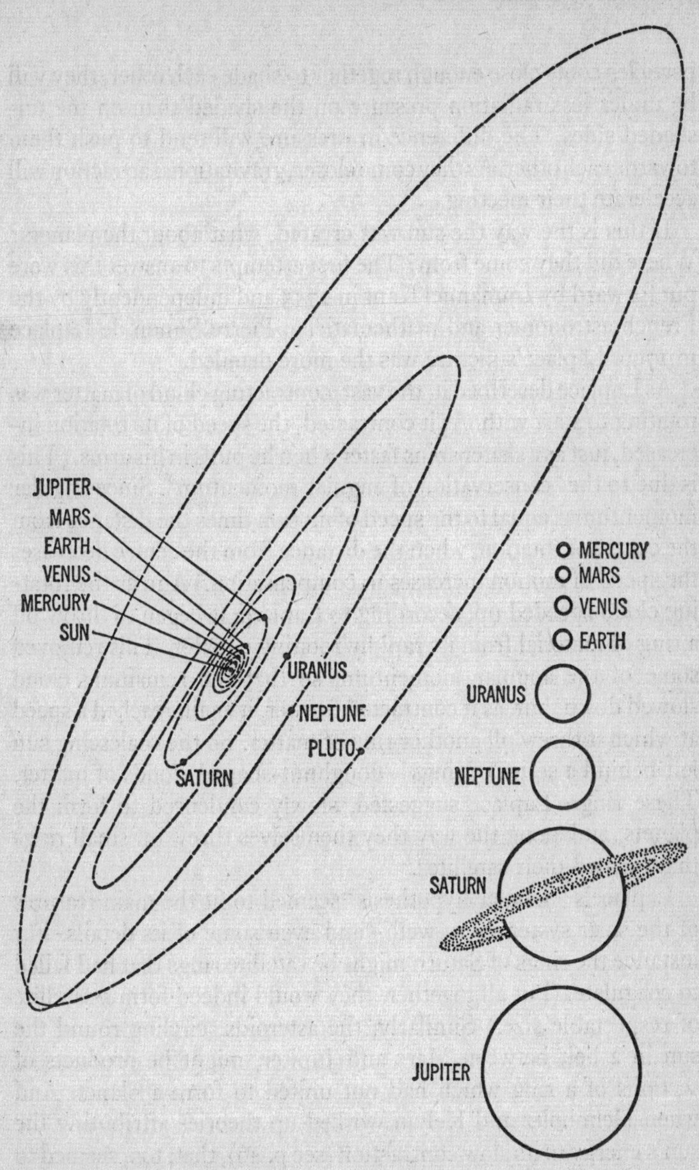

The solar system, drawn schematically, with an indication of the hierarchy of planets according to relative size.

particles come close enough together to shade each other, they will be under less radiation pressure on the shaded than on the unshaded sides. The difference in pressure will tend to push them towards each other. As they come closer, gravitational attraction will accelerate their meeting.

If this is the way the sun was created, what about the planets? Where did they come from? The first attempts to answer this were put forward by Immanuel Kant in 1755 and independently by the French astronomer and mathematician Pierre Simon de Laplace in 1796. Laplace's picture was the more detailed.

As Laplace described it, the vast, contracting cloud of matter was rotating to start with. As it contracted, the speed of its rotation increased, just as a skater spins faster when he pulls in his arms. (This is due to the 'conservation of angular momentum'. Since angular momentum is equal to the speed of motion times the distance from the centre of rotation, when the distance from the centre decreases the speed of motion increases in compensation.) And as the rotating cloud speeded up, according to Laplace, it began to throw off a ring of material from its rapidly rotating equator. This removed some of the angular momentum, so that the remaining cloud slowed down, but, as it contracted further, it again reached a speed at which it threw off another ring of matter. So the coalescing sun left behind a series of rings – doughnut-shaped clouds of matter. These rings, Laplace suggested, slowly condensed to form the planets, and along the way they themselves threw off small rings that formed their satellites.

Laplace's 'nebular hypothesis' seemed to fit the main features of the solar system very well – and even some of its details. For instance the rings of Saturn might be satellite rings that had failed to coagulate. (Put all together, they would indeed form a satellite of respectable size.) Similarly, the asteroids, circling round the sun in a belt between Mars and Jupiter, might be products of sections of a ring which had not united to form a planet. And when Helmholtz and Kelvin worked up theories attributing the sun's energy to its slow contraction (see p. 46), that, too, seemed to fit right in with Laplace's picture.

The nebular hypothesis held the field through most of the nineteenth century. But fatal flaws began to appear well before its end. In 1859, James Clark Maxwell, analysing Saturn's rings mathematically, showed that a ring of gaseous matter thrown off by any body could only condense to a collection of small particles like the rings of Saturn; it would never form a solid body, because gravitational forces would pull the ring apart before such a condensation materialized.

The problem of angular momentum also arose. It turned out that the planets, making up only a little more than 0·1 per cent of the mass of the whole solar system, carried 98 per cent of its total angular momentum! Jupiter alone possesses 60 per cent of all the angular momentum of the solar system. The sun, then, retained only a tiny fraction of the angular momentum of the original cloud. How did almost all of the angular momentum get shoved into the small rings split off the nebula? The problem is all the more puzzling since, in the case of Jupiter and Saturn which have satellite systems that seem like miniature solar systems and have, presumably, been formed in the same way, the central planetary body retains most of the angular momentum.

By 1900, the nebular hypothesis was so dead that the idea of any evolutionary process at all seemed discredited. The stage was set for the revival of a catastrophic theory. In 1905, two American scientists, Thomas Chrowder Chamberlin and Forest Ray Moulton, proposed a new one, this time explaining the planets as the result of a near collision between our sun and another star. The encounter pulled gaseous matter out of both suns, and the clouds of material left in the vicinity of our sun afterwards condensed into small 'planetesimals' and these into planets. This is the 'planetesimal hypothesis'. As for the problem of angular momentum, the British scientists James Hopwood Jeans and Harold Jeffreys proposed, in 1918, a 'tidal hypothesis', suggesting that the passing sun's gravitational attraction had given the dragged-out masses of gas a kind of sideways yank and thus imparted angular momentum to them. If such a catastrophic theory were true, then planetary systems would have to be extremely scarce. Stars are so widely spaced that stellar collisions are 10,000 times less common than are supernovae, which are themselves not common. It is estimated that in the lifetime of the galaxy there has been time for only ten encounters of the type that would produce solar systems by this theory.

However, these initial attempts at designing catastrophes failed when put to the test of mathematical analysis. Russell showed that in any such near collision the planets would have to end up thousands of times as far from the sun as they actually are. Furthermore, attempts to patch up the theory by imagining a variety of actual collisions, rather than near misses, had little success. During the 1930s, Lyttelton speculated about the possibility of a three-star collision, and later Hoyle had suggested that the sun had had a companion that had gone supernova and left planets as a legacy. In 1939, however, the American astronomer Lyman Spitzer showed that any material ejected from the sun under any circumstances would be so hot that it would not condense into planetesimals, but would merely expand into a thin gas. That seemed to end all thought of catastrophe (although, in 1965, a British astronomer, M. M. Woolfson, got around this by suggesting that the sun may have drawn its planetary material from a very diffuse, cool star, so that extreme temperatures need not be involved).

And so, after the planetesimal theory had come to a dead end, astronomers returned to the evolutionary idea and took another look at Laplace's nebular hypothesis.

By that time, their view of the universe had expanded enormously. They now had to account for the formation of galaxies. This called, of course, for much bigger clouds of gas and dust than Laplace had envisaged as the parent of the solar system. And it now appeared that such vast collections of matter would experience turbulence and would break up into eddies, each of which could condense into a separate system.

In 1944, the German astronomer Carl F. von Weizsäcker made a thorough analysis of this idea. He calculated that the largest eddies would contain enough matter to form galaxies. During the turbulent contraction of such an eddy, sub-eddies would develop. Each sub-eddy would be large enough to give birth to a solar system (with one or more suns). On the outskirts of the solar eddy itself, sub-sub-eddies might give rise to planets. This would take place at junctions where sub-sub-eddies met, moving against each other like meshing gears; at such places dust particles would collide and coalesce. As a result of these collisions, first planetesimals and then planets would form.

The Weizsäcker theory, in itself, did not solve the matter of the angular momentum of the planets any more than the much simpler Laplacian version did. The Swedish astrophysicist Hannes Alfven took into account the magnetic field of the sun. As the young sun whirled rapidly, its magnetic field acted as a brake, slowing it up, and the angular momentum was passed on to the planets. Hoyle elaborated on this notion so that the Weizsäcker theory, modified to include magnetic as well as gravitational forces, seems the best one yet to account for the origin of the solar system.

There remain irregularities in the solar system, however, that no overall theory for the general formation can easily account for, and which would probably require sub-theories, so to speak. For instance, there are the comets: small bodies in vastly elongated orbits that circle the sun in periods of dozens, hundreds, even thousands of years. Their orbits are completely un-planetlike; they enter the inner solar system from all angles; and they are made up in part of light, low-melting substances, which vaporize and stream away as the temperature goes up when they approach the neighbourhood of the sun.

In 1950, Oort suggested the existence of a vast shell of small, icy bodies, slowly circling the sun at the distance of a light-year or more. There might be as many as 100,000 million of them altogether – material left over from the original cloud of dust and gas that condensed to form the solar system, material too far out to be effectively captured by gravitational forces and remaining as an outermost shell, not drawn inward.

On the whole, they would remain undisturbed in their orbit. But every once in a while, some chance combination of gravitational pulls from nearby stars, might slow up one body or another sufficiently to allow it to fall towards the inner solar system, move round the sun, and go shooting out to the cloud. In doing so, they approach from all possible directions. If they pass near one of the large outer planets, the planetary gravitational pull may further alter their orbits to keep them permanently within the bounds of the planetary system. Once within the planetary system, the heating and vaporizing effect of the sun breaks up their substance in a very short time, geologically speaking. However, there are many more where they come from, since Oort estimates that only 20 per cent of the total supply of cometary bodies have been sent hurtling in towards the sun in all the thousands of millions of years of the existence of the solar system.

A second irregularity is that represented by the planetoids, a group of tens of thousands of tiny planetary bodies (the largest is less than five hundred miles in diameter; the smallest, under a mile) that, for the most part, are to be found between the orbits of Mars and Jupiter. If the spacing of the planets were absolutely regular, astronomers would expect to find a planet about where the largest of the planetoids is. Did a planet once actually exist there? Did it explode for some reason, sending fragments scattering? Were there sub-explosions, which would account for some planetoids having elongated orbits, while others have unusually tilted ones (though all rotate more or less counter-clockwise)? Or is it that, thanks to the overriding effect of the gravitational field of nearby giant Jupiter, the cloud in the region between the orbits of Mars and Jupiter, coalesced into planetesimals but never into a single planet? The problem of the origin of the planetoids remains open.

Pluto, the outermost planet, first discovered in 1930 by the American astronomer Clyde William Tombaugh, is another problem. The other outer planets – Jupiter, Saturn, Uranus, and Neptune – are large, gaseous, speedily rotating giants; Pluto is small, dense, and rotates once in 6·4 days. Furthermore, its orbit is elongated more than any other planet, and it is tipped at a greater angle to the general plane of revolution than any other planet. Its orbit is so elongated that when it approaches that part which is closest to the sun, it is actually closer, for twenty years or so, than Neptune can ever approach.

Some astronomers wonder if Pluto might once have been a satellite of Neptune. It is a little large for the role, but that hypothesis would account for its slow rotation, since 6·4 days might have been the time of its revolution about Neptune – a revolution equal to rotation, as in the case of our moon. Perhaps a vast cosmic accident freed Pluto of Neptune's grip and sent it hurtling into an elongated orbit. The same accident may have forced Triton, Neptune's large satellite, to circle in an orbit far removed from that of Neptune's equator, and moved Neptune closer to the sun, for its orbit ought to be distinctly farther out if the evenly increasing separation of successive planets were to be obeyed. Unfortunately, astronomers haven't the faintest notion of the kind of cosmic accident that could have resulted in all this.

The rotation of the planets also offers its problems. Ideally, all the planets ought to rotate in a counter-clockwise direction (as viewed from a point high above the earth's north Pole) with their axes of rotation perpendicular to the plane of their revolution about the sun. This is reasonably true for the sun itself and for Jupiter, the two major bodies of the solar system, but there is a puzzling variation in the others whose plane of rotation we can measure.

The earth's axis is tipped about 23½° to the vertical, while the axes of Mars, Saturn, and Neptune are tipped by 25°, 27°, and 29° respectively. Uranus represents an even more extreme case, for its axis is tipped by 98° – or a little more than a right angle – so that in effect its axis is lined up with its plane of rotation and it rolls along its orbit like a top rolling on its side, instead of standing upright (or leaning a little) on its peg. Uranus has five small satellites whose orbits are all tipped with the planet's axis so they remain in Uranus's equatorial plane.

What has tipped so many of the planets, and what has tipped Uranus so drastically, is still a puzzle, yet not nearly as much a puzzle as that posed by the planet Venus. For a long time now, astronomers have understood that when a small body circles a larger one, tidal pulls slow the rotation of the small body until it faces one side constantly to the larger one, rotating exactly once every revolution. This is true of the moon with respect to the earth, for instance, so that it rotates about its axis (and revolves about the Earth) once in 29½ days.

It was thought very likely that Mercury and Venus, so close to the sun, were likewise slowed, and that both planets rotated once per revolution – Mercury in 88 days and Venus in 225 days. Mercury is hard to observe because it is small, distant, and extraordinarily close to the masking light of the sun. However, as long ago as the 1880s, the Italian astronomer Giovanni Virginio Schiaparelli noted vague markings on Mercury's surface and used them to measure its period of rotation. He decided Mercury *did* rotate in 88 days, once per revolution.

The case of Venus was much more difficult. An eternal layer of clouds totally and permanently obscured the surface of Venus as far as ordinary vision was concerned. No markings could be seen and, down to the 1960s, no direct determination of the period of rotation of the nearest planet was possible (though that of distant Pluto was known).

In the 1960s, however, it became possible to 'see' astronomical bodies by something other than the reflection of light waves. Tight beams of short radio waves could be sent out in the direction of such a body, and the reflected beam could be detected on earth. In 1946, this had been done in the case of the moon. The short radio waves were of the type used in radar, and so 'radar astronomy' was born.

Radar reflections from the moon were of minor importance, however, since the moon's surface could be seen very well by the reflection of ordinary sunlight. What about Venus, however? Radar waves could slip past the cloud layer and touch the actual surface before being reflected. In 1961, this was accomplished by groups of scientists in the United States, in Great Britain, and in the Soviet Union. From the time it took the radar waves to reach Venus and return, a more accurate measure could be made of Venus's distance at that moment (and therefore of all the distances in the solar system). It was not long afterwards that radar contact was made with Mercury, too; a Soviet team was first to succeed, in 1962.

The nature of the reflected radar beam varies according to whether the surface it has touched is rough or smooth, and whether it is rotating or not. Roughness tends to broaden the reflected beam, while rotation tends to expand the wavelength range. The extent of change depends upon the degree of roughness or the rapidity of rotation.

In 1965, the nature of the reflected radar beam from Mercury made it clear to the Americans Rolf Buchanan Dyce and G. H. Pettengill that Mercury had to be rotating faster than had been thought. The period of rotation was not 88 days but 59 days! This discovery, which was optically confirmed in 1968, was a considerable surprise, but astronomers quickly recovered. The period of Mercury's rotation was just two thirds its period of revolution, which meant that it presented alternate faces to the sun at each perihelion (point of closest approach). Tidal effects were studied, and it was found that this situation was stable and could be accounted for.

Venus was something else again. There was considerable satisfaction over the fact that the radar beam could bring back information concerning the planet's solid surface – something that could not be done by light waves.

The surface was, for instance, rough. Late in 1965 it was decided that there were at least two huge mountain ranges on Venus. One of them runs from north to south for about 2,000 miles and is several hundred miles wide. The other, even larger, runs east to west. The two ranges are named after the first two letters of the Greek alphabet and are the 'Alpha Mountains' and the 'Beta Mountains'.

But earlier than that, in 1964, it turned out that Venus was rotating slowly. So far, so good, for the period of rotation was thought to be (a matter of pure speculation) 225 days. The period turned out to be 243 days, and the axis of rotation was just about perpendicular to the plane of revolution. It was disappointing that the period of rotation was not exactly 225 days (equal to the period of revolution), for that would have been expected and easily explained. What really astonished the astronomers was that the rotation was in the 'wrong' direction. Venus rotated clockwise (as viewed from high above earth's North Pole), east to west, instead of west to east as did all the other planets, except Uranus. It was as though Venus were standing on its head, with its North Pole pointing downwards and its South Pole pointing upwards.

Why? Nobody can yet offer any explanation.

Furthermore, the time of rotation is such that every time Venus makes its closest approach to earth, it presents the same (cloud-hidden) side to us. Can there be some gravitational influence of earth over Venus? But how can small and distant earth compete with the more distant but enormously larger sun? This is puzzling, too.

In short, in the latter part of the 1960s, Venus emerged as the puzzle-planet of the solar system.

And yet there are puzzles even closer to home. The moon is extraordinarily large in some ways. It is 1/81 as massive as the earth. No other satellite in the solar system is nearly as large in comparison to the planet it circles. What's more, the moon does not circle earth in the plane of earth's Equator, but has an orbit markedly tipped to that plane – an orbit that is more nearly in the plane within which the planets generally revolve about the sun. Is it possible that the moon originally was not a satellite of the earth, but was an independent planet somehow captured by earth? Are earth and the moon twin planets?

Burning curiosity about the origin of the moon and the past history of the earth–moon system is one of the motives that have led scientists to such an excited study of the moon's surface, up to and including manned landings on our satellite.

Of Shape and Size

One of the major inspirations of the ancient Greeks was their decision that the earth has the shape of a sphere. They conceived this idea originally (tradition credits Pythagoras of Samos with being the first to suggest it, about 525 B.C.) on philosophical grounds – e.g., that a sphere was the perfect shape. But the Greeks also verified it with observations. Around 350 B.C., Aristotle marshalled conclusive evidence that the earth was not flat but round. His most telling argument was that as one travelled north or south, new stars appeared over the horizon and visible ones disappeared below the horizon behind. Then, too, ships sailing out to sea vanished hull first in whatever direction they travelled, while the cross-section of the earth's shadow on the moon, during a lunar eclipse, was always a circle, regardless of the position of the moon. Both these latter facts could be true only if the earth were a sphere.

Among scholars, at least, the notion of the spherical earth never entirely died out, even during the Dark Ages. The Italian poet Dante Alighieri assumed a spherical earth in that epitome of the medieval view, *The Divine Comedy*.

It was another thing entirely when the question of a *rotating* sphere arose. As long ago as 350 B.C., the Greek philosopher Heraclides of Pontus suggested that it was far easier to suppose that the earth rotated on its axis than that the entire vault of the heavens revolved round the earth. This, however, most ancient and medieval scholars refused to accept, and, as late as 1632, Galileo was condemned by the Inquisition at Rome and forced to recant his belief in a moving earth.

Nevertheless, the Copernican theory made a stationary earth completely illogical, and slowly its rotation was accepted by everyone. It was only in 1851, however, that this rotation was actually demonstrated experimentally. In that year, the French physicist Jean Bernard Léon Foucault set a huge pendulum swinging from the dome of a Parisian church. According to the conclusions of physicists, such a pendulum ought to maintain its swing in a fixed plane, regardless of the rotation of the earth. At the North Pole, for instance, the pendulum would swing in a fixed plane, while the earth rotated under it, counter-clockwise, in twenty-four hours. Since a person watching the pendulum would be carried with the earth (which would seem motionless to him), it would seem to that person that the pendulum's plane of swing was turning clockwise through one full revolution every twenty-four hours. At the South Pole, the same thing would happen except that the pendulum's plane of swing would turn counter-clockwise.

At latitudes below the poles, the plane of the pendulum would still turn (clockwise in the Northern Hemisphere and counter-clockwise in the Southern), but in longer and longer periods as one moved farther and farther from the poles. At the Equator, the pendulum's plane of swing would not alter at all.

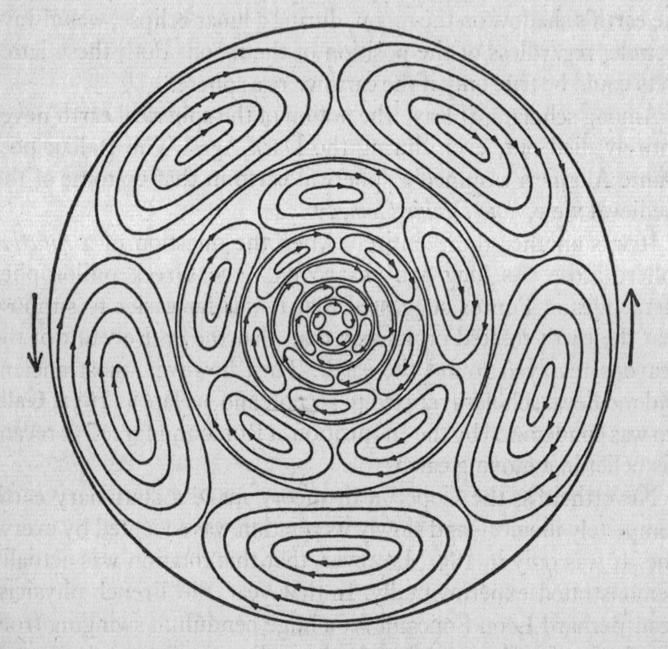

Carl F. von Weizsäcker's model of the origin of the solar system. His theory holds that the great cloud from which it was formed broke up into eddies and sub-eddies that then coalesced into the sun, the planets, and their satellites.

During Foucault's experiment, the pendulum's plane of swing turned in the proper direction and at just the proper rate. The observer, so to speak, could see with his own eyes the earth turn under the pendulum.

The rotation of the earth brings with it many consequences. The surface moves faster at the Equator, where it must make a circle of 25,000 miles in twenty-four hours, at a speed of just over 1,000 miles an hour. As one travels north (or south) from the Equator, a spot on the earth's surface need travel more slowly, since it must make a smaller circle in the same twenty-four hours. Near the poles, the circle is small indeed, and, at the poles, the surface is motionless.

The air partakes of the motion of the surface of the earth over which it hovers. If an air mass moves northwards from the Equator, its own speed (matching that of the Equator) is faster than that of the surface it travels towards. It overtakes the surface in the west-to-east journey and drifts eastwards. This drift is an example of a 'Coriolis effect', named after the French mathematician Gaspard Gustave de Coriolis, who first studied it in 1835.

The effect of such Coriolis effects on air masses is to set them to turning with a clockwise twist in the Northern Hemisphere. In the Southern Hemisphere, the effect is reversed, and a counter-clockwise twist is produced. In either case, 'cyclonic disturbances' are set up. Massive storms of this type are called 'hurricanes' in the North Atlantic and 'typhoons' in the North Pacific. Smaller but more intense storms of this sort are 'cyclones' or 'tornadoes'. Over the sea, such violent twisters set up dramatic 'sea spouts'.

However, the most exciting deduction obtained from the earth's rotation was made two centuries before Foucault's experiment, in Isaac Newton's time. At that time, the notion of the earth as a perfect sphere had already held sway for nearly 2,000 years, but then Newton took a careful look at what happens to such a sphere when it rotates. He noted the difference in the rate of motion of the earth's surface at different latitudes and considered what it must mean.

The faster the rotation, the stronger the centrifugal effect – that is, the tendency to push material away from the centre of rotation. It follows, therefore, that the centrifugal effect increases steadily from zero at the stationary poles to a maximum at the rapidly whirling equatorial belt. This means that the earth should be pushed out around its middle. In other words, it should be an 'oblate spheroid', with an 'equatorial bulge' and flattened poles. It must have roughly the shape of a tangerine rather than that of a golf ball. Newton even calculated that the polar flattening should be about 1/230 of the total diameter, which is surprisingly close to the truth.

The earth rotates so slowly that the flattening and bulging are too slight to be readily detected. But at least two astronomical observations supported Newton's reasoning, even in his own day. First, Jupiter and Saturn were clearly seen to be markedly flattened at the poles, as was first pointed out by the Italian-born French astronomer Giovanni Domenico Cassini in 1687. Both planets are much larger than the earth and rotate much faster, so that Jupiter's surface, for instance, is speeding at 27,000 miles per hour at its equator. With centrifugal effects born of such speeds, no wonder it is flattened.

Second, if the earth really bulges at the Equator, the varying gravitational pull on the bulge by the moon, which most of the time is either north or south of the Equator in its circuit round the earth, should cause the earth's axis of rotation to mark out a double cone, so that each pole points to a steadily changing point in the sky. The points mark out a circle about which the pole makes a complete revolution every 26,000 years. In fact, Hipparchus of Nicaea had noted this shift about 150 B.C. when he compared the position of the stars in his day with those recorded a century and a half earlier. The shift of the earth's axis has the effect of causing the sun to reach the point of equinox about fifty seconds of arc eastwards each year (that is, in the direction of morning). Since the equinox thus comes to a preceding (i.e., earlier) point each year, Hipparchus named this shift the 'precession of the equinoxes', and it is still known by that name.

Naturally scientists set out in search of more direct proof of the earth's distortion. They resorted to a standard device for solving geometrical problems – trigonometry. On a curved surface, the angles of a triangle add up to more than 180 degrees. The greater the curvature, the greater the excess over 180 degrees. Now if the earth was an oblate spheroid, as Newton had said, the excess should be greater on the more sharply curved surface of the equatorial bulge than on the less curved surface towards the poles. In the 1730s, French scientists made the first test by doing some large-scale surveying at separate sites in the north and the south of France. On the basis of these measurements, the French astronomer Jacques Cassini (son of Giovanni Domenico, who had pointed out the flattening of Jupiter and Saturn) decided that the earth bulged at the poles, not at the Equator! To use an exaggerated analogy, its shape was more like that of a cucumber than of a tangerine.

But the difference in curvature between the north and the south of France obviously was too small to give conclusive results. Consequently, in 1735 and 1736 a pair of French expeditions went forth to more widely separated regions – one to Peru, near the Equator, and the other to Lapland, approaching the Arctic. By 1744, their surveys had given a clear answer: the earth was distinctly more curved in Peru than in Lapland.

Today the best measurements show that the diameter of the earth is 26·68 miles longer through the Equator than along the axis through the poles (i.e., 7,926·36 miles against 7,899·78 miles).

Perhaps the most important scientific result of the eighteenth-century inquiry into the shape of the earth was that it made the scientific community dissatisfied with the state of the art of measurement. No decent standards for precise measurement existed. This dissatisfaction was partly responsible for the adoption, during the French Revolution half a century later, of the logical and scientifically worked-out 'metric' system based on the metre. The metric system now is used by scientists all over the world, to their great satisfaction, and it is the system in general public use in every civilized country except the English-speaking nations, chiefly Great Britain and the United States.

The importance of accurate standards of measure cannot be overestimated. A good percentage of scientific effort is continually being devoted to improvement in such standards. The standard metre and standard kilogram were made of platinum–iridium alloy (virtually immune to chemical change) and were kept in a Paris suburb under conditions of great care; in particular, under constant temperature to prevent expansion or contraction.

New alloys such as 'Invar' (short for invariable), composed of nickel and iron in certain proportions, were discovered to be almost unaffected by temperature change. These could be used in forming better standards of length and the Swiss-born French physicist Charles Édouard Guillaume, who developed Invar, received the Nobel Prize for physics in 1920 for this discovery.

In 1960, however, the scientific community abandoned material standards of length. The General Conference of Weights and Measures adopted as standard the length of a tiny wave of light produced by the rare gas krypton. Exactly 1,650,763·73 of these waves (far more unchanging than anything man-made could be) equal one metre, a length which is now a thousand times as exact as it had been before.

The smoothed-out, sea-level shape of the earth is called the 'geoid'. Of course, the earth's surface is pocked with irregularities – mountains, ravines, and so on. Even before Newton raised the question of the planet's over-all shape, scientists had tried to measure the magnitude of these minor deviations from a perfect sphere (as they thought). They resorted to the device of a swinging pendulum. Galileo, in 1581, as a seventeen-year-old boy, had discovered that a pendulum of a given length always completed its swing in just about the same time, whether the swing was short or long; he is supposed to have made the discovery while watching the swinging chandeliers in the cathedral of Pisa during services. There is a lamp in the cathedral still called 'Galileo's lamp', but it was not hung until 1584. (Huygens hooked a pendulum to the gears of a clock and used the constancy of its motion to keep the clock going with even accuracy. In 1656, he devised the first modern clock in this way – the 'grandfather clock' – and at once increased the accuracy of timekeeping tenfold.)

The period of the pendulum depends both on its length and on the gravitational force. At sea level, a pendulum with a length of 39·1 inches makes a complete swing in just one second, a fact worked out in 1644 by Galileo's pupil, the French mathematician Marin Mersenne. The investigators of the earth's irregularities made use of the fact that the period of a pendulum's swing depends on the strength of gravity at any given point. A pendulum that swings perfect seconds at sea level, for instance, will take slightly longer than a second to complete a swing on a mountain top, where gravity is slightly weaker because the mountain top is farther from the centre of the earth.

In 1673, a French expedition to the north coast of South

America (near the Equator) found that at that location the pendulum was slowed even at sea level. Newton later took this as evidence for the existence of the equatorial bulge, since that lifted the camp farther from the earth's centre, and weakened the force of gravity. After the expedition to Peru and Lapland had proved his theory, a member of the Lapland expedition, the French mathematician Alexis Claude Clairault, worked out methods of calculating the oblateness of the earth from pendulum swings. Thus the geoid, or sea-level shape of the earth, can be determined, and it turns out to vary from the perfect oblate spheroid by less than 300 feet at all points. Nowadays, gravitational force is also measured by a 'gravimeter', a weight suspended from a very sensitive spring. The position of the weight against a scale in the background indicates the force with which it is pulled downwards and hence measures variations in gravity with great delicacy.

Gravity at sea level varies by about 0·6 per cent, being least at the Equator, of course. The difference is not noticeable in ordinary life, but it can affect sports records. Achievements at the Olympic Games depend to some extent on the latitude (and altitude) of the city in which they are conducted.

A knowledge of the exact shape of the geoid is essential for accurate map-making, and only 7 per cent of the earth's land surface can really be said to be accurately mapped. As late as the 1950s, the distance between New York and London, for instance, was not known to better than a mile or so, and the location of some islands in the Pacific were known only within a possible error of several miles. In these days of air travel and (alas!) potential missile-aiming, this is inconvenient. But truly accurate mapping has now been made possible – oddly enough, not by surveys of the earth's surface, but by astronomical measurements of a new kind. The first instrument of these new measurements was the man-made satellite called Vanguard I, launched by the United States on 17 March 1958. Vanguard I makes a revolution round the earth in two and a half hours, and in the first couple of years of its lifetime it had already made more revolutions than the moon had in all the centuries it has been observed with the telescope. By observations of Vanguard I's position at specific times from specific points of the earth, the distances between these observing points can be calculated precisely. In this way, positions and distances not known to within a matter of miles were, in 1959, determined to within a hundred yards or so. (Another satellite, named Transit I-B, launched by the United States on 13 April 1960, was the first of a series specifically intended to extend this into a system for the accurate location of position on the earth's surface, something which could greatly improve and simplify air and sea navigation.)

Like the moon, Vanguard I circles the earth in an ellipse which is not in the earth's equatorial plane. As in the case of the moon, the perigee (closest approach) of Vanguard I shifts because of the attraction of the equatorial bulge. Because Vanguard I is far closer to the bulge and far smaller than the moon, it is affected to a greater extent, and because of its many revolutions, the effect of the bulge can be well studied. By 1959, it was certain that the perigee shift of Vanguard I was not the same in the Northern Hemisphere as in the Southern. This showed that the bulge was not quite symmetrical with respect to the Equator. The bulge seemed to be twenty-five feet higher (that is, twenty-five feet more distant from the earth's centre) at spots south of the Equator than at spots north of it. Further calculations showed that the South Pole was fifty feet closer to the centre of the earth (counting from sea level) than was the North Pole.

Further information, obtained in 1961, based on the orbits of Vanguard I and Vanguard II (the latter having been launched on 17 February 1959) indicates that the sea-level Equator is not a perfect circle. The equatorial diameter is 1,400 feet (nearly a quarter of a mile) longer in some places than in others.

Newspaper stories have described the earth as 'pear-shaped' and the Equator as 'egg-shaped'. Actually, these deviations from the perfectly smooth curve are perceptible only to the most refined measurements. No one looking at the earth from space would see anything resembling a pear or an egg; he would see only what would seem a perfect sphere. Besides, detailed studies of the geoid have shown so many regions of very slight flattening and very slight humping that, if the earth must be described dramatically, it had better be called 'lumpy shaped'.

A knowledge of the exact size and shape of the earth makes it possible to calculate its volume, about 260,000 million cubic miles. Calculating the earth's mass, however, is a more complex matter, but Newton's law of gravitation gives us something to begin with. According to Newton, the gravitational force (f) between any two objects in the universe can be expressed as follows:

$$f = \frac{gm_1m_2}{d^2}$$

where m_1 and m_2 are the masses of the two bodies concerned and d is the distance between them, centre to centre. As for g, that represents the 'gravitational constant'.

What the value of the constant was, Newton could not say. If we can learn the values of the other factors in the equation, however, we can find g, for by transposing the terms we get:

$$g = \frac{fd^2}{m_1m_2}$$

To find the value of g, therefore, all we need to do is to measure the gravitational force between two bodies of known mass at the separation of a known distance. The trouble is that gravitational force is the weakest force we know, and the gravitational attraction between two masses of any ordinary size that we can handle is almost impossible to measure.

Nevertheless, in 1798 the English physicist Henry Cavendish, a wealthy, neurotic genius who lived and died in almost complete seclusion but performed some of the most astute experiments in the history of science, managed to make the measurement. Cavendish attached a ball of known mass to each end of a long rod and suspended this dumb-bell-like contraption on a fine thread. Then he placed a larger ball, also of known mass, close to each ball on the rod – on opposite sides, so that gravitational attraction between the fixed large balls and the suspended small balls would cause the horizontally hung dumb-bell to turn, thus twisting the thread. The dumb-bell did indeed turn slightly. Cavendish now measured how much force was needed to produce this amount of twist of the thread. This told him the value of f. He also knew m_1

and m_2, the masses of the balls, and d, the distance between the attracted balls. So he was able to compute the value of g. Once he had that, he could calculate the mass of the earth, because the

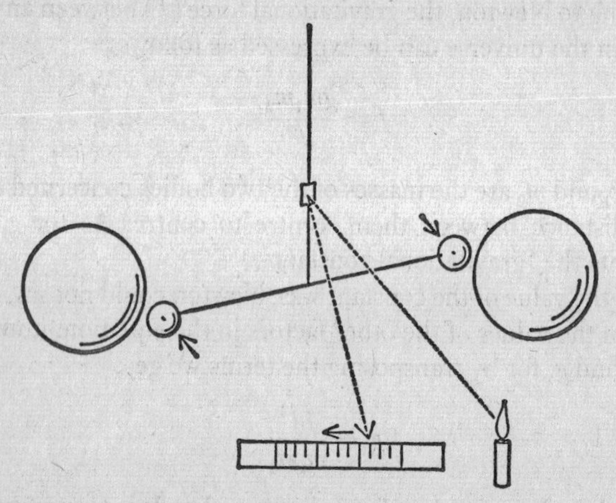

Henry Cavendish's apparatus for measuring gravity. The two small balls are attracted by the larger ones, causing the thread on which they are suspended to twist. The mirror shows the amount of this slight twist by the deflection of reflected light on the scale.

earth's gravitational pull (f) on any given body can be measured. Thus Cavendish 'weighed' the earth for the first time.

The measurements have since been greatly refined. In 1928, the American physicist Paul R. Heyl at the United States Bureau of Standards determined the value of g to be 0·00000006673 dyne centimetre squared per gram squared. You need not be concerned about those units, but note the smallness of the figure. It is a measure of the weakness of gravitational force. Two one-pound weights placed a foot apart attract each other with a force of only one half of one thousand millionth of an ounce.

The fact that the earth itself attracts such a weight with the force of one pound even at a distance of 4,000 miles from its centre emphasizes how massive the earth is. In fact, the mass of the earth turns out to be 6,595,000,000,000,000,000,000 tons or, in metric units, 5,983,000,000,000,000,000,000 kilograms.

From the mass and volume of the earth, its average density is easily calculated. In metric units, the answer comes out to 5·522 grams per cubic centimetre (5·522 times the density of water). The density of the earth's surface rocks averages only 2·8 grams per cubic centimetre, so the density of the interior must be much greater. Does it increase smoothly all the way down to the centre? The first proof that it does not – that the earth is made up of a series of different layers – came from the study of earthquakes.

The Layers of the Planet

On 1 November 1755, a great earthquake, possibly the most violent of modern times, struck the city of Lisbon, demolishing every house in the lower part of the city. Then a tidal wave swept in from the ocean. Sixty thousand people were killed, and the city was left a scene of devastation.

The shock was felt over an area of one and a half million square miles, doing substantial damage in Morocco as well as in Portugal. Because it was All Soul's Day, people were in church, and it is said that all over southern Europe those in the cathedrals saw the chandeliers dance and sway.

The Lisbon disaster made a great impression on the scholars of the day. It was an optimistic time when many thinkers felt that the new science of Galileo and Newton would place in man's hands the means of making the earth a human paradise. This blow showed that there were still giant, unpredictable, and apparently malicious forces beyond man's control. The earthquake inspired Voltaire, the great literary figure of the time, to write his famous pessimistic satire *Candide*, with its ironical refrain that all was for the best in this best of all possible worlds.

We are accustomed to thinking of dry land as shaking with the effect of an earthquake, but the earth beneath the ocean floor may be set to quivering, too, with even more devastating effects. The vibration sets up long, gentle swells in the ocean which, on reaching the shallow shelves in the neighbourhood of land, pile up into towers of water, sometimes fifty to one hundred feet high. If the waves hit with no warning, thousands are drowned. The popular name for such earthquake-generated waves are 'tidal waves', but this is a misnomer. They may resemble monstrous tides, but they have entirely different causes. Nowadays, they are referred to by the Japanese name 'tsunami'. Japan's coastline is particularly vulnerable to such waves, so this nomenclature is justified.

After the Lisbon disaster, to which a tsunami had added its share of destruction, scientists began turning their thoughts earnestly to what the causes of earthquakes might be. The best the ancient Greeks were able to do had been Aristotle's suggestion that it was caused by masses of air, imprisoned underground and trying to escape. Modern scientists, however, suspected that it might be the effect of earth's internal heat on stresses within the solid rock itself.

The English geologist John Michell (who had studied the forces involved in 'torsion', or twisting, later used by Cavendish to measure the mass of the earth) suggested in 1760 that earthquakes were waves set up by the shifting of masses of rock miles below the surface. To study earthquakes properly, an instrument for detecting and measuring these waves had to be developed, and this did not come to pass until one hundred years after the Lisbon quake. In 1855, the Italian physicist Luigi Palmieri devised the first 'seismograph' (from Greek words meaning 'earthquake-writing').

In its simplest form, the seismograph consists of a massive block suspended by a comparatively weak spring from a support firmly fixed in bedrock. When the earth moves, the suspended block remains still, because of its inertia. However, the spring attached to the bedrock stretches or contracts a little with the earth's motion. This motion is recorded on a slowly rotating drum by means of a pen attached to the stationary block, writing on smoked paper. Actually, two blocks are used, one oriented to record the earthquake waves travelling north and south, the other, east and west. Nowadays, the most delicate seismographs, such as the one at Fordham University, use a ray of light in place of a pen, to avoid the frictional drag of the pen on the paper. This ray shines on sensitized paper, making tracings that are developed as a photograph.

The English engineer John Milne, using seismographs of his own design, showed conclusively in the 1890s that Michell's description of earthquakes as waves propagated through the body of the earth was correct. Milne was instrumental in setting up stations for the study of earthquakes and related phenomena in various parts of the world, particularly in Japan. By 1900, thirteen seismograph stations were in existence, and today there are over 500, spread over every continent including Antarctica.

The earth suffers a million quakes a year, including at least ten disastrous ones and a hundred serious ones. Some 15,000 people are killed by these tremors each year. The most murderous quake is supposed to have taken place in northern China in 1556, with an estimated 830,000 dead. As recently as 1923, a quake that shook Tokyo devastated the city and left 143,000 dead.

The largest earthquakes are estimated to release a total energy equal to 100,000 ordinary atomic bombs or, if you prefer, one hundred large H-bombs. It is only because their energies are dissipated over a large area that they are not more destructive than they are. They can make the earth vibrate as though it were a gigantic tuning fork. The Chilean earthquake of 1960 caused our planet to vibrate at a frequency of just under once an hour (20 octaves below middle C and quite inaudible).

Earthquake intensity is measured on a scale from 0 up through 9, where each number represents an energy release ten times that of the number below. (No quake of intensity greater than 9 has ever been recorded, but the Good Friday quake in Alaska in 1964 recorded an intensity of 8·5.) This is called the 'Richter scale' because it was introduced in 1935 by the American seismologist Charles Francis Richter.

About 80 per cent of earthquake energy is released in the areas bordering the vast Pacific Ocean. Another 15 per cent is released in an east–west band sweeping across the Mediterranean. These earthquake zones (see map on p. 130) are closely associated with volcanic areas, which is one reason why the effect of internal heat was associated with earthquakes.

Volcanoes are a natural phenomenon that are as frightening as earthquakes and longer-lasting, though, of course, in most cases their effects are confined to a smaller area. About 500 volcanoes are known to have been active in historical times, two thirds of them along the rim of the Pacific.

On rare occasions, when a volcano traps and overheats huge quantities of water, appalling catastrophes can take place. On 26–27 August 1883, the small East Indian volcanic island Krakatoa, situated in the strait between Sumatra and Java, exploded with a roar that has been described as the loudest sound ever formed on earth during historic times. The sound was heard by human ears as far away as 3,000 miles and could be picked up by instruments all over the globe. The sound waves travelled several times completely round the planet. Five cubic miles of rock were fragmented and hurled into the air; ash fell over an area of 300,000 square miles. Ashes darkened the sky over hundreds of square miles, leaving in the stratosphere dust that brightened sunsets for years. Tsunamis a hundred feet in height killed 36,000 people on the shores of Java and Sumatra, and their waves could be detected easily in all parts of the world.

A similar event, with even greater consequences, may have taken place over three thousand years before in the Mediterranean Sea. In 1967, American archaeologists discovered the ash-covered remains of a city on the small island of Thera, eighty miles north of Crete. About 1400 B.C., apparently, it exploded as Krakatoa did. The tsunami that resulted struck the island of Crete, then the home of a long-developed and admirable civilization, a crippling blow from which that civilization never recovered. The Cretan control of the seas vanished, and a period of turmoil and darkness eventually followed; recovery took many centuries. The dramatic disappearance of Thera lived on in the minds of survivors, and its tale passed down the line of generations with embellishments. It may very well have given rise to Plato's tale of Atlantis, which was told about eleven centuries after the death of Thera and of Cretan civilization.

And yet perhaps the most famous single volcanic eruption in the history of the world was a minute one, compared to Krakatoa or

Thera. It was the eruption of Vesuvius in A.D. 79 (at that time it had been considered a dead volcano), which buried the Roman resort cities of Pompeii and Herculaneum. The famous encyclopaedist Gaius Plinius Secundus (better known as Pliny) died in that catastrophe, which was described by his nephew, Pliny the Younger, an eyewitness.

Excavations of the buried cities began in serious fashion after 1763. These offered an unusual opportunity to study relatively complete remains of a city that had existed during the most prosperous period of ancient times.

Another unusual phenomenon is the actual birth of a new volcano. Such an awesome event was witnessed in Mexico on 20 February 1943, when in the village of Paricutin, 200 miles west of Mexico City, a volcano began to appear in what had been a quiet cornfield. In eight months, it had built itself up to an ashy cone 1,500 feet high. The village had to be abandoned, of course.

Modern research in volcanoes and their role in forming much of the earth's crust began with the French geologist Jean Étienne Guettard in the mid eighteenth century. For a while, in the late eighteenth century, the single-handed efforts of the German geologist Abraham Gottlob Werner popularized the false notion that most rocks were of sedimentary origin, from an ocean that had once been world-wide ('neptunism'). The weight of the evidence, particularly that presented by Hutton, made it quite certain however that most rocks were formed through volcanic action ('plutonism'). Both volcanoes and earthquakes would seem the expression of the earth's internal energy, originating for the most part from radioactivity (Chapter 6).

Once seismographs allowed the detailed study of earthquake waves, it was found that those most easily studied came in two general varieties: 'surface waves' and 'bodily waves'. The surface waves follow the curve of the earth; the bodily waves go through the interior – and by virtue of this short cut usually are the first to arrive at the seismograph. These bodily waves in turn are of two types: primary ('P waves') and secondary ('S waves'). The primary waves, like sound waves, travel by alternate compression and expansion of the medium (to visualize them, think of the pushing together and pulling apart of an accordion). Such waves can pass through any medium – solid or fluid. The secondary waves, on the other hand, have the familiar form of snake-like

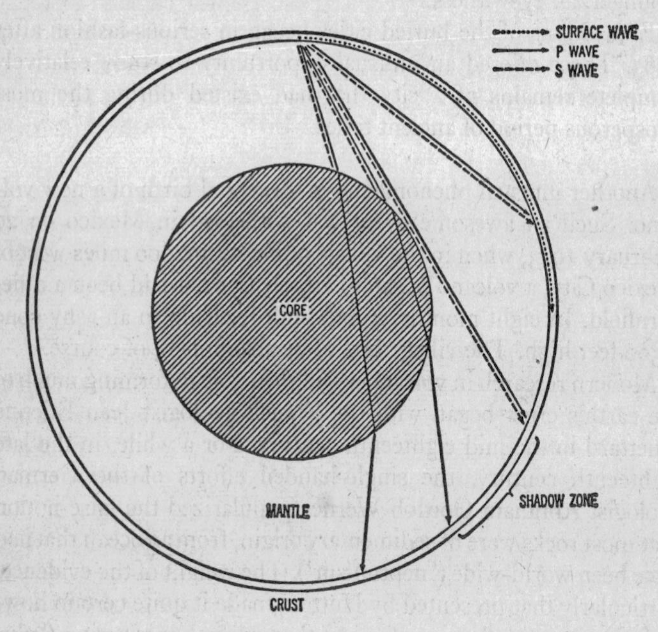

Earthquake waves' routes in the earth. Surface waves travel along the crust. The earth's liquid core refracts the P-type bodily waves. S waves cannot travel through the core.

wiggles at right angles to the direction of travel, and they cannot travel through liquids or gases.

The primary waves move faster than secondary waves and consequently reach a seismograph station sooner. From the time lag of the secondaries, it is possible to estimate the distance of the earthquake. And its location or 'epicentre' (the spot on the earth's surface directly above the rock disturbance) can be pinpointed by getting distance bearings at three or more stations: the three radii trace out three circles that will intersect at a single point.

The speed of both the P and S types of wave is affected by the kind of rock, the temperature, and the pressure, as laboratory studies have shown. Therefore earthquake waves can be used as probes to investigate conditions deep under the earth's surface.

A primary wave near the surface travels at 5 miles per second; 1,000 miles below the surface, judging from the arrival times, its velocity must be nearly 8 miles per second. Similarly, a secondary wave has a velocity of less than 3 miles per second near the surface and 4 miles per second at a depth of 1,000 miles. Since increase in velocity is a measure of increase in density, we can estimate the density of the rock beneath the surface. At the surface of the earth, as I have mentioned, the average density is 2·8 grams per cubic centimetre; 1,000 miles down it amounts to 5 grams per cubic centimetre, 1,800 miles down, nearly 6 grams per cubic centimetre.

At the depth of 1,800 miles, there is an abrupt change. Secondary waves are stopped cold. The British geologist R. D. Oldham maintained, in 1906, that this must mean that the region below is liquid: the waves have reached the boundary of the earth's 'liquid core'. And primary waves on reaching this level change direction sharply; apparently they are refracted by entering the liquid core.

The boundary of the liquid core is called the 'Gutenberg discontinuity', after the American geologist Beno Gutenberg, who in 1914 defined the boundary and showed that the core extended to 2,160 miles from the earth's centre. The density of the various deep layers of the earth were worked out in 1936 from earthquake data by the Australian mathematician Keith Edward Bullen. His results were confirmed by the data yielded by the huge Chilean earthquake of 1960. We can therefore say that at the Gutenberg discontinuity the density of the material jumps from 6 to 9, and thereafter it increases smoothly to 11·5 grams per cubic centimetre at the centre.

What is the nature of the liquid core? It must be composed of a substance which has a density of from 9 to 11·5 grams per cubic centimetre under the conditions of temperature and pressure in the core. The pressure is estimated to range from 10,000 tons per square inch at the top of the liquid core to 25,000 tons per square inch at the centre of the earth. The temperature is less certain. On the basis of the rate at which temperature is known to increase with depth in deep mines and of the rate at which rocks can conduct heat, geologists estimate (rather roughly) that temperatures in the liquid core must be as high as 5,000° C. (The centre of the much larger planet Jupiter may be as high as 500,000° C.)

The substance of the core must be some common element – common enough to be able to make up a sphere half the diameter of the earth and one third its mass. The only heavy element that is at all common in the universe is iron. At the earth's surface its density is only 7·86 grams per cubic centimetre, but under the enormous pressures of the core, it would have a density in the correct range – 9 to 12 grams per cubic centimetre. What is more, under centre-of-the-earth conditions it would be liquid.

If more evidence is needed, meteorites supply it. These fall into two broad classes: 'stony' meteorites, composed chiefly of silicates, and 'iron' meteorites, made up of about 90 per cent iron, 9 per cent nickel, and 1 per cent other elements. Many scientists believe that the meteorites are remnants of a shattered planet; if so, the iron meteorites may be pieces from the liquid core of that planet and the stony meteorites fragments of its mantle. (Indeed, in 1866, long before seismologists had probed the earth's core, the composition of the iron meteorites suggested to the French geologist Gabriel Auguste Daubrée that the core of our planet was made of iron.)

Today most geologists accept the liquid nickel–iron core as one of the facts of life as far as the earth's structure is concerned. One major refinement, however, has been introduced. In 1936, the Danish geologist Inge Lehmann, seeking to explain the puzzling fact that some primary waves show up in a 'shadow zone' on the surface from which most such waves are excluded, proposed that a discontinuity within the core about 800 miles from the centre introduced another bend in the waves and sent a few careering into the shadow zone. Gutenberg supported this view, and now many geologists differentiate between an 'outer core' that is liquid nickel–iron and an 'inner core' that differs from the outer core in some way, perhaps in being solid or slightly different chemically. As a result of the great Chilean earthquake of 1960, the entire globe was set into slow vibrations at rates matching those predicted by taking the inner core into account. This is strong evidence in favour of its existence.

The portion of the earth surrounding the nickel–iron core is called the 'mantle'. It seems to be composed of silicates, but judging from the velocity of earthquake waves passing through them, these silicates are different from the typical rocks of the earth's surface – something first shown in 1919 by the American physical chemist Leason Heberling Adams. Their properties suggest that they are rocks of the so-called 'olivine' type (olive-green in colour), which are comparatively rich in magnesium and iron and poor in aluminium.

The mantle does not quite extend to the surface of the earth. A Croation geologist named Andrija Mohorovičić, while studying the waves produced by a Balkan earthquake in 1909, decided that there was a sharp increase in wave velocity at a point about twenty miles beneath the surface. This 'Mohorovičić discontinuity' (known as 'Moho' for short) is now accepted to be the boundary of the earth's 'crust'.

The nature of the crust and of the upper mantle is best explored by means of the 'surface waves' I mentioned earlier. Like the 'bodily waves', the surface waves come in two varieties. One kind are called 'Love waves' (after their discoverer A. E. H. Love). The Love waves are horizontal ripples, like the shape of a snake moving over the ground. The other variety is the 'Rayleigh waves' (named after the English physicist John William Strutt, Lord Rayleigh); these ripples are vertical, like the path of a sea serpent moving through the water.

Analysis of these surface waves (notably by Maurice Ewing of Columbia University) shows that the crust is of varying thickness. It is thinnest under the ocean basins, where the Moho discontinuity in some places is only 8 to 10 miles below sea level. Since the oceans themselves are 5 to 7 miles deep in spots, the solid crust

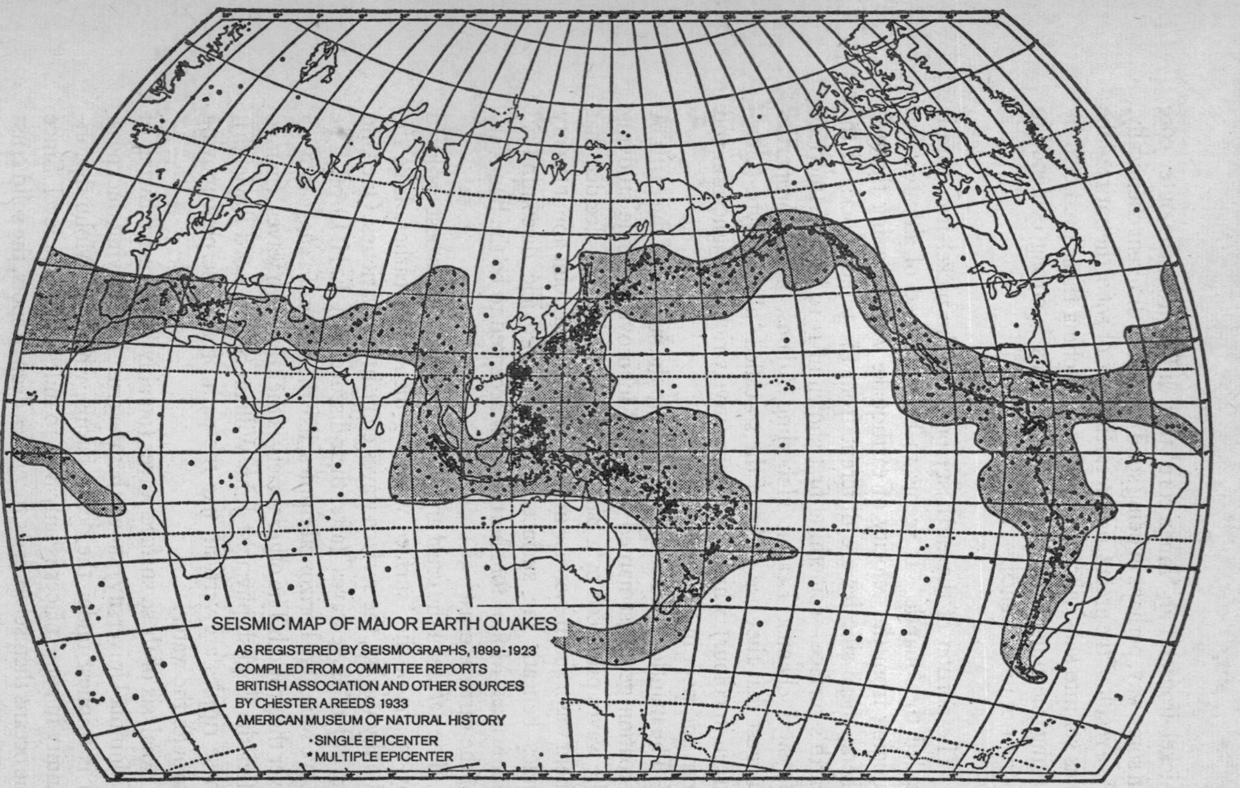

The earth's earthquake belts. They follow the principal zones of new mountain building.

may be as thin as 3 miles under the ocean deeps. Under the continents, on the other hand, the Moho discontinuity lies at an average depth of about 20 miles below sea level (it is about 22 miles under New York City, for instance), and it plunges to a depth of nearly 40 miles beneath mountain ranges. This fact, combined with evidence from gravity measurements, shows that the rock in mountain ranges is less dense than the average.

The general picture of the crust is that of a structure composed of two main types of rock – basalt and granite – with the less dense granite riding buoyantly on the basalt, forming continents and, in places where the granite is particularly thick, mountains (just as a large iceberg rises higher out of the water than a small one). Young mountains thrust their granite roots deep into the basalt, but, as the mountains are worn down by erosion, they adjust by floating slowly upwards (to maintain the equilibrium of mass called 'isostasy', a name suggested in 1889 by the American geologist Clarence Edward Dutton). In the Appalachians, a very ancient mountain chain, the root is about gone.

The basalt beneath the oceans is covered with a quarter to a half mile of sedimentary rock, but little or no granite – the Pacific basin is completely free of granite. The thinness of the crust under the oceans has suggested a dramatic project: Why not drill a hole through the crust down to the Moho discontinuity and tap the mantle to see what it is made of? It will not be an easy task, for it will mean anchoring a ship over an abyssal section of the ocean, lowering drilling gear through miles of water, and then drilling through a greater thickness of rock than anyone has yet drilled. Early enthusiasm for the project evaporated, however, and the matter now lies in abeyance.

The 'floating' of the granite in the basalt inevitably suggests the possibility of 'continental drift'. In 1912, the German geologist Alfred Lothar Wegener suggested that the continents were originally a single piece of granite, which he called 'Pangaea' ('All-Earth'). At some early stage of the earth's history this fractured and the continents drifted apart. He argued that they were still drifting – Greenland, for instance, moving away from Europe at the rate of a yard a year. What gave him (and others, dating back to Francis Bacon about 1620) the idea was mainly the fact that the eastern coastline of South America seemed to fit like a jigsaw piece into the shape of the western coast of Africa.

For a half-century, Wegener's theory was looked upon with hard disfavour. As late as 1960, when the first edition of this book was published, I felt justified, in view of the state of geophysical opinion at that time, in categorically dismissing it. The most telling argument against it was that the basalt underlying both oceans and continents was simply too stiff to allow the continental granite to drift sideways.

And yet evidence in favour of the supposition that the Atlantic Ocean once did not exist, and that the separate continents once formed a single land-mass, grew massively impressive. If the continents were matched, not by their actual shore line (an accident of the current sea-level) but by the central point of the continental slope (the shallow floor of the ocean neighbouring the continents which is exposed during ages of low sea-level), then the fit is excellent all along the Atlantic, in the north as well as the south. Then, too, rock formations in parts of western Africa match the formations in parts of eastern South America in fine detail. Past wanderings of the magnetic poles look less startling if one considers that the continents, not the poles, wandered.

Perhaps the most devastating piece of evidence arrived in 1968 when a 2½-inch fossilized bone from an extinct amphibian was found in Antarctica. Such a creature could not possibly have lived so close to the South Pole, so Antarctica must once have been farther from the pole, or at least milder in temperature. The amphibian could not have crossed even a narrow stretch of salt water, so Antarctica must have been part of a larger body of land, containing warmer areas.

That still leaves the question of what it was that caused the original super-continent to break up and drift apart. About 1960, the American geologist Harry Hammond Hess suggested that molten mantle material might be welling up – along certain fracture-lines running the length of the Atlantic Ocean, for instance – be forced sideways near the top of the mantle, cool, and harden. The ocean floor is, in this way, pulled apart and stretched.

It is not, then, that the continents drift, but that they are pushed apart by a spreading sea-floor.

As the story seems now, Pangaea did exist, after all, and was intact as recently as 225 million years ago, when the dinosaurs were coming into prominence. Judging from the evolution and distribution of plants and animals, the break-up must have be-

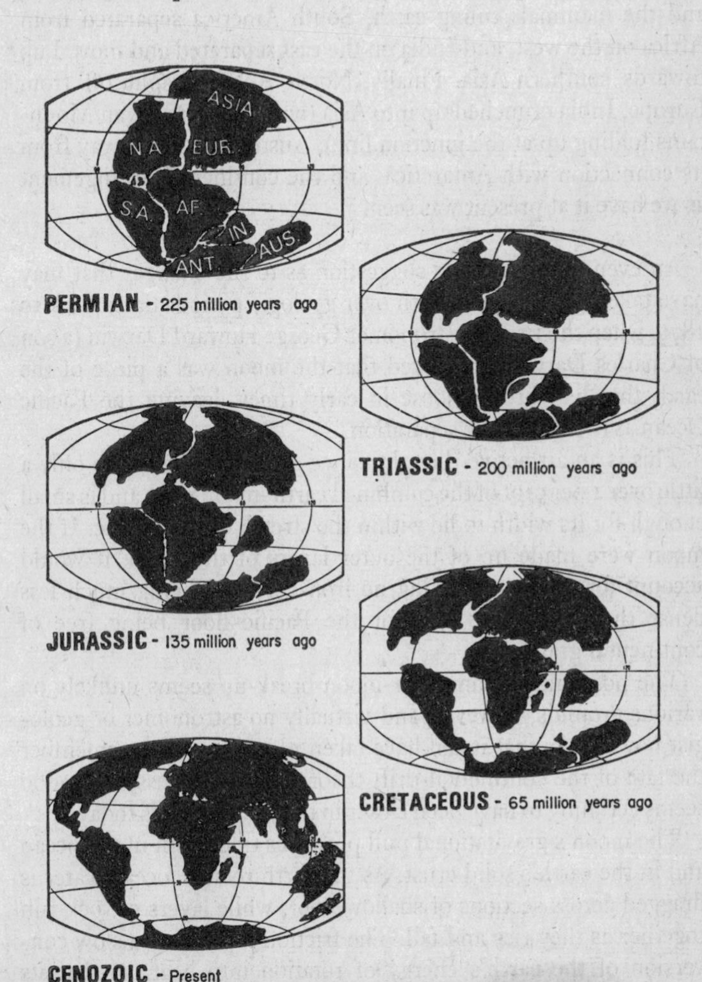

PERMIAN - 225 million years ago

TRIASSIC - 200 million years ago

JURASSIC - 135 million years ago

CRETACEOUS - 65 million years ago

CENOZOIC - Present

come pronounced about 200 million years ago. Pangaea then broke into three parts. The northern part (North America, Europe, and Asia) is called 'Laurasia'; the southern part (South America, Africa, and India) is called 'Gondwana', from an Indian province. Antarctica plus Australia formed a third part.

Some 65 million years ago, with the dinosaurs already extinct and the mammals ruling earth, South America separated from Africa on the west, and India on the east separated and moved up towards southern Asia. Finally, North America split off from Europe, India crunched up into Asia (with the Himalayan Mountains folding up at the junction line), Australia moved away from its connection with Antarctica, and the continental arrangement as we have it at present was seen.

An even more startling suggestion as to the changes that may have taken place on the earth over geologic periods dates back to 1879, when the British astronomer George Howard Darwin (a son of Charles Darwin) suggested that the moon was a piece of the earth that had broken loose in early times, leaving the Pacific Ocean as the scar of the separation.

This is an attractive thought, since the moon makes up only a little over 1 per cent of the combined earth–moon mass and is small enough for its width to lie within the stretch of the Pacific. If the moon were made up of the outer layers of the earth, it would account for the moon having no iron core and being much less dense than the earth, and for the Pacific floor being free of continental granite.

The possibility of an earth–moon break-up seems unlikely on various grounds, however, and virtually no astronomer or geologist now thinks that it can have taken place (however, remember the fate of the continental-drift theory). Nevertheless, the moon seems certainly to have been closer in the past than it is today.

The moon's gravitational pull produces tides both in the ocean and in the earth's solid crust. As the earth rotates, ocean water is dragged across sections of shallow floor, while layers of rock rub together as they rise and fall. The friction represents a slow conversion of the earth's energy of rotation into heat, so that its rotational period gradually increases. The effect is not great in human terms, for the day lengthens by one second in about a hundred thousand years. As the earth loses rotational energy, the angular momentum must be conserved. What the earth loses, the moon gains. Its speed increases as it revolves about the earth, which means it drifts farther away very slowly.

If one works backwards in time towards the far geologic past, we see that the earth's rotation must speed up, the day be significantly shorter, the moon significantly closer, and the whole effect more rapid. Darwin calculated backwards to find out when the moon was close enough to earth to form a single body, but even if we don't go that far we ought to find evidence of a shorter day in the past. For instance, about 570 million years ago – the time of the oldest fossils – the day may have been only a little over 20 hours long, and there may have been 428 of them in a year.

Nor is this only theory now. Certain corals lay down bands of calcium carbonate more actively at some seasons than others, so that you can count annual bands just as in tree trunks. It is also suggested that some lay down calcium carbonate more actively by day than by night, so that there are very fine daily bands. In 1963, the American palaeontologist John West Wells counted the fine bands in fossil corals and reported there were, on the average, 400 daily bands per annual band in coral dating back 400 million years and 380 daily bands per annual band in corals dating back only 320 million years.

Of course, the question is, if the moon was much closer to the earth then, and the earth rotated more rapidly, what happened in still earlier periods? If Darwin's theory of an earth–moon separation is not so, what *is* so?

One suggestion is that the moon was captured at some time in the past. If it were captured 600 million years ago, for instance, that might account for the fact that at about that time we begin to find numerous fossils in rocks, whereas earlier rocks have nothing but uncertain traces of carbon. Perhaps the earlier rocks were washed clean by the vast tides that accompanied the capture of the moon. (There was no land life at the time; if there had been, it would have been destroyed.) If the moon were captured, it would have been closer then than now, and there would be a lunar recession and a lengthening of the day since, but nothing of the sort before.

Another suggestion is that the moon was formed in the neighbourhood of the earth, out of the same gathering dust-cloud, and has been receding ever since, but that it never was actually part of the earth. The truth is that astronomers still don't know, but they hope to find out through a continued exploration of the moon's surface either by men or machines landed on our companion.

The fact that the earth consists of two chief portions – the silicate mantle and the nickel–iron core (in about the same proportions as the white and yolk of an egg) – had persuaded most geologists that the earth must have been liquid at some time in its early history. It might then have consisted of two mutually insoluble liquids. The silicate liquid, being the lighter, would float to the top and cool by radiating its heat into space. The underlying iron liquid, insulated from direct exposure to space, would give up its heat far more slowly and would thus remain liquid to the present day.

There are at least three ways in which the earth could have become hot enough to melt, even from a completely cold start as a collection of planetesimals. These bodies, on colliding and coalescing, would give up their energy of motion ('kinetic energy') in the form of heat. Then, as the growing planet was compressed by gravitational force, still more energy would be liberated as heat. Third, the radioactive substances of the earth – uranium, thorium, and potassium – have delivered large quantities of heat over the ages as they have broken down; in the early stages, when there was a great deal more radioactive material than now, radioactivity itself might have supplied enough heat to liquefy the earth.

Not all scientists are willing to accept a liquid stage as an absolute necessity. The American chemist Harold Clayton Urey, in particular, believes that most of the earth was always solid. He argues that in a largely solid earth an iron core could still be formed by a slow separation of iron; even now, he suggests, iron may be migrating from the mantle into the core at the rate of 50,000 tons a second.

Cooling of the earth from an original molten or near-molten state would help to explain its wrinkled exterior. As the cooling earth shrank, its crust would occasionally buckle. Minor buckling would give rise to earthquakes. Larger buckling, or a steady accumulation of smaller adjustments, would eventually produce mountain ranges. The mountain-building eras, however, would be relatively brief. After mountains were formed, they would be worn down by erosion in fairly short order (on the geological time scale), and then would come a long period of stability before compressional forces built up great enough strains to start a new crust-buckling stage. Thus during most of its lifetime the earth would be a rather drab and featureless planet, with low continents and shallow seas.

The trouble with this view is that the earth does not seem really to be cooling off. The thought that it must be doing so arises from the natural assumption that a hot body *must* cool down if there is no source of continuing heat. True! But in earth's case, there *is* a source of continuing heat, one that was not understood prior to the twentieth century. This new source became apparent with the discovery of radioactivity in 1896, when it appeared that a hitherto utterly unsuspected form of energy lay hidden deep within the recesses of the atom.

It appears that over the last several hundred million years radioactivity has been generating enough heat in the crust and mantle at least to keep the earth's internal temperature from falling; if anything, the earth may be very slowly heating up. Yet, despite that, we are now living at the tag end of a mountain-building era (fortunately for those of us who are fond of rugged scenery). If the earth has not been cooling and shrinking during that period, how were our present mountains built?

A couple of decades ago a theory was put forward by the Israeli physicist Chaim L. Perkeris and elaborated by the American geologist D. T. Briggs. This theory, which resembles the later notion of the spreading of the ocean floor, begins by supposing that heat coming from the core periodically sets up a series of vertical eddies in the mantle. The eddies of heated material rise towards the crust and sink again after they cool there. Since the mantle is not liquid, merely plastic, this motion is very slow – perhaps not more than two inches a year.

Now, where two neighbouring eddies move downwards, a portion of crust is sucked downwards too, forming a root of light crustal material in the heavier mantle. This root is converted by the mantle's heat into granite. Afterwards, isostasy causes the root and its overlay of light material to rise and form a mountain chain. The period of mountain-building, lasting perhaps 60 million years, is followed by a quiescent period of 500 million years during which enough heat accumulates in the mantle to start a new cycle. It may be, then, that mountain-building and continental drift are interrelated.

The Ocean

The earth is unusual among the planets of the solar system in possessing a surface temperature that permits water to exist in all three states: liquid, solid, and gas. The earth is also the only body in the solar system, as far as we know, to have oceans. Actually I should say 'ocean', because the Pacific, Atlantic, Indian, Arctic, and Antarctic oceans all comprise one connected body of salt water in which the Europe–Asia–Africa mass, the American continents, and smaller bodies such as Antarctica and Australia can be considered islands.

The statistics of this ocean are impressive. It has a total area of 140 million square miles and covers 71 per cent of the earth's surface. Its volume, reckoning the average depth of the oceans as two and one third miles, is about 326 million cubic miles – 0·15 per cent of the total volume of our planet. It contains 97·2 per cent of all the H_2O on the earth and is the source of the earth's fresh-water supply as well, for 80,000 cubic miles of it are evaporated each year to fall again as rain or snow. As a result of such precipitation, there is some 200,000 cubic miles of fresh water under the continents' surface and about 30,000 cubic miles of fresh water gathered into the open as lakes and rivers.

The ocean is of peculiar importance to life. Almost certainly the first forms of life originated there, and, from the standpoint of sheer quantity, the oceans still contain most of our planet's life. On land, life is confined to within a few feet of the surface (though birds and aeroplanes do make temporary sorties from this base); in the oceans, life permanently occupies the whole of a realm as deep as seven miles or more in some places.

And yet, until recent years mankind has been as ignorant of the oceans, and particularly of the ocean floor, as of another planet. Even today, astronomers know more about the surface of the moon than geologists know about the surface of the earth under the oceans.

The founder of modern oceanography was an American naval officer named Matthew Fontaine Maury. In his early thirties, he was lamed in an accident that, however unfortunate for himself, brought benefits to humanity. Placed in charge of the depot of charts and instruments (undoubtedly intended as a sinecure), he threw himself into the task of charting ocean currents. In particular, he studied the course of the Gulf Stream, which had first been investigated as early as 1769 by the American scholar Benjamin Franklin. Maury gave it a description that has become a classic remark in oceanography: 'There is a river in the ocean.' It is certainly a much larger river than any on land. It transports a thousand times as much water each second as does the Mississippi. It is 50 miles wide at the start, nearly ½ mile deep, and moves at speeds of up to 4 miles an hour. Its warming effect is felt even in the far northern island of Spitzbergen.

Maury also initiated international cooperation in studying the ocean; he was the moving figure behind a historic international conference held in Brussels in 1853. In 1855, he published the first textbook in oceanography, entitled *Physical Geography of the Sea*. The Naval Academy at Annapolis honoured his achievements by naming Maury Hall after him.

Since Maury's time, the ocean currents have been thoroughly mapped. They move in large clockwise circles in the oceans of the Northern Hemisphere and in large counter-clockwise circles in those of the Southern, thanks to the Coriolis effect (see p. 113). A current moving directly along the Equator is not subjected to a Coriolis effect and may move in a straight line. Such a thin, straight current was located in the Pacific Ocean, moving due east for several thousand miles along the Equator. It is called the 'Cromwell current' after its discoverer, the American oceanographer Townsend Cromwell. A similar current, somewhat slower, was discovered in the Atlantic, in 1961, by the American oceanographer Arthur D. Voorhis.

Furthermore, oceanographers have even begun to explore the more sluggish circulation of the ocean depths. That the deeps cannot maintain a dead calm is clear from several indirect forms of evidence. For one thing, the life at the top of the sea is continually consuming its mineral nutrients – phosphate and nitrate – and carrying this material down to the depths with itself after death; if there were no circulation to bring it up again, the surface would become depleted of these minerals. For another thing, the oxygen supplied to the oceans by absorption from the air would not percolate down to the depths at a sufficient rate to support life there if there were no conveying circulation. Actually oxygen is found in adequate concentration down to the very floor of the abyss. This can be explained only by supposing that there are regions in the ocean where oxygen-rich surface waters sink.

The engine that drives this vertical circulation is temperature difference. The ocean's surface water is cooled in arctic regions, and it therefore sinks. This continual flow of sinking water spreads out all along the ocean floor, so that even in the tropics the bottom water is very cold – near the freezing point. Eventually the cold water of the depths wells up towards the surface, for it has no other place to go. After rising to the surface, the water warms and drifts off towards the Arctic or the Antarctic, there to sink again. The resulting circulation, it is estimated, would bring about complete mixing of the Atlantic Ocean, if something new were added to part of it, in about 1,000 years. The larger Pacific Ocean would undergo complete mixing in perhaps 2,000 years.

The continental barriers complicate this general picture. To follow the actual circulations, oceanographers have resorted to oxygen as a tracer. Cold water absorbs more oxygen than warm water can. The arctic surface water, therefore, is particularly rich in oxygen. After it sinks, it steadily loses oxygen to organisms feeding in it. So by sampling the oxygen concentration in deep water at various locations, it is possible to plot the direction of the deep-sea currents.

Such mapping has shown that one major current flows from the Arctic Ocean down the Atlantic under the Gulf Stream and in the opposite direction, another from the Antarctic up the south Atlantic. The Pacific Ocean gets no direct flow from the Arctic to speak of because the only outlet into it is the narrow and shallow Bering Strait. This is why it is the end of the line for the deep-sea flow. That the north Pacific is the dead end of the global flow is shown by the fact that its deep waters are poor in oxygen. Large parts of this largest ocean are therefore sparsely populated with life forms and are the equivalent of desert areas on land. The same may be said of nearly land-locked seas like the Mediterranean, where full circulation of oxygen and nutrients is partly choked off.

More direct evidence for this picture of the deep-sea currents was obtained in 1957 during a joint British–American oceanographic expedition. The investigators used a special float, invented by the British oceanographer John C. Swallow, which is designed to keep its level at a depth of a mile or more and is equipped with a device for sending out short-wave sound waves. By means of these signals the float can be tracked as it moves with the deep-sea current. The expedition thus traced the deep-sea current down the Atlantic along its western edge.

All this information will acquire practical importance when the world's expanding population turns to the ocean for more food. Scientific 'farming of the sea' will require knowledge of these fertilizing currents, just as land farming requires knowledge of river courses, ground water, and rainfall. The present harvest of sea-food – some 55 million tons per year – can, with careful and efficient management, be increased (it is estimated) to something over 200 million tons per year, while leaving sea life enough leeway to maintain itself adequately. (This, of course, presupposes that we do not continue our present course of heedlessly damaging and polluting the ocean, particularly those portions of the ocean –

nearest the continental shores – that contain and offer man the major portion of sea organisms. So far, we are not only failing to rationalize a more efficient use of the sea for food, but are decreasing its ability to yield us the quantity of food we harvest now.)

Food is not the only important resource of the ocean. Sea water contains in solution vast quantities of almost every element. As much as 4,000 million tons of uranium, 300 million tons of silver, and 4 million tons of gold are contained in the oceans, but in dilution too great for practical extraction. However, both magnesium and bromine are now obtained from sea water on a commercial scale. By the end of the 1960s, the value of the magnesium obtained from the ocean was $70 million per year, while 75 per cent of all the bromine produced in the world came from the sea. Moreover, an important source of iodine is dried seaweed, the living plants having previously concentrated the element out of sea water to an extent that man cannot yet profitably duplicate.

Much more prosaic material is dredged up from the sea. From the relatively shallow waters bordering the United States, some 20 million tons of oyster shells are obtained each year to serve as a valuable source of limestone. In addition, 50 million cubic yards of sand and gravel are obtained in similar fashion.

Scattered over the deeper portions of the ocean floor are metallic nodules that have precipitated out about some nucleus that may be a pebble or a shark tooth. (It is the oceanic analogue of the formation of a pearl about a sand-grain inside an oyster.) These are usually referred to as manganese nodules because they are richest in that metal. It is estimated that there are 31,000 tons of these nodules per square mile of the Pacific floor. Obtaining these in quantity would be difficult indeed and the manganese content alone would not make it worthwhile under present conditions. However, the nodules contain 1 per cent nickel, 0·5 per cent of copper, and 0·5 per cent cobalt. These minor constituents make the nodules far more attractive than they would otherwise be.

Even the 97 per cent of the ocean substance that is actually water is important. Mankind presses ever harder on the limited fresh-water supplies of the planet; eventually more and more use will have to be made of ocean water from which the salts have been removed, a process known as 'desalination'. Already some 700 desalination plants, with a capacity of up to 30,000 gallons of fresh water per day, exist throughout the world. On the whole, such sea-borne fresh water cannot yet compete with rain-borne fresh water in most parts of the world, but the technology involved is as yet young.

It is only within the last century that man has plumbed the great deeps of the ocean. The sea bottom first became a matter of practical interest to mankind (rather than one of intellectual curiosity to a few scientists) when it was decided to lay a telegraph cable across the Atlantic. In 1850, Maury had worked up a chart of the Atlantic sea bottom for purposes of cable-laying. It took fifteen years, punctuated by many breaks and failures, before the Atlantic cable was finally laid – under the incredibly persevering drive of the United States financier Cyrus West Field, who lost a fortune in the process. (More than twenty cables now span the Atlantic.)

Systematic exploration of the sea bottom began with the famous around-the-world expedition of the British *Challenger* in the 1870s. To measure the depth of the oceans the *Challenger* had no better device than the time-honoured method of paying out four miles of cable with a weight on the end until it reached the bottom. Over 360 soundings were made in this fashion. This procedure is not only fantastically laborious (for deep sounding) but also of low accuracy. Ocean-bottom exploration was revolutionized in 1922 with the introduction of echo-sounding by means of sound waves; in order to explain how this works, a digression on sound is in order.

Mechanical vibrations set up longitudinal waves in matter (in air, for instance), and we can detect some of these as sound. We hear different wavelengths as sounds of different pitch. The deepest sound we hear has a wavelength of 22 metres and a frequency of 15 cycles per second. The shrillest sound a normal adult can hear has a wavelength of 2·2 centimetres and a frequency of 15,000 cycles per second. (Children can hear somewhat shriller sounds.)

The absorption of sound by the atmosphere depends on the wavelength. The longer the wavelength, the less sound is absorbed by a given thickness of air. For this reason, fog-horn blasts are far in the bass register so that they can penetrate as great a distance as possible. The fog-horn of the *Queen Mary* sounded at twenty-seven vibrations per second, about that of the lowest note on the piano. It could be heard at a distance of 10 miles, and instruments could pick up the sound at a distance of 100 to 150 miles.

Sounds deeper in pitch than the deepest we can hear also exist. Some of the sounds set up by earthquakes or volcanoes are in this 'infrasonic' range. Such vibrations can encircle the earth, sometimes several times, before being completely absorbed.

The efficiency with which sound is reflected depends on the wavelength in the opposite way. The shorter the wavelength, the more efficient the reflection. Sound waves with frequencies higher than those of the shrillest sounds we hear are even more efficiently reflected. Some animals can hear shriller sounds than we can and make use of this. Bats squeak to emit sound waves with 'ultrasonic' frequencies as high as 130,000 cycles per second and listen for the reflections. From the direction in which reflections are loudest and from the time lag between squeak and echo they can judge the location of insects to be caught and twigs to be avoided. They can thus fly with perfect efficiency if they are blinded, but not if they are deafened. (The Italian biologist Lazzaro Spallanzani, who first observed this in 1793, wondered if bats could see with their ears, and, of course, in a sense, they do.)

Porpoises, as well as guacharos (cave-dwelling birds of Venezuela), also use sounds for 'echo-location' purposes. Since they are interested in locating larger objects, they can use the less efficient sound waves in the audible region for the purpose. (The complex sounds emitted by the large-brained porpoises and dolphins may even, it is beginning to be suspected, be used for purposes of general communication – for talking, to put it bluntly. The American biologist John C. Lilly has been investigating this possibility exhaustively.)

To make use of the properties of ultrasonic sound waves, men must first produce them. Small-scale production and use are exemplified by the 'dog whistle' (first constructed in 1883). It produces sound in the near ultrasonic range that can be heard by dogs, but not by humans.

A route whereby much more could be done was opened by the French chemist Pierre Curie and his brother, Jacques, who in 1880 discovered that pressures on certain crystals produced an electric potential ('piezoelectricity'). The reverse was also true. Applying an electric potential to a crystal of this sort produced a slight constriction as though pressure were being applied ('electrostriction'). When the technique for producing a very rapidly fluctuating potential was developed, crystals could be made to vibrate quickly enough to form ultrasonic waves. This was first done in 1917 by the French physicist Paul Langevin, who immediately applied the excellent reflective powers of this short-wave sound to the detection of submarines. During the Second World War, this method was perfected and became 'sonar' ('*so*und *n*avigation *a*nd *r*anging', 'ranging' meaning 'determining distance').

The determination of the distance of the sea bottom by the reflection of ultrasonic sound waves was what replaced the sounding line. The time interval from the sending of the signal (a sharp pulse) and the return of its echo measures the distance to the bottom. The only thing the operator has to worry about is whether the reading signals a false echo from a school of fish or some other obstruction. (Obviously the instrument is useful to fishing fleets.)

The echo-sounding method not only is swift and convenient, but also makes it possible to trace a continuous profile of the bottom over which the vessel moves, so that oceanographers are obtaining a picture of the topography of the ocean bottom. It turns out to be more rugged than the land surface, and its features have a grander scale. There are plains of continental size and mountain ranges longer and higher than any on land. The island of Hawaii is the top of an underwater mountain 33,000 feet high – higher than anything in the Himalayas – so that Hawaii may fairly be called the tallest mountain on the earth. There are also numerous flat-topped cones called 'sea-mounts' or 'guyots'. The latter name honours the Swiss-American geographer Arnold Henry Guyot, who brought scientific geography to the United States when he emigrated to America in 1848. Sea-mounts were first discovered during the Second World War by the American geologist Harry Hammond Hess, who located nineteen in quick succession. At least 10,000 exist, mostly in the Pacific. One of these, discovered in 1964 just south of Wake Island, is over 14,000 feet high.

Moreover, there are deep abysses (trenches) in which the Grand Canyon would be lost. The trenches, all located alongside island archipelagoes, have a total area amounting to nearly 1 per cent of the ocean bottom. This may not seem much, but it is actually equal to one half the area of the United States, and the trenches contain fifteen times as much water as all the rivers and lakes in the world. The deepest of them are in the Pacific; they are found there alongside the Philippines, the Marianas, the Kuriles, the Solomons, and the Aleutians. There are other great abysses in the Atlantic off the West Indies and the South Sandwich Islands, and there is one in the Indian Ocean off the East Indies.

Besides the trenches, oceanographers have traced on the ocean bottom canyons, sometimes thousands of miles long, which look like river channels Some of them actually seem to be extensions of rivers on land, notably a canyon extending from the Hudson River into the Atlantic. At least twenty such huge gouges have been located in the Bay of Bengal alone, as a result of oceanographic studies of the Indian Ocean during the 1960s. It is tempting to suppose that these were once river beds on land, when the ocean was lower than now. But some of the undersea channels are so far below the present sea level that it seems altogether unlikely they could ever have been above the ocean. In recent years, various oceanographers, notably Maurice Ewing and Bruce C. Heezen, have developed another theory: that the undersea canyons were gouged out by turbulent flows ('turbidity currents') of soil-laden water in an avalanche down the off-shore continental slopes at speeds of up to sixty miles an hour. One turbidity current, which focused scientific attention on the problem, took place in 1929 after an earthquake off Newfoundland. The current snapped a number of cables, one after the other, and made a great nuisance of itself.

The most dramatic find concerning the sea bottom, though, was foreshadowed in 1853, when the Atlantic-cable project was in progress. Soundings of the ocean depth were taken, and it was reported that there seemed signs of an undersea plateau in the middle of the ocean. The Atlantic seemed shallower in the middle than on either side.

Naturally, it was only practical to make a few soundings by actual line, but in 1922 the German oceanographic vessel *Meteor* began to make soundings in the Atlantic with ultrasonic devices. By 1925, they were able to report a vast undersea mountain range winding down the Atlantic. The highest peak broke through the water surface and appeared as islands such as the Azores, Ascension, and Tristan da Cunha.

Later soundings elsewhere showed that the mountain range was not confined to the Atlantic. At its southern end it curves around Africa and moves up the western Indian Ocean to Arabia. In mid-Indian Ocean, it branches so that the range continues south of Australia and New Zealand and then works northwards in a vast circle all around the Pacific Ocean. What began (in men's minds) as the Mid-Atlantic ridge became the Mid-Oceanic Ridge. And in one rather basic fashion, the Mid-Oceanic Ridge is not like the mountain ranges on the continent. The continental highlands are of folded sedimentary rocks, while the vast oceanic ridge is of basalt squeezed up from the hot lower depths.

After the Second World War, the details of the ocean floor were probed with new energy by Ewing and Heezen. Detailed soundings in 1953 showed, rather to their astonishment, that a deep canyon ran the length of the Ridge and right along its centre. This was eventually found to exist in all portions of the Mid-Oceanic Ridge, so that sometimes it is called the 'Great Global Rift'. There are places where the Rift comes quite close to land: it runs up the Red Sea between Africa and Arabia, and it skims the borders of the Pacific through the Gulf of California and up the coast of the state of California.

At first it seemed that the Rift might be continuous, a 40,000-mile crack in the earth's crust. Closer examination, however, showed that it consisted of short, straight sections that were set off from each other as though earthquake shocks had displaced one section from the next. And, indeed, it was along the Rift that the earth's quakes and volcanoes tended to occur.

The Rift was a weak spot up through which heated molten rock, 'magma', welled slowly from the interior – cooling, piling up to form the ridge, and spreading out farther still. The spreading can be as rapid as 16 centimetres per year, and the entire Pacific Ocean floor could be covered with a new layer in 100 million years. Indeed, sediment drawn up from the ocean floor is rarely found to be older, which would be remarkable in a planetary life forty-five times as long, were it not for the concept of 'sea-floor spreading'.

The Rift and its branches seem to divide the earth's crust into six large plates and some smaller ones. As a result of the activity along the Rift, these plates move, but as units; there is no motion to speak of among the surface features of a given plate. It is the movement of these plates that accounts for the break-up of Pangaea and the continental drifting since. There is nothing to show that the drifting may not eventually bring the continents together again, perhaps in a new arrangement. There may have been many Pangaeas formed and broken up in the earth's lifetime, with the latest break-up most clearly seen in the records only because it is the latest.

This concept of the motion of the plates may serve to explain many features of earth's crust whose origin was obscure earlier. When two plates come together slowly, and crust buckles and bulges both up and down, forming mountains and their 'roots'. Thus, the Himalayan Mountains seem to have been formed when the plate bearing India made slow contact with the plate bearing the rest of Asia.

On the other hand, when two plates come together too rapidly to allow buckling, the surface of one plate may gouge its way under the other, forming a trench, a line of islands, and a disposition towards volcanic activity. Such trenches and islands are found in the western Pacific, for instance.

Plates pull apart under the influence of sea-floor spreading, as well as come together. The rift passes right through western Iceland, which is (very slowly) pulling apart. Another place of division is at the Red Sea, which is rather young and exists only because Africa and Arabia have already pulled apart somewhat. (The opposite shores of the Red Sea fit closely if put together.) This process is continuing, so that the Red Sea is, in a sense, a new ocean in the process of formation. Active upwelling in the Red Sea is indicated by the fact that at the bottom of that body of water there are, as discovered in 1965, sections with a temperature of 56°C and a salt concentration at least five times normal.

The existence of the Rift is of greatest immediate importance, naturally, to those people who live on those parts of earth's land surface that happen to be in its neighbourhood. The San Andreas Fault in California is actually part of the Rift, for instance, and it was the yielding of that fault which caused the San Francisco earthquake of 1906 and the Good Friday earthquake in Alaska in 1964.

The deep sea, surprisingly enough, contains life. Until nearly a century ago, life in the ocean was thought to be confined to the surface region. The Mediterranean, long the principal centre of civilization, is indeed rather barren of life in its lower levels. But though this sea is a semi-desert – warm and low in oxygen – the English naturalist Edward Forbes dredged up living starfish from a depth of a quarter of a mile in the 1840s. Then, in 1860, a telegraph cable was brought up from the Mediterranean bottom, a mile deep, and was found to be encrusted with corals and other forms of life.

In 1872, the *Challenger*, under the direction of the British naturalist Charles Wyville Thomson, in a voyage spanning 69,000 miles, made the first systematic attempt to dredge up life forms from the ocean bottom; he found plenty. Nor is the world of underwater life a region of eerie silence by any means. An underwater listening device, the 'hydrophone', has, in recent years, shown that sea creatures click, grunt, snap, moan, and, in general, make the ocean depths as maddeningly noisy as ever the land is.

Since the Second World War, numerous expeditions have explored the abyss. A new *Challenger* probed the Marianas Trench in the western Pacific in 1951 and found that it (and not one off the Philippine Islands) was one of the deepest gashes in the earth's

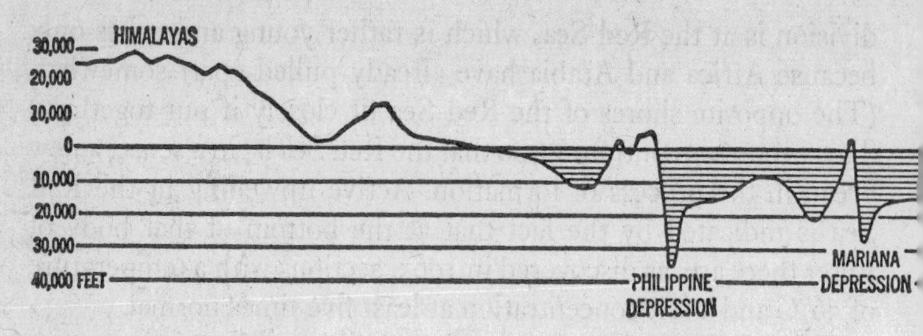

Profile of the Pacific bottom. The great trenches in the sea floor go deeper below sea level than the height of the Himalayas, and the Hawaiian peak stands higher from the bottom than the tallest land mountain.

crust. The deepest portion is now called the 'Challenger Deep'. It is over 36,000 feet deep. If Mount Everest were placed in it, a mile of water would roll over its topmost peak. Yet the *Challenger* brought up bacteria from the floor of the abyss. They look much like bacteria of the surface, but cannot live at a pressure of less than a thousand atmospheres!

The creatures of the trenches are so adapted to the great pressures of these bottoms that they are unable to rise out of their trench; in effect, they are imprisoned in an island. They have experienced a segregated evolution. Yet they are in many respects related to other organisms closely enough, so that it seems their evolution in the abyss has not gone on for a very long time. One can visualize some groups of ocean creatures being forced into lower and lower depths by the pressure of competition, just as other groups were forced higher and higher up the continental shelf until they emerged on to the land. The first group had to become adjusted to higher pressures, the second to the absence of water. On the whole, the latter adjustment was probably the more difficult, so we should not be amazed that life exists in the abyss.

To be sure, life is not as rich in the depths as nearer the surface. The mass of living matter below four and one half miles is only a tenth as great per unit volume of ocean as it is estimated to be at two miles. Furthermore, there are few, if any, carnivores below four and a half miles, since there are insufficient prey to support

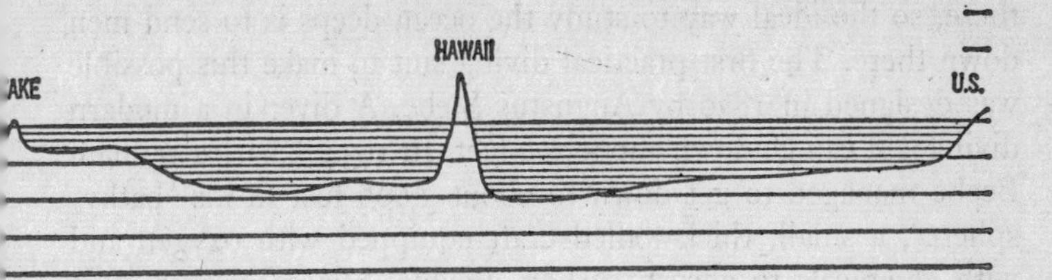

them. They are scavengers instead, eating anything organic that they can find. The recentness with which the abyss has been colonized is brought out by the disclosure that no species of creature found there has been developed earlier than 200 million years ago, and most have histories of no more than 50 million years. It is only at the beginning of the age of the dinosaurs that the deep sea, hitherto bare of organisms, was finally invaded by life.

Nevertheless, some of the organisms that invaded the deep survived there, whereas their relatives nearer the surface died out. This was demonstrated, most dramatically, in the late 1930s. On 25 December 1938, a trawler fishing off South Africa brought up an odd fish about five feet long. What was odd about it was that its fins were attached to fleshy lobes rather than directly to the body. A South African zoologist, J. L. B. Smith, who had the chance of examining it, recognized it as a matchless Christmas present. It was a coelacanth, a primitive fish that zoologists had thought extinct for 70 million years. Here was a living specimen of an animal that was supposed to have disappeared from the earth before the dinosaurs reached their prime.

The Second World War halted the hunt for more coelacanths, but in 1952 another of a different genus was fished up off Madagascar. By now numbers have been found. Because it is adapted to fairly deep waters, the coelacanth dies soon after being brought to the surface.

Evolutionists have been particularly interested in studying the coelacanth specimens because it was from this fish that the first amphibians developed, in other words, the coelacanth is a rather direct descendant of our fishy ancestors.

Just as the ideal way to study outer space is to send men out there, so the ideal way to study the ocean deeps is to send men down there. The first practical diving suit to make this possible was designed in 1830 by Augustus Siebe. A diver in a modern diving suit can go down about 300 feet. In 1934, Charles William Beebe managed to get down to about 3,000 feet in his 'bathysphere', a small, thick-walled craft equipped with oxygen and with chemicals to absorb carbon dioxide. His co-worker, Otis Barton, plumbed to a depth of 4,500 feet in 1948, using a modified bathysphere called a 'benthoscope'.

The bathysphere was an inert object suspended from a surface vessel by a cable (a snapped cable meant the end). What was needed was a manoeuvrable ship of the abyss. Such a ship, the 'bathyscaphe', was invented in 1947 by the Swiss physicist Auguste Piccard. Built to withstand great pressures, it used a heavy ballast of iron pellets (which are automatically jettisoned in case of emergency) to take it down and a 'balloon' containing petrol (which is lighter than water) to provide buoyancy and stability. In its first test off Dakar, West Africa, in 1948, the bathyscaphe (unmanned) descended 4,500 feet.

Later, Piccard and his son Jacques built an improved version of the bathyscaphe and named the new vessel *Trieste* because the then Free City of Trieste had helped finance its construction. In 1953, Piccard plunged two and a half miles into the depths of the Mediterranean.

The *Trieste* was bought by the United States Navy for research. On 14 January 1960, Jacques Piccard and a Navy man, Don Walsh, took it to the bottom of the Marianas Trench, plumbing seven miles to the deepest part of any abyss. There, at the ultimate ocean depth, where the pressure was 1,100 atmospheres, they found water currents and living creatures. In fact, the first creature seen was a vertebrate, a one-foot-long, flounder-like fish, with eyes.

In 1964, the French-owned bathyscaphe *Archimède* made ten trips to the bottom of the Puerto Rico Trench, which, with a depth of five and one quarter miles, is the deepest abyss in the Atlantic. There, too, every square foot of the ocean floor had its life form.

Oddly enough, the bottom did not descend smoothly into the abyss; rather it seemed terraced, like a giant, spread-out staircase.

The Ice Caps

The extremities of our planet have always fascinated mankind, and one of the most adventurous chapters in the history of science has been the exploration of the polar regions. Those regions are charged with romance, spectacular phenomena, and elements of man's destiny – the strange auroras in the sky, the extreme cold, and especially the immense ice-caps or glaciers, which hold the key to the world climate and man's way of life.

The actual push to the poles came rather late in human history. It began during the great age of exploration following the discovery of the Americas by Christopher Columbus. The first Arctic explorers went chiefly to find a sea route around the top of North America. Pursuing this will-o'-the-wisp, the English navigator Henry Hudson (in the employ of Holland) found Hudson Bay and his death in 1610. Six years later, another English navigator, William Baffin, discovered what came to be called Baffin Bay and penetrated to within 800 miles of the North Pole. Eventually, in the years 1846 to 1848, the British explorer John Franklin worked his way over the northern coast of Canada and discovered the 'North-West Passage' (and a most impractical passage for ships it then was). He died on the voyage.

There followed a half century of efforts to reach the North Pole, motivated in large part by sheer adventure and the desire to be the first to get there. In 1873, the Austrian explorers Julius Payer and Carl Weyprecht reached within 600 miles of the Pole and named a group of islands they found Franz Josef Land, after the Austrian emperor. In 1896, the Norwegian explorer Fridtjof Nansen drifted on the Arctic ice to within 300 miles of the Pole. At length, on 6 April 1909, the American explorer Robert Edwin Peary arrived at the Pole itself.

By now, the North Pole has lost much of its mystery. It has been explored on the ice, from the air, and under water. Richard Evelyn Byrd and Floyd Bennett were the first to fly over it, in 1926, and submarines have traversed its waters.

Meanwhile, the largest nothern ice cap, which is centred in Greenland, has drawn a number of scientific expeditions. The Greenland glacier has been found to cover about 640,000 of that island's 840,000 square miles, and its ice is known to reach a thickness of a mile in some places.

As the ice accumulates, it is pushed down to the sea, where the edges break off or 'calve' to form icebergs. Some 16,000 icebergs are thus formed in the Northern Hemisphere each year, 90 per cent of them breaking off the Greenland ice cap. The icebergs work slowly southwards, particularly down the west Atlantic. About 400 icebergs per year pass Newfoundland and threaten shipping lanes; between 1870 and 1890, fourteen ships were sunk and forty damaged by collision with icebergs.

The climax came in 1912, when the luxury liner *Titanic* collided with an iceberg and sank on her maiden voyage. An international watch over the positions of these inanimate monsters has been maintained ever since. During the years since this Ice Patrol has come into existence, not one ship has been sunk by an iceberg.

Far larger than Greenland is the South Pole's great continental glacier. The Antarctic ice cap covers seven times the area of the Greenland glacier and has an average thickness of one and a half miles, with three-mile depths in spots. This is due to the great size of the Antarctic continent – some 5 million square miles, though how much of this is land and how much ice-covered sea is still uncertain. Some explorers believe that Antarctica is a group of large islands bound together by ice, but at the moment the continent theory seems to have the upper hand.

The famous English explorer James Cook (better known as Captain Cook) was the first European to cross the Antarctic Circle. In 1773, he circumnavigated the Antarctic regions. (It was perhaps this voyage that inspired Samuel Taylor Coleridge's *The Rime of the Ancient Mariner*, published in 1798, which described a voyage from the Atlantic to the Pacific by way of the icy regions of Antarctica.)

In 1819, the British explorer William Smith discovered the South Shetland Islands, just fifty miles off the coast of Antarctica; in 1821, a Russian expedition sighted a small island (Peter I Island) within the Antarctic Circle; and, in the same year, the Englishman George Powell and the American Nathaniel B. Palmer first laid eyes on a peninsula of the Antarctic continent itself – now called Palmer Peninsula.

In the following decades, explorers inched towards the South Pole. By 1840, the American naval officer Charles Wilkes announced that the land strikes added up to a continental mass, and, subsequently, he was proved right. The Englishman James Weddell penetrated an ocean inlet east of Palmer Peninsula (now called Weddell Sea) to within 900 miles of the Pole. Another British explorer, James Clark Ross, discovered the other major ocean inlet into Antarctica (now called the Ross Sea) and got within 710 miles of the Pole. In 1902–4, a third Briton, Robert Falcon Scott, travelled over the Ross ice shelf (a section of ice-covered ocean as large as the state of Texas) to within 500 miles of it. And, in 1909, still another Englishman, Ernest Shackleton, crossed the ice to within about 100 miles of the Pole.

On 16 December 1911, the goal was finally reached by the Norwegian explorer Roald Amundsen. Scott, making a second dash of his own, got to the South Pole just three weeks later, only to find Amundsen's flag already planted there. Scott and his men perished on the ice on their way back.

In the late 1920s, the aeroplane helped to make good the conquest of Antarctica. The Australian explorer George Hubert Wilkins flew over 1,200 miles of its coastline, and Richard Evelyn Byrd, in 1929, flew over the South Pole. By that time the first base, Little America I, had been established in the Antarctic.

The north and south polar regions became focal points of the greatest international project in science of modern times. This had its origin in 1882–3, when a number of nations joined in an 'International Polar Year' of exploration and scientific investigation of

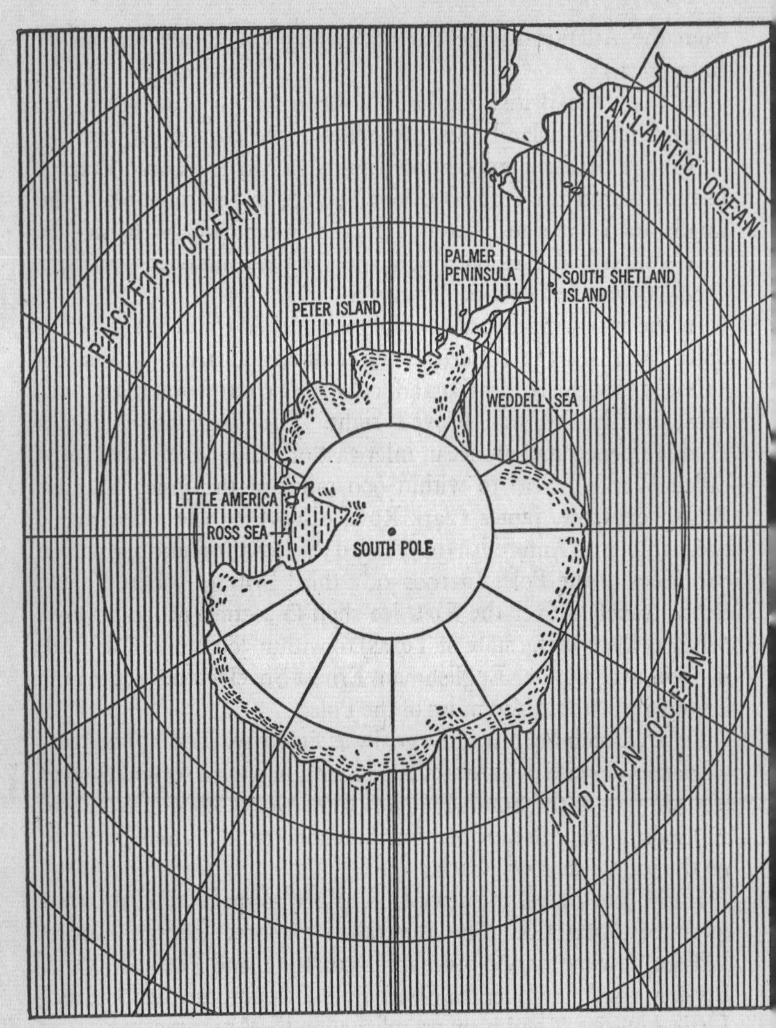

The major continental glaciers are today largely restricted to Greenland and Antarctica. At the height of the last ice age, the glaciers extended over most of northern and western Europe and south of the Great Lakes on the North American continent.

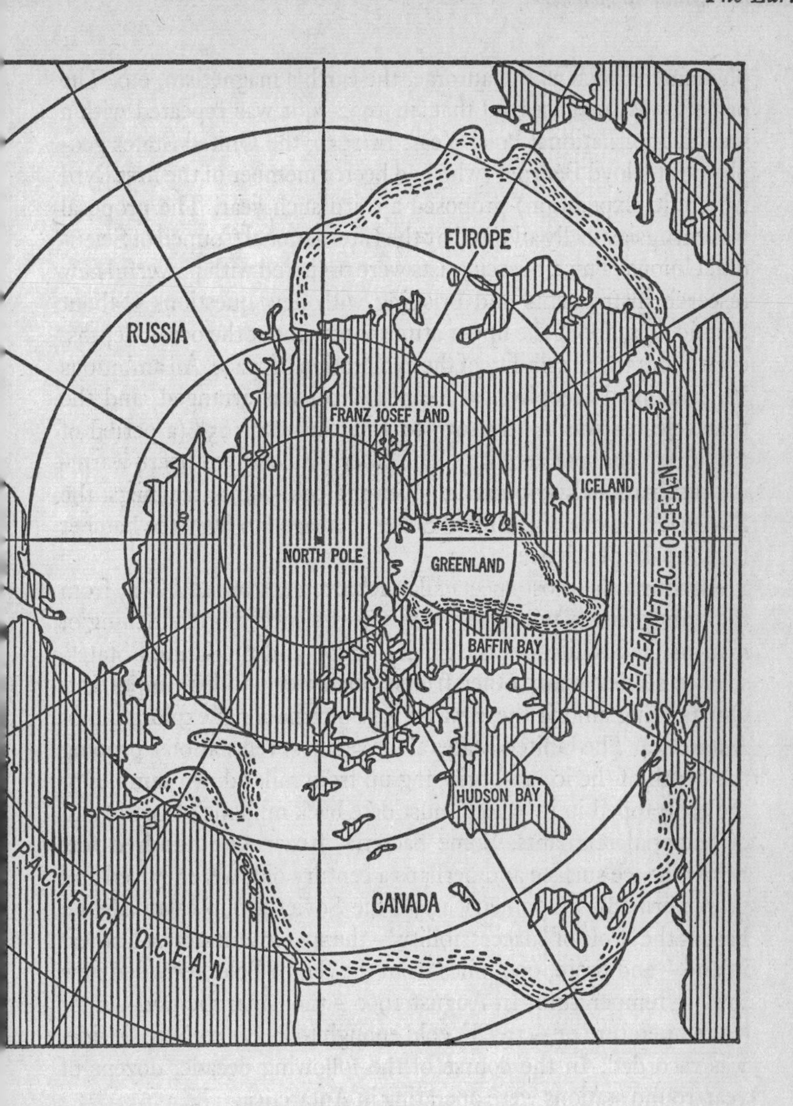
EUROPE
RUSSIA
FRANZ JOSEF LAND
ICELAND
NORTH POLE
GREENLAND
BAFFIN BAY
ATLANTIC OCEAN
HUDSON BAY
PACIFIC OCEAN
CANADA

phenomena such as the aurorae, the earth's magnetism, etc. The project was so successful that, in 1932–3, it was repeated with a second International Polar Year. In 1950, the United States geophysicist Lloyd Berkner (who had been a member of the first Byrd Antarctic Expedition) proposed a third such year. The proposal was enthusiastically adopted by the International Council of Scientific Unions. This time scientists were prepared with powerful new research instruments and bristling with new questions – about cosmic rays, about the upper atmosphere, about the ocean depths, even about the possibility of the exploration of space. An ambitious 'International Geophysical Year' (IGY) was arranged, and the time selected was 1 July 1957 to 31 December 1958 (a period of maximum sunspot activity). The enterprise enlisted heart-warming international cooperation; even the cold-war antagonists, the Soviet Union and the United States, managed to bury the hatchet for the sake of science.

Although the most spectacular achievement of the IGY, from the standpoint of public interest, was the successful launching of man-made satellites by the Soviet Union and the United States, science reaped many other fruits which were no less important. Outstanding among these was a vast international exploration of Antarctica. The United States alone set up seven stations, probing the depth of the ice and bringing up from miles down samples of the air trapped in it (which must date back millions of years) and of bacterial remnants. Some bacteria, frozen one hundred feet below the ice surface and perhaps a century old, were revived and grew normally. In January 1958, the Soviet group established a base at the 'Pole of Inaccessibility' – the spot in Antarctica farthest inland – and there, 600 miles from the South Pole, recorded new lows in temperature. In August 1960 – the Antarctic mid-winter – a temperature of −127°F, cold enough to freeze carbon dioxide, was recorded. In the course of the following decade, dozens of year-round stations were operating in Antarctica.

In the most dramatic Antarctic feat, a British exploring team headed by Vivian Ernest Fuchs and Edmund Percival Hillary crossed the continent by land for the first time in history (with special vehicles and all the resources of modern science at their disposal, to be sure). Hillary, by the way, had also been the first, along with the Sherpa mountaineer Tenzing Norgay, to climb Mount Everest, the highest mountain on earth, in 1953.

The success of the IGY and the warmth generated by this demonstration of cooperation in the midst of the cold war led to an agreement in 1959 among twelve nations to bar all military activities (including nuclear explosions and the dumping of radioactive wastes) from the Antarctic. Thus Antarctica will be reserved for scientific activities.

The earth's load of ice, amounting to nearly 9 million cubic miles, covers about 10 per cent of its land area. About 86 per cent of the ice is piled up in the Antarctic continental glacier and 10 per cent in the Greenland glacier. The remaining 4 per cent makes up the small glaciers in Iceland, Alaska, the Himalayas, the Alps, and a few other locations.

The Alpine glaciers have been under study for a long time. In the 1820s, two Swiss geologists, J. Venetz and Jean de Charpentier, noticed that rocks characteristic of the central Alps were scattered over the plains to the north. How had they got there? The geologists speculated that the mountain glaciers had once covered a much larger area and had left boulders and piles of debris behind when they retreated.

A Swiss zoologist, Jean Louis Rodolphe Agassiz, looked into this notion. He drove lines of stakes into the glaciers and waited to see whether they moved. By 1840, he had proved beyond doubt that glaciers flowed like very slow rivers at a rate of about 225 feet per year. Meanwhile, he had travelled over Europe and found marks of glaciers in France and England. He found boulders foreign to their surroundings in other areas and scoured marks on rock that could only have been made by the grinding of glaciers, carrying pebbles encrusted along their bottoms.

Agassiz went to the United States in 1846 and became a Harvard professor. He found signs of glaciation in New England and the Midwest. By 1850, it seemed quite obvious that there must have been a time when a large part of the Northern Hemisphere was under a large continental glacier. The deposits left by the glacier have been studied in detail since Agassiz' time. These studies have shown that the ice advanced and retreated four times. They were as far south as Cincinnati a mere 18,000 years ago. When they advanced, the climate to the south was wetter and colder; when they retreated (leaving lakes behind, of which the largest still in existence are the Canadian–American Great Lakes), the climate to the south grew warmer and drier.

The last retreat of the ice took place between 8,000 and 12,000 years ago. Before the ice ages, there was a period of mild climate on the earth lasting at least 100 million years. There were no continental glaciers, even at the poles. Coal beds in Spitzbergen and signs of coal even in Antarctica testify to this, because coal marks the site of ancient lush forests.

The coming and going of glaciers leaves its mark, not only on the climate of the rest of the earth, but on the very shape of the continents. For instance, if the now-shrinking glaciers of Greenland and Antarctica were to melt completely, the ocean level would rise nearly 200 feet. It would drown the coastal areas of all the continents, including many of the world's largest cities, with the water level reaching the twentieth storey of the Manhattan skyscrapers. On the other hand, Alaska, Canada, Siberia, Greenland, and even Antarctica would become more habitable.

The reverse situation takes place at the height of an ice age. So much water is tied up in the form of land-based ice caps (up to three or four times the present amount) that the sea-level mark is as much as 440 feet lower than it now is. When this is so, the continental shelves are exposed.

The continental shelves are relatively shallow portions of the ocean adjoining the continents. The sea floor slopes more or less gradually until a depth of about 130 metres is achieved. After this the slope is much steeper, and considerably greater depths are achieved rapidly. The continental shelves are, structurally, part of the continents they adjoin; it is the edge of the shelf that is the true boundary of the continent. What it amounts to is that at the present moment there is enough water in the ocean basins to flood the borders of the continent.

Nor is the continental shelf small in area. It is much broader in some places than others; there is considerable shelf area off the east coast of the United States, but little off the west coast (which is at the edge of a crustal plate). On the whole, though, the continental shelf is some fifty miles wide on the average and makes up a total area of 10 million square miles. In other words, a potential continental area rather greater than the Soviet Union in size is drowned under ocean waters.

It is this area that is exposed during periods of maximum glaciation and was indeed exposed in the last great Ice Ages. Fossils of land animals (such as the teeth of elephants) have been dredged up from the continental shelves, miles from land and under yards of water. What's more, with the northern continental sections ice-covered, rain was more common than now farther south, so that the Sahara Desert was then grassland. The drying of the Sahara as the ice caps receded took place not long before the beginning of historic times.

There is thus a pendulum of habitability. As the sea-level drops, large continental areas become deserts of ice, but the continental shelves become habitable, as do present-day deserts. As the sea-level rises, there is further flooding of the lowlands, but the polar regions become habitable, and again deserts retreat.

The major question regarding the Ice Ages involves their cause. What makes the ice advance and retreat, and why is it that the glaciations have been relatively brief, the present one having occupied only 1 million of the last 100 million years?

It takes only a small change in temperature to bring on or to terminate an ice age – just enough fall in temperature to accumulate a little more snow in the winter than melts in the summer or enough rise to melt a little more snow in the summer than falls in the winter. It is estimated that a drop in the earth's average annual temperature of only 3·5°C is sufficient to make glaciers grow, whereas a rise of the same amount would melt Antarctica and Greenland to bare rock in a matter of centuries.

Such changes in the temperature of the earth have indeed taken place in the past. A method has now been evolved by which primeval temperatures can be measured with amazing accuracy. The

American chemist Jacob Bigeleisen, working with H. C. Urey, showed in 1947 that the ratio of the common variety of oxygen (oxygen 16) to its rarer isotopes (e.g., oxygen 18), present in compounds, would vary with temperature. Consequently, if one

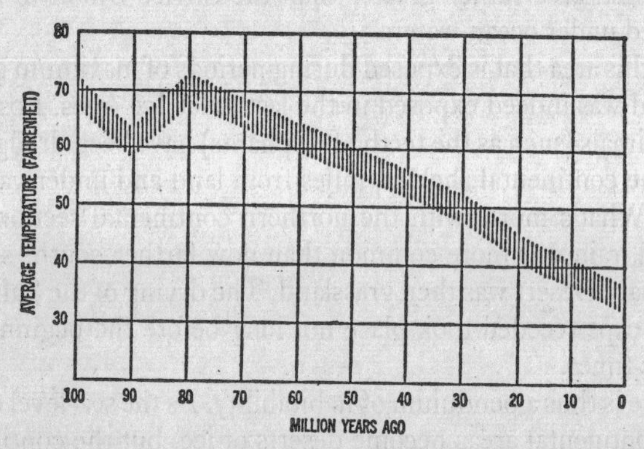

The record of the ocean temperatures during the last 100 million years.

measured the ratio of oxygen 16 to oxygen 18 in an ancient fossil of a sea animal, one could tell the temperature of the ocean water at the time the animal lived. By 1950, Urey and his group had developed the technique to so fine a point that by analysing the shell layers of a millions-of-years-old fossil (an extinct form of squid), they could determine that the creature was born during a summer, lived four years, and died in the spring.

This 'thermometer' has established that 100 million years ago the average world-wide ocean temperature was about 70°F. It cooled slowly to 61° 10 million years later and then rose to 70° again 10 million years after that. Since then, the ocean temperature has declined steadily. Whatever triggered this decline may also be a factor in the extinction of the dinosaurs (who were probably adapted to mild and equable climates) and put a premium on the warm-blooded birds and mammals, who can maintain a constant internal temperature.

Cesare Emiliani, using the Urey technique, studied the shells of foraminifera brought up in cores from the ocean floor. He found that the over-all ocean temperature was about 50°F 30 million years ago, 43° 20 million years ago, and is now 35°.

What caused these long-term changes in temperature? One possible explanation is the so-called 'greenhouse effect' of carbon dioxide. Carbon dioxide absorbs infra-red radiation rather strongly. This means that when there are appreciable amounts of it in the atmosphere, it tends to block the escape of heat at night from the sun-warmed earth. The result is that heat accumulates. On the other hand, when the carbon dioxide content of the atmosphere falls, the earth steadily cools.

If the current concentration of carbon dioxide in the air should double (from 0·03 per cent of the air to 0·06 per cent), that small change would suffice to raise the earth's over-all temperature by three degrees and would bring about the complete and quick melting of the continental glaciers. If the carbon dioxide dropped to half the present amount, the temperature would drop sufficiently to bring the glaciers down to New York City again.

Volcanoes discharge large amounts of carbon dioxide into the air; the weathering of rocks absorbs carbon dioxide (thus forming limestone). Here, then, is a possible pair of mechanisms for long-term climatic changes. A period of greater than normal volcanic action might release a large amount of carbon dioxide into the air and initiate a warming of the earth. Contrariwise, an era of mountain-building, exposing large areas of new and unweathered rock to the air, could lower the carbon dioxide concentration in the atmosphere. This is what may have happened at the close of the Mesozoic (the age of reptiles) some 80 million years ago, when the long decline in the earth's temperature began.

But what about the comings and goings of the four ice ages within the last million years? Why was there this rapid alternation of glaciation and melting in comparatively short spells of tens of thousands of years?

In 1920, a Serbian physicist named Milutin Milankovich suggested slight variations in the earth's relation to the sun might explain the situation. Sometimes the earth's tilt changes a little; sometimes its perihelion (closest approach to the sun in its orbit) is slightly closer than at other times. A combination of these factors, Milankovich argued, could so affect the amount of heat received from the sun by, say, the Northern Hemisphere as to cause a cyclic rise and fall of its average temperature. He thought that such a cycle might last 40,000 years, giving the earth a 'Great Spring', 'Great Summer', 'Great Autumn', and 'Great Winter', each some 10,000 years in length. Precise dating of coral reefs and deep-sea sediments has shown such temperature shifts.

The difference between Great Summer and Great Winter is small, and the theory implies that only after a long period of overall temperature decline did the additional small temperature fall of the Great Winter suffice to reduce the Northern Hemisphere's temperature to the point where the ice ages began a million years ago. According to the Milankovich theory, we are now in a Great Summer and, in 10,000 years or so, will begin to enter another Great Winter.

The Milankovich theory has disturbed some geologists, mainly because it implies that the ice ages of the Northern and Southern Hemispheres have come at different times, which has not been demonstrated. In recent years, several other theories have been proposed: that the sun has cycles of slight fluctuation in its output of heat; that dust from volcanic eruptions, rather than carbon dioxide, has produced the 'greenhouse' warming; and so on. An alternative hypothesis is one advanced by Maurice Ewing of the Lamont Geological Observatory and a colleague, William Donn.

Ewing and Donn ascribe the succession of ice ages in the Northern Hemisphere to the geographical conditions around the North Pole. The Arctic Ocean is nearly surrounded by land. In the mild aeons before the recent ice ages began, when this ocean was open water, winds sweeping across it picked up water vapour and dropped snow on Canada and Siberia. As glaciers grew on the land, according to the Ewing–Donn theory, the earth absorbed less heat from the sun, because the cover of ice, as well as clouds resulting from stormier weather, reflected away part of the sunlight. Consequently, the general temperature of the earth dropped.

But as it did so, the Arctic Ocean froze over, and, consequently, the winds picked up less moisture from it. Less moisture in the air meant less snow each winter. So the trend was reversed: with less snowy winters, summer melting took the upper hand over winter snowfall. The glaciers retreated until the earth warmed sufficiently to melt the Arctic Ocean to open water again – at which point the cycle started anew with a rebuilding of the glaciers.

It seems a paradox that the melting of the Arctic Ocean, rather than its freezing, should bring on an ice age. Geophysicists, however, find the theory plausible and capable of explaining many things. The main problem about the theory is that it makes the absence of ice ages up to a million years ago more mysterious than ever. But Ewing and Donn have an answer for that. They suggest that during the long period of mildness before the ice ages the North Pole may have been located in the Pacific Ocean. In that case, most of the snow would have fallen in the ocean instead of on land, and no important glaciers could have got started.

The North Pole, of course, has a constant small motion, moving in thirty-foot irregular circles in a period of 435 days or so, as was discovered at the beginning of the twentieth century by the American astronomer Seth Carlo Chandler. It has also drifted thirty feet towards Greenland since 1900. However, such changes – caused perhaps by earthquakes and consequent shifts in the mass-distribution in the globe – are small potatoes.

What is needed for the Ewing–Donn theory are large sweeps, and these might possibly be brought to pass by continental drift. As the crustal plates shift about, the North Pole may at times be enclosed by land or be left in open sea. However, can such changes produced by drift be matched with the occurrence or non-occurrence of periods of glaciation?

Whatever the cause of the ice ages may have been, it seems now that man himself may be changing the climate in store for the future. The American physicist Gilbert N. Plass has suggested that we may be seeing the last of the ice ages, because the furnaces of civilization are loading the atmosphere with carbon dioxide. A hundred million chimneys are ceaselessly pouring carbon dioxide into the air; the total amount is about 6,000 million tons a year – 200 times the quantity coming from volcanoes. Plass pointed out that, since 1900, the carbon dioxide content of our atmosphere has increased about 10 per cent and may increase as much again by the year 2000. This addition to the earth's 'greenhouse' shield against the escape of heat, he calculated, should raise the average temperature by about 1·1°C per century. During the first half of the twentieth century, the average temperature has indeed risen at this rate, according to the available records (mostly in North America and Europe). If the warming continues at the same rate, the continental glaciers may disappear in a century or two.

Investigations during the IGY seemed to show that the glaciers are indeed receding almost everywhere. One of the large glaciers in the Himalayas was reported in 1959 to have receded 700 feet since 1935. Others had retreated 1,000 or even 2,000 feet. Fish adapted to frigid waters are migrating northwards, and warm-climate trees are advancing in the same direction. The sea level is rising slightly each year, as would be expected if the glaciers are melting. The sea level is already so high that, at times of violent storms at high tide, the ocean is not far from threatening to flood the New York subway system.

And yet there seems to be a slight downturn in temperature since the early 1940s, so that half the temperature increase between 1880 and 1940 has been wiped out. This may be due to increasing dust and smog in the air since 1940: particles that cut off sunlight and, in a sense, shade the earth. It would seem that two different types of man-made atmospheric pollution are currently cancelling each other's effect, at least in this respect, and at least temporarily.

4 The Atmosphere

The Shells of Air

Aristotle supposed the world to be made up of four shells, constituting the four elements of matter: earth (the solid ball), water (the ocean), air (the atmosphere), and fire (an invisible outer shell that occasionally became visible in the flashes of lightning). The universe beyond these shells, he said, was composed of an unearthly, perfect fifth element that he called 'aether' (from a Latin derivative the name became 'quintessence', which means 'fifth element').

There was no room in this scheme for nothingness: where earth ended, water began; where both ended, air began; where air ended, fire began; and where fire ended, aether began and continued to the end of the universe. 'Nature,' said the ancients, 'abhors a vacuum' (Latin for 'nothingness').

The suction pump, an early invention to lift water out of wells, seemed to illustrate this abhorrence of a vacuum admirably. A piston is fitted tightly within a cylinder. When the pump handle is pushed down, the piston is pulled upwards, leaving a vacuum in the lower part of the cylinder. But since nature abhors a vacuum, the surrounding water opens a one-way valve at the bottom of the cylinder and rushes into the vacuum. Repeated pumping lifts the water higher and higher in the cylinder, until it pours out of the pump spout.

According to Aristotelian theory, it should have been possible in this way to raise water to any height. But miners who had to pump water out of the bottoms of mines found that no matter how hard and long they pumped, they could never lift the water higher than thirty-three feet above its natural level.

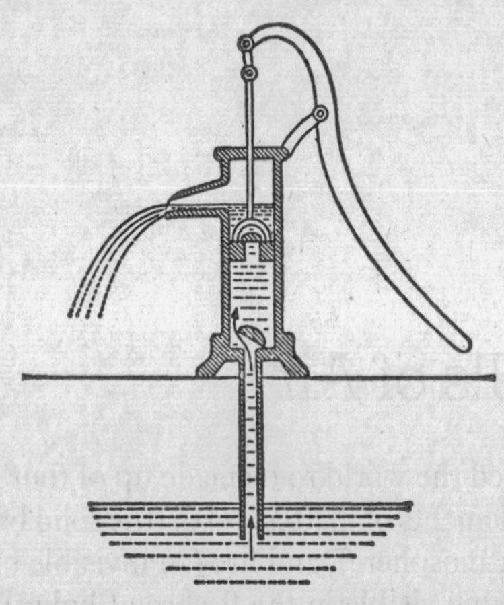

Principle of the water pump. When the handle raises the piston, a partial vacuum is created in the cylinder, and water rises into it through a one-way valve. After repeated pumping, the water level is high enough for the water to flow out of the spout.

Galileo got interested in this puzzle towards the end of his long and inquisitive life. He could come to no conclusion except that apparently nature abhorred a vacuum only up to certain limits. He wondered whether the limit would be lower if he used a liquid denser than water, but he died before he could try this experiment.

Galileo's students Evangelista Torricelli and Vincenzo Viviani did perform it in 1644. Selecting mercury (which is thirteen and a half times as dense as water), they filled a yard-long glass tube with mercury, stoppered the open end, upended the tube in a dish of mercury, and removed the stopper. The mercury began to run out of the tube into the dish, but, when its level had dropped to thirty inches above the level in the dish, it stopped pouring out of the tube and held at that level.

Thus was constructed the first 'barometer'. Modern mercury barometers are not essentially different. It did not take long to discover that the height of the mercury column was not always the same. The English scientist Robert Hooke pointed out, in the 1660s, that the height of the mercury column decreased before a storm, thus pointing the way to the beginning of scientific weather forecasting or 'meteorology'.

What was holding the mercury up? Viviani suggested that it was the weight of the atmosphere pressing down on the liquid in the dish. This was a revolutionary thought, for the Aristotelian notion had been that air had no weight, being drawn only to its proper sphere above the earth. Now it became plain that a thirty-three-foot column of water, or a thirty-inch column of mercury, measured the weight of the atmosphere – that is, the weight of a column of air of the same cross-section from sea level up to as far as the air went.

The experiment also showed that nature did not necessarily abhor a vacuum under all circumstances. The space left in the closed end of the tube after the mercury fell was a vacuum, containing nothing but a very small quantity of mercury vapour. This 'Torricellian vacuum' was the first decent vacuum produced by man.

The vacuum was pressed into the service of science almost at once. In 1650, the German scholar Athanasius Kircher demonstrated that sound could not be transmitted through a vacuum, thus upholding an Aristotelian theory (for once). In the next decade, Robert Boyle showed that very light objects will fall as rapidly as heavy ones in a vacuum, thus upholding Galileo's theories of motion against the views of Aristotle.

If air had a finite weight, it must have some finite height. The weight of the atmosphere turned out to be fourteen and seven tenths pounds per square inch; on this basis the atmosphere was just about five miles high – if it was evenly dense all the way up. But, in 1662, Boyle showed that it could not be, because pressure increased air's density. He stood up a tube shaped like the letter 'J' and poured some mercury into the mouth of the tube, on the tall side of the J. The mercury trapped a little air in the closed end on the short side. As he poured in more mercury, the air pocket shrank. At the same time its pressure increased, Boyle discovered, for it shrank less and less as the mercury grew weightier. By actual measurement, Boyle showed that reducing the volume of gas to one half doubled its pressure; in other words, the volume varied in inverse ratio to the pressure. This historic discovery, known as 'Boyle's law', was the first step in the long series of discoveries about matter that eventually led to the atomic theory.

Since air contracted under pressure, it must be densest at sea level and become steadily thinner as the weight of the overlying air declined towards the top of the atmosphere. This was first demonstrated by the French mathematician Blaise Pascal, who sent his brother-in-law Florin Périer nearly a mile up a mountain-side in 1648 and had him carry a barometer and note the manner in which the mercury level dropped as altitude increased.

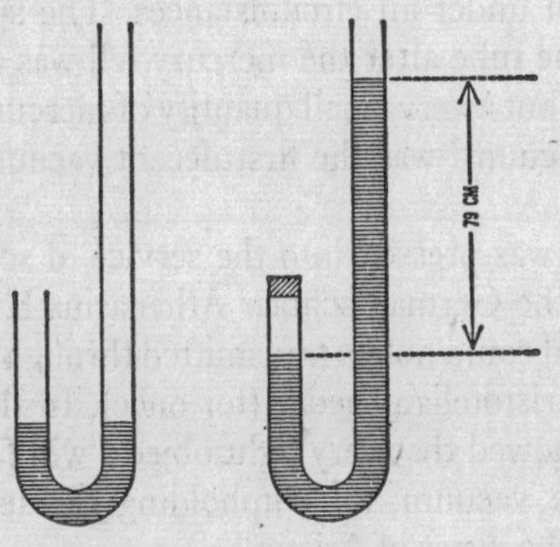

Diagram of Boyle's experiment. When the left arm of the tube is stoppered and more mercury is poured into the right arm, the trapped air is compressed. Boyle showed that the volume of the trapped air varied inversely with the pressure. That is 'Boyle's law'.

Theoretical calculations showed that, if the temperature were the same all the way up, the air pressure would decrease tenfold with every 12 miles of rise in altitude. In other words, at 12 miles the column of mercury it could support would have dropped from 30 inches to 3 inches; at 24 miles it would be ·3 of an inch; at 36 miles, ·03 of an inch and so on. At 108 miles, the air pressure would amount to only 0·000000003 of an inch of mercury. This may not sound much, but over the whole earth the weight of the air above 108 miles would still total 6 million tons.

Actually all these figures are only approximations, because the air temperature changes with height. Nevertheless, they do clarify the picture, and we can see that the atmosphere has no definite boundary; it simply fades off gradually into the near-emptiness of space. Meteor trails have been detected as high as one hundred miles, where the air pressure is only a millionth what it is on the earth's surface, and the air density only a thousand millionth. Yet that is enough to burn these tiny bits of matter to incandescence by friction. And the aurora borealis (Northern Lights), formed of glowing wisps of gas bombarded by particles from outer space, has been located as high as 500 to 600 miles above sea-level.

Until the late eighteenth century, it seemed that man would never be able to get any closer to the upper atmosphere than the top of the mountains. The highest mountain close to the centres of scientific research was Mont Blanc in south-eastern France, and that was only three miles high. An interesting effort to substitute technology for mountain-climbing came in 1749 when the Scottish astronomer Alexander Wilson attached thermometers to kites, hoping thus to measure atmospheric temperatures at a height. The real breakthrough, however, came in 1782, when the two French brothers Joseph Michel and Jacques Étienne Montgolfier lit a fire under a large bag with an opening underneath and thus filled the bag with hot air. The bag rose slowly; the Montgolfiers had successfully launched man's first balloon! Within a few months balloons were being made with hydrogen, a gas only one fourteenth as dense as air, so that each pound of hydrogen could carry aloft a payload of thirteen pounds. Now gondolas went up carrying animals and, soon, men.

Within a year of the launching of the first balloon, an American named John Jeffries made a balloon flight over London with a barometer and other instruments, plus an arrangement to collect air at various heights. By 1804, the French scientist Joseph Louis

Gay-Lussac had ascended nearly four and a half miles and brought down samples of the rarefied air. Such adventures were made a little safer by the French balloonist Jean Pierre Blanchard, who, in 1785, at the very onset of the 'balloon age', invented the parachute.

This was nearly the limit for men in an open gondola; three men rose to six miles in 1875, but only one, Gaston Tissandier, survived the lack of oxygen. He was able to describe the symptoms of air deficiency, and that was the birth of 'aviation medicine'. Unmanned balloons carrying instruments were designed and put into action in 1892, and these could be sent higher and bring back information on temperature and pressure from hitherto unexplored regions.

In the first few miles of altitude rise, the temperature dropped, as was expected. At seven miles or so, it was −55°C. But then came a surprise. Above this level, the temperature did not decrease. In fact, it even rose slightly.

The French meteorologist Léon Phillippe Teisserenc de Bort suggested in 1902 that the atmosphere might have two layers: (1) a turbulent lower layer containing clouds, winds, storms, and all the familiar weather changes (in 1908 he called this layer the 'troposphere', from the Greek for 'sphere of change') and (2) a quiet upper layer containing sub-layers of lighter gases, helium and hydrogen (he named this the 'stratosphere', meaning 'sphere of layers). Teisserenc de Bort called the level at which the temperature ceased to decline the 'tropopause' – 'end of change', or the boundary between the troposphere and the stratosphere. The tropopause has since been found to vary from an altitude of about ten miles above sea level at the Equator to only five miles above at the poles.

During the Second World War, high-flying United States bombers discovered a dramatic phenomenon just below the tropopause: the 'jet stream', consisting of very strong, steady, west-to-east winds blowing at speeds up to 500 miles per hour. Actually there are two jet streams, one in the Northern Hemisphere at the general latitude of the United States, the Mediterranean, and north China, and one in the Southern at the latitude of New Zealand

and Argentina. The streams meander, often debouching into eddies far north or south of their usual course. Aeroplanes now take advantage of the opportunity to ride on these swift winds. But far more important is the discovery that the jet streams have a powerful influence on the movement of air masses at lower levels. This knowledge at once helped to advance the art of weather forecasting.

But man did not resign his desire for personal exploration to instruments. One could not survive in the thin, cold atmosphere of great heights – but why expose oneself to that atmosphere? Why not a sealed cabin, within which the pressures and temperatures of earth's surface air could be maintained?

In the 1930s, thanks to sealed cabins, man reached the stratosphere. In 1931, the Piccard brothers (Auguste and Jean Félix) – the former later invented the bathyscaphe – rose to 11 miles in a balloon carrying a sealed gondola. Then new balloons of plastic material, lighter and less porous than silk, made it possible to go higher and remain up longer. In 1938, a balloon named Explorer II went to 13 miles, and by the 1960s manned balloons have gone as high as 21½ miles and unmanned balloons almost to 29 miles.

These higher flights showed that the zone of nearly constant temperature did not extend indefinitely upwards. The stratosphere came to an end at a height of about twenty miles, for above that the temperature started to rise!

This 'upper atmosphere', above the stratosphere, containing only 2 per cent of the earth's total air mass, was in turn penetrated in the 1940s. This time man needed a new type of vehicle altogether – the rocket.

The Chinese, as long ago as the thirteenth century, invented and used small rockets for psychological warfare – to frighten the enemy. Modern Western civilization adapted rockets to a bloodier purpose. In 1801, a British artillery expert, William Congreve, having learned about rockets in the Orient, where Indian troops used them against the British in the 1780s, devised a number of deadly missiles. Some were used against the United States in the

War of 1812, notably at the bombardment of Fort McHenry in 1814, which inspired Francis Scott Key to write the 'Star-Spangled Banner', singing of 'the rockets' red glare'. Rocket weapons faded out in the face of improvements in range, accuracy, and power of conventional artillery. However, the Second World War saw the development of the American bazooka and the Soviet 'Katusha', both of which are essentially rocket-propelled packets of explosives. Jet planes, on a much larger scale, also make use of the rocket principle of action and reaction.

Around the beginning of the twentieth century, two men independently conceived a new and finer use of rockets – exploring the upper atmosphere and space. They were a Russian, Konstantin Eduardovich Tsiolkovsky, and an American, Robert Hutchings Goddard. (It is odd indeed, in view of later developments, that a Russian and an American were the first heralds of the age of rocketry, though an imaginative German inventor, Hermann Ganswindt, also advanced even more ambitious, though less systematic and scientific, speculations at this time.)

The Russian was the first in print; he published his speculations and calculations in 1903 to 1913, whereas Goddard did not publish until 1919. But Goddard was the first to put speculation into practice. On 16 March 1926, from a snow-covered farm in Auburn, Massachusetts, he fired a rocket 200 feet into the air. The remarkable thing about his rocket was that it was powered by a liquid fuel, instead of gunpowder. Then, too, whereas ordinary rockets, bazookas, jet planes, and so on make use of the oxygen in the surrounding air, Goddard's rocket, designed to work in outer space, had to carry its own oxidizer in the form of liquid oxygen ('lox', as it is now called in missile-man slang).

Jules Verne, in his nineteenth-century science fiction, had visualized a cannon as a launching device for a trip to the moon, but a cannon expends all its force at once and at the start, when the atmosphere is thickest and offers the greatest resistance. Goddard's rockets moved upwards slowly at first, gaining speed and expending final thrust high in the thin atmosphere, where resistance is low. The gradual attainment of speed means that acceleration is kept at bearable levels, an important point for manned vessels.

Unfortunately, Goddard's accomplishment got almost no recognition except from his outraged neighbours, who managed to have him ordered to take his experiments elsewhere. Goddard went off to shoot his rockets in greater privacy, and, between 1930 and 1935, his vehicles attained speeds of as much as 550 miles an hour and heights of a mile and a half. He developed systems for steering a rocket in flight and gyroscopes to keep a rocket headed in the proper direction. Goddard also patented the idea of multi-stage rockets. Because each successive stage sheds part of the original weight and starts at a high velocity imparted by the preceding stage, a rocket divided into a series of stages can attain much higher speeds and greater heights than could a rocket with the same quantity of fuel all crammed into a single stage.

During the Second World War, the United States Navy half-heartedly supported further experiments by Goddard. Meanwhile, the German government threw a major effort into rocket research, using as its corps of workers a group of youngsters who had been inspired, primarily, by Hermann Oberth, a Rumanian mathematician who, in 1923, wrote on rockets and spacecraft independently of Tsiolkovsky and Goddard. German research began in 1935 and culminated in the development of the V-2. Under the guidance of the rocket expert Wernher von Braun (who, after the Second World War, placed his talents at the disposal of the United States), the first true rocket missile was shot off in 1942. The V-2 came into combat use in 1944, too late to win the war for the Nazis, although they fired 4,300 of them altogether, of which 1,230 hit London. Von Braun's missiles killed 2,511 Englishmen and seriously wounded 5,869 others.

On 10 August 1945, almost on the very day of the war's end, Goddard died – just in time to see his spark blaze into flame at last. The United States and the Soviet Union, stimulated by the success of the V-2, plunged into rocket research, each carrying off as many German experts in rocketry as could be lured to its side.

By 1949, the United States had fired a captured German V-2 to a height of 128 miles, and, in the same year, its rocket experts sent a WAC Corporal, the second stage of a two-stage rocket, to 250 miles. The exploration of the upper atmosphere had begun.

Rockets alone would have accomplished little in that exploration had it not been for a companion invention – 'telemetering'. Telemetering was first applied to atmospheric research, in a balloon, in 1925 by a Russian scientist named Pyotr A. Molchanoff. Essentially, this technique of 'measuring at a distance' entails translating the conditions to be measured (e.g., temperature) into electrical impulses that are transmitted back to earth by radio. The observations take the form of changes in the intensity or spacing of the pulses. For instance, a temperature change affects the electrical resistance of a wire and so changes the nature of the pulse; a change in air pressure similarly is translated into a certain kind of pulse by the fact that air cools the wire, the extent of the cooling depending on the pressure; radiation sets off pulses in a detector, and so on. Nowadays, telemetering has become so elaborate that the rockets seem to do everything but talk, and their intricate messages have to be interpreted by rapid computers.

Rockets and telemetering, then, showed that above the stratosphere the temperature rose to a maximum of some −10°C at a height of 30 miles and then dropped again to a low of −90°C at a height of 50 miles. This region of rise and fall in temperature is called the 'mesosphere', a word coined in 1950 by the British geophysicist Sydney Chapman.

Beyond the mesosphere what is left of the thin air amounts to only a few thousandths of 1 per cent of the total mass of the atmosphere. But this scattering of air atoms steadily increases in temperature to an estimated 1,000° at 300 miles and probably to still higher levels above that height. It is therefore called the 'thermosphere' ('sphere of heat') – an odd echo of Aristotle's original sphere of fire. Of course, temperature here does not signify heat in the usual sense: it is merely a measure of the speed of the particles.

Above 300 miles we come to the 'exosphere', a term first used by Lyman Spitzer in 1949, which may extend to as high as 1,000 miles and gradually merges into interplanetary space.

Increasing knowledge of the atmosphere may enable man to do something about the weather some day and not merely talk about it. Already, a small start has been made. In the early 1940s, the

American chemists Vincent Joseph Schaefer and Irving Langmuir noted that very low temperatures could produce nuclei about which raindrops would form. In 1946, an aeroplane dropped powdered carbon dioxide into a cloud bank in order to form first nuclei and then raindrops ('cloud seeding'). Half an hour later, it was raining. Bernard Vonnegut later improved the technique when he discovered that powdered silver iodide generated on the ground and directed upwards worked even better. Rainmakers, of a new scientific variety, are now used to end droughts – or to attempt to end them, for clouds must first be present before they can be seeded. In 1961, Soviet astronomers used cloud seeding to clear a patch of sky through which an eclipse might be glimpsed. It was partially successful.

Those interested in rocketry strove constantly for new and better results. The captured German V-2s were used up by 1952, but by then larger and more advanced rocket-boosters were being built in both the United States and the Soviet Union, and progress continued.

A new era began when, on 4 October 1957 (within a month of the hundredth anniversary of Tsiolkovsky's birth), the Soviet Union put the first man-made satellite in orbit. Sputnik I travelled around the earth in an elliptical orbit – 156 miles above the surface (or 4,100 miles from the earth's centre) at perigee and 560 miles at apogee. An elliptical orbit is something like the course of a switchback. In going from apogee (the highest point) to perigee, the satellite slides downhill, so to speak, and loses gravitational potential. This brings an increase in velocity, so that at perigee the satellite starts uphill again at top speed, as a switchback does. The satellite loses velocity as it climbs (as does the switchback) and is moving at its slowest speed at apogee, before it turns downhill again.

Sputnik I at perigee was in the mesosphere, where the air resistance, though slight, was sufficient to slow the satellite a bit on each trip. On each successive revolution, it failed to attain its previous apogee height. Slowly, it spiralled inward. Eventually, it lost so much energy that it yielded to the earth's pull sufficiently to dive into the denser atmosphere, there to be burned up by friction with the air.

The rate at which a satellite's orbit decays in this way depends partly on the mass of the satellite, partly on its shape, and partly on the density of the air through which it passes. Thus the density of the atmosphere at that level can be calculated. The satellites have given man the first direct measurements of the density of the upper atmosphere. The density proved to be higher than had been thought, but at the altitude of 150 miles, for instance, it is still only one ten-millionth of that at sea level, and, at 225 miles, only one billionth.

These wisps of air ought not to be dismissed too readily, however. Even at a height of 1,000 miles, where the atmospheric density is only one thousand billionth the sea-level figure, that faint breath of air is a thousand million times as dense as are the gases in outer space itself. The earth's envelope of gases spreads far outwards.

The Soviet Union did not remain alone in this field, of course. Within four months it was joined by the United States, which, on 30 January 1958, launched its first satellite in orbit, Explorer I. Since then each nation has put hundreds of satellites whirling about the earth, for a variety of purposes. The upper atmosphere, the portion of space in the vicinity of the earth, has been studied by satellite-borne instruments and, in addition, the earth itself has been the target of studies. For one thing, satellites made it possible for the first time in man's history to see our planet (or at least half of it at any one time) as a unit, and to study the air circulation as a whole.

On 1 April 1960, the United States launched the first 'weather-eye' satellite, Tiros I ('Tiros' standing for '*T*elevision *I*nfra-*R*ed *O*bservation *S*atellite'), then Tiros II in November, which, for ten weeks, sent down over 20,000 pictures of vast stretches of the earth's surface and its cloud cover, including pictures of a cyclone in New Zealand and a patch of clouds in Oklahoma that was apparently spawning tornadoes. Tiros III, launched in July 1961, photographed eighteen tropical storms and, in September, showed Hurricane Esther developing in the Caribbean two days before it was located by more orthodox methods. The more sensitive

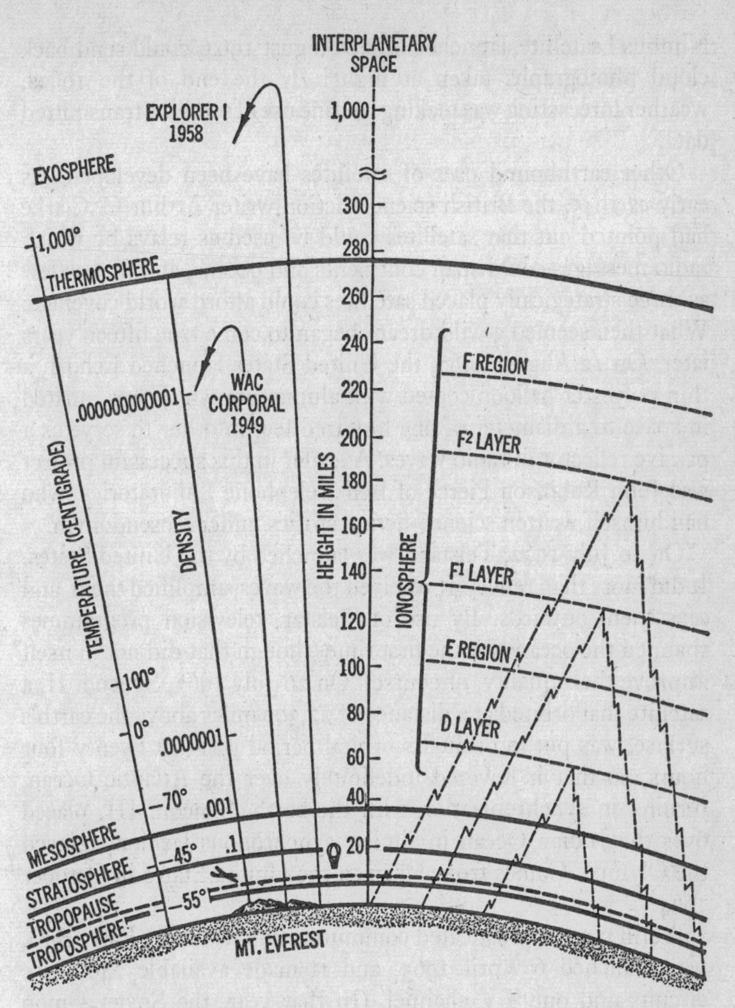

Profile of the atmosphere. The jagged lines indicate the reflection of radio signals from the Kennelly–Heaviside and Appleton layers of the ionosphere. Air density decreases with height and is expressed in percentages of barometric pressure at sea level.

Nimbus I satellite, launched on 28 August 1964, could send back cloud photographs taken at night! By the end of the 1960s, weather forecasting was making routine use of satellite-transmitted data.

Other earthbound uses of satellites have been developed. As early as 1945, the British science-fiction writer Arthur C. Clarke had pointed out that satellites could be used as relays by which radio messages could span continents and oceans, and that as few as three strategically placed satellites could afford world coverage. What then seemed a wild dream began to come true fifteen years later. On 12 August 1960, the United States launched Echo I, a thin polyester balloon coated with aluminium which was inflated in space to a diameter of one hundred feet in order to serve as a passive reflector of radio waves. A leader in this successful project was John Robinson Pierce of Bell Telephone Laboratories, who had himself written science-fiction stories under a pseudonym.

On 10 July 1962, Telstar I was launched by the United States. It did more than reflect. It received the waves, amplified them, and sent them onwards. By use of Telstar, television programmes spanned the oceans for the first time (though that did not in itself improve their quality, of course). On 26 July 1963, Syncom II, a satellite that orbited at a distance of 22,300 miles above the earth's surface, was put in orbit. Its orbital period was just twenty-four hours, so that it hovered indefinitely over the Atlantic Ocean, turning in synchronization with the earth. Syncom III, placed over the Indian Ocean in similar synchronous fashion, relayed the Olympic Games from Japan to the United States in October 1964.

A still more sophisticated communications satellite, Early Bird, was launched 6 April 1965, and it made available 240 voice circuits and one TV channel. (In that year, the Soviet Union began to send up communications satellites as well.) Earth seems on the threshold of becoming 'one world', at least as far as communications are concerned.

Satellites have also been launched for the specific purpose of being used to determine position on earth. The first satellite, Transit 1B, was launched on 13 April 1960.

The Gases in Air

Up to modern times, air was considered a simple homogeneous substance. In the early seventeenth century, the Flemish chemist Jan Baptista van Helmont began to suspect that there were a number of chemically different gases. He studied the vapour given off by fermenting fruit juice (carbon dioxide) and recognized it as a new substance. Van Helmont was, in fact, the first to use the term 'gas' – a word he is supposed to have coined from 'chaos', the ancients' word for the original substance out of which the universe was made. In 1756, the Scottish chemist Joseph Black studied carbon dioxide thoroughly and definitely established it as a gas other than air. He even showed that small quantities of it existed in the air. Ten years later, Henry Cavendish studied a flammable gas not found in the atmosphere. It was eventually named hydrogen. The multiplicity of gases was thus clearly demonstrated.

The first to realize that air was a mixture of gases was the French chemist Antoine-Laurent Lavoisier. In experiments conducted in the 1770s, he heated mercury in a closed vessel and found that the mercury combined with part of the air, forming a red powder (mercuric oxide), but four fifths of the air remained a gas. No amount of heating would consume any of this remaining gas. A candle would not burn in it, nor could mice live in it.

Lavoisier decided that air was made up of two gases. The one fifth that combined with mercury in his experiment was the portion of the air that supported life and combustion. This he called 'oxygen'. The remainder he called 'azote', from Greek words meaning 'no life'. Later it became known as 'nitrogen', because the substance was present in sodium nitrate, commonly called 'nitre'. Both gases had been discovered in the previous decade. Nitrogen had been discovered in 1772 by the Scottish physician Daniel Rutherford, and oxygen in 1774 by the English Unitarian minister Joseph Priestley.

By the mid nineteenth century, the French chemist Henri Victor Regnault had analysed air samples from all over the world and discovered the composition of the air to be the same everywhere. The oxygen content was 20·9 per cent, and it was assumed that all the rest (except for a trace of carbon dioxide) was nitrogen.

Nitrogen is a comparatively inert gas; that is, it does not readily combine with other substances. It can, however, be forced into combination, for instance, by heating it with magnesium metal, forming the solid magnesium nitride. Some years after Lavoisier's discovery, Henry Cavendish tried to exhaust the nitrogen by combining it with oxygen under the influence of an electric spark. He failed. No matter what he did, he could not get rid of a small bubble of remaining gas, amounting to less than 1 per cent of the original quantity. Cavendish thought this might be an unknown gas, even more inert than nitrogen. But not all chemists are Cavendishes, and the puzzle was not followed up, so the nature of this residue of air was not discovered for another century.

In 1882, the British physicist Robert John Strutt, Lord Rayleigh, compared the density of nitrogen obtained from air with the density of nitrogen obtained from certain chemicals and found, to his surprise, that the air nitrogen was definitely denser. Could it be that nitrogen obtained from air was not pure but contained small quantities of another, heavier gas? A Scottish chemist, Sir William Ramsay, helped Lord Rayleigh look further into the matter. By this time, they had the aid of spectroscopy. When they heated the small residue of gas left after exhaustion of nitrogen from air and examined its spectrum, they found a new set of bright lines – lines that belonged to no known element. To their newly discovered, very inert element they gave the name 'argon' (from a Greek word meaning 'inert').

Argon accounted for nearly all of the approximately 1 per cent of unknown gas in air – but not quite all. There were still several 'trace constituents' in the atmosphere, each constituting only a few parts per million. During the 1890s Ramsay went on to discover four more inert gases: 'neon' (new), 'krypton' (hidden), 'xenon' (stranger), and helium, which had been discovered more than thirty years before in the sun. In recent decades, the infra-red spectroscope has turned up three others: nitrous oxide ('laughing gas'), whose origin is unknown; methane, a product of the decay of organic matter; and carbon monoxide. Methane is released by bogs, and some 45 million tons of the same gas, it has been calculated, are added to the atmosphere each year by the venting of intestinal gases by cattle and other large animals. The carbon monoxide is probably man-made, resulting from the incomplete combustion of wood, coal, petrol, and so on.

All this, of course, refers to the composition of the lowest reaches of the atmosphere. What about the stratosphere? Teisserenc de Bort believed that helium and hydrogen might exist in some quantity up there, floating on the heavier gases underneath. He was mistaken. In the middle 1930s, Russian balloonists brought down samples of air from the upper stratosphere, and it proved to be made up of oxygen and nitrogen in the same one-to-four mixture as the air of the troposphere.

But there were reasons to believe some unusual gases existed still higher in the upper atmosphere, and one of the reasons was the phenomenon called the 'air glow'. This is the very feeble general illumination of all parts of the night sky, even in the absence of the moon. The total light of the air glow is considerably greater than that of the stars, but is so diffuse that it is not noticeable except to the delicate light-gathering instruments of the astronomer.

The source of the light had been a mystery for many years. In 1928, the astronomer V. M. Slipher succeeded in detecting in the air glow some mysterious spectral lines that had been found in nebulae in 1864 by William Huggins and were thought to represent an unfamiliar element, named 'nebulium'. In 1927, through experiments in the laboratory, the American astronomer Ira Sprague Bowen showed that the lines came from 'atomic oxygen', that is, oxygen existing as single atoms and not combined in the normal form of the two-atom molecule. Similarly, other strange spectral lines from the aurora turned out to represent atomic nitrogen. Both atomic oxygen and atomic nitrogen in the upper atmosphere are produced by energetic radiation from the sun, which breaks down the molecules into single atoms, something first suggested in 1931 by Sydney Chapman. Fortunately the high-energy radiation is, in this way, absorbed or weakened before it reaches the lower atmosphere.

The air glow, Chapman maintained, comes from the re-combination at night of the atoms that are split apart by solar energy during the day. In recombining, the atoms give up some of the energy they absorbed in splitting, so that the air glow is a kind of delayed and very feeble return of sunlight in a new and specialized form. The rocket experiments of the 1950s supplied direct evidence of this. Spectroscopes carried by the rockets recorded the green lines of atomic oxygen most strongly at a height of sixty miles. A smaller proportion of the nitrogen was in the atomic form, because nitrogen molecules hold together more strongly than do oxygen molecules; nevertheless, the red light of atomic nitrogen was strong at a height of ninety-five miles.

Slipher had also found lines in the air glow that were suspiciously like well-known lines emitted by sodium. The presence of sodium seemed so unlikely that the matter was dropped in embarrassment. What would sodium, of all things, be doing in the upper atmosphere? It is not a gas, after all, but a very reactive metal that does not occur alone anywhere on the earth. It is always combined with other elements, most commonly in sodium chloride (table salt). But, in 1938, French scientists established that the lines were indeed identical with the sodium lines. Unlikely or not, sodium had to be in the upper atmosphere. Again rocket experiments clinched the matter: their spectroscopes recorded the yellow light of sodium unmistakably, and most strongly at a height of fifty-five miles. Where the sodium comes from is still a mystery; it may come from ocean salt spray or perhaps from vaporized meteors. Still more puzzling is the fact that lithium – a rarer relative of sodium – was also found, in 1958, to be contributing to the air glow.

In 1956, a team of United States scientists under the leadership of Murray Zelikoff produced an artificial air glow. They fired a rocket that at sixty miles released a cloud of nitric oxide gas. This accelerated the re-combination of oxygen atoms in the upper atmosphere. Observers on the ground easily sighted the bright glow that resulted. A similar experiment with sodium vapour also was successful: it created a clearly visible, yellow glow. When

Soviet scientists sent Lunik III in the direction of the moon in October 1959, they arranged for it to expel a cloud of sodium vapour as a visible signal that it had gone into orbit.

At lower levels in the atmosphere, atomic oxygen disappears, but the solar radiation is still energetic enough to bring about the formation of the three-atom variety of oxygen called 'ozone'. The ozone concentration is greatest at a height of fifteen miles. Even there, in what is called the 'ozonosphere' (first discovered in 1913 by the French physicist Charles Fabry), it makes up only one part in 4 million of the air, but that is enough to absorb ultra-violet light sufficiently to protect life on the earth. The ozone is formed by the action of ultra-violet light on ordinary two-atom oxygen. The ultra-violet radiation is, in this way, consumed: that is why the ozonosphere is earth's shield against it. It is the absorption of ultra-violet by oxygen that raises the temperature of the mesosphere above that of the stratosphere. Near the earth's surface, the concentration of ozone is very low, although it may rise high enough to form an irritating component of 'smog'.

Further rocket experiments showed that Teisserenc de Bort's speculations concerning layers of helium and hydrogen were not wrong – merely misplaced. From 200 to 600 miles upwards, where the atmosphere has thinned out to near-vacuum, there is a layer of helium, now called the 'heliosphere'. The existence of this layer was first deduced in 1961 by the Belgian physicist Marcel Nicolet from the frictional drag on the Echo I satellite. This was confirmed by actual analysis of the thin-gas surroundings by Explorer XVII, launched on 2 April 1963.

Above the heliosphere is an even thinner layer of hydrogen, the 'protonosphere', which may extend upwards some 40,000 miles before quite fading off into the general density of interplanetary space.

High temperatures and energetic radiation can do more than force atoms apart or into new combinations. They can chip electrons away from atoms and so 'ionize' the atoms. What remains of the atom is called an 'ion' and differs from ordinary atoms in carrying an electric charge. The word 'ion' comes from a Greek word meaning 'traveller'. Its origin lies in the fact that when an electric current passes through a solution containing ions, the positively charged ions travel in one direction and the negatively charged ions in the other.

A young Swedish student of chemistry named Svante August Arrhenius was the first to suggest that the ions were charged atoms, as the only means of explaining the behaviour of certain solutions that conducted an electric current. His notions, advanced in the thesis he presented for his degree of Doctor of Philosophy in 1884, were so revolutionary that his examiners could scarcely bring themselves to pass him. The charged particles within the atom had not yet been discovered, and the concept of an electrically charged atom seemed ridiculous. Arrhenius got his degree, but with only a minimum passing grade.

When the electron was discovered in the late 1890s, Arrhenius's theory suddenly made startling sense. He was awarded the Nobel Prize in chemistry in 1903 for the same thesis that nineteen years earlier had nearly lost him his doctoral degree. (This sounds like an improbable film scenario, I admit, but the history of science contains many episodes that make Hollywood seem unimaginative.)

The discovery of ions in the atmosphere did not emerge until after Guglielmo Marconi started his experiments with wireless. When, on 12 December 1901, he sent signals from Cornwall to Newfoundland, across 2,100 miles of the Atlantic Ocean, scientists were startled. Radio waves travel only in a straight line. How had they managed to go round the curvature of the earth and get to Newfoundland?

A British physicist, Oliver Heaviside, and an American electrical engineer, Arthur Edwin Kennelly, soon suggested that the radio signals might have been reflected back from the sky by a layer of charged particles high in the atmosphere. The 'Kennelly–Heaviside layer', as it has been called ever since, was finally located in the 1920s. The British physicist Edward Victor Appleton discovered it by paying attention to a curious fading phenomenon in radio transmission. He decided that the fading was the result of interference between two versions of the same signal, one coming directly from the transmitter to his receiver, the other by a roundabout route via reflection from the upper atmosphere. The delayed wave was out of phase with the first, so the two waves partly cancelled each other; hence the fading.

It was a simple matter then to find the height of the reflecting layer. All he had to do was to send signals at such a wavelength that the direct signal completely cancelled the reflected one – that is, the two signals arrived at directly opposite phases. From the wavelength of the signal used and the known velocity of radio waves, he could calculate the difference in the distances the two trains of waves had travelled. In this way, he determined that the Kennelly–Heaviside layer was some sixty-five miles up.

The fading of radio signals generally occurred at night. Appleton found that shortly before dawn radio waves were not reflected back by the Kennelly–Heaviside layer but were reflected from still higher layers (now sometimes called the 'Appleton layers') which begin at a height of 140 miles.

For all these discoveries Appleton received the Nobel Prize in physics in 1947. He had defined the important region of the atmosphere called the 'ionosphere', a word introduced in 1930 by the Scottish physicist Robert Alexander Watson-Watt. It includes the later-named mesosphere and thermosphere, and is now divided into a number of layers. From the stratopause up to sixty-five miles or so is the 'D region'. Above that is the Kennelly–Heaviside layer, called the 'D layer'. Above the D layer, to a height of 140 miles, is the 'E region' – an intermediate area relatively poor in ions. This is followed by the Appleton layers: the 'F_1 layer' at 140 miles and 'F_2 layer' at 200 miles. The F_1 layer is the richest in ions, the F_2 layer being significantly strong only in the daytime. Above these layers is the 'F region'.

These layers reflect and absorb only the long radio waves used in ordinary radio broadcasts. The shorter waves, such as those used in television, pass through, for the most part. That is why television broadcasting is limited in range – a limitation which can be remedied by satellite relay stations in the sky, most notably, so far, by the Early Bird satellite, launched in 1965, which allows live television to span oceans and continents. The radio waves from space (e.g., from radio stars) also pass through the ionosphere, fortunately; if they did not, there would be no radio astronomy.

The ionosphere is strongest at the end of the day, after the day-long effect of the sun's radiation, and weakens by dawn because many ions and electrons have recombined. Storms on the sun, intensifying the streams of particles and high-energy radiation sent to the earth, cause ionized layers to strengthen and thicken. The regions above the ionosphere also flare up into auroral displays. During these electric storms long-distance transmission of radio waves on the earth is disrupted and sometimes blacked out altogether.

It has turned out that the ionosphere is only one of the belts of radiation surrounding the earth. Outside the atmosphere, in what used to be considered 'empty' space, man's satellites in 1958 disclosed a startling surprise. To understand it we must make an excursion into the subject of magnetism.

Magnets

Magnets got their name from the ancient Greek town of Magnesia, near which the first 'lodestones' were discovered. The lodestone is an iron oxide with natural magnetic properties. Tradition has it that Thales of Miletus, about 550 B.C., was the first philosopher to describe it.

Magnets became something more than a curiosity when it was discovered that a steel needle stroked by a lodestone was magnetized and that, if the needle was allowed to pivot freely in a horizontal plane, it would end up laying along a north–south line. Such a needle was, of course, of tremendous use to mariners; in fact, it became indispensable to ocean navigation, though the Polynesians did manage to cross the Pacific without a compass.

It is not known who first put such a magnetized needle on a pivot and enclosed it in a box to make a compass. The Chinese are supposed to have done it first and passed it on to the Arabs, who, in turn, passed it on to the Europeans. This is all very doubtful and may be only legend. At any rate, in the twelfth century the compass came into use in Europe and was described in detail in 1269 by a French scholar best known by his Latinized name of Peter Peregrinus. Peregrinus named the end of the magnet that pointed north the 'north pole' and the other the 'south pole'.

Naturally, people speculated as to why a magnetized needle should point north. Because magnets were known to attract other magnets, some thought there was a gigantic lodestone mountain in the far north towards which the needle strained. Others were even more romantic and gave magnets a 'soul' and a kind of life.

The scientific study of magnets began with William Gilbert, the court physician to Queen Elizabeth I. It was Gilbert who discovered that the earth itself was a giant magnet. He mounted a magnetized needle so that it could pivot freely in a vertical direction (a 'dip needle'), and its north pole then dipped towards the ground ('magnetic dip'). Using a spherical lodestone as a model of the earth, he found that the needle behaved in the same way when it was placed over the 'northern hemisphere' of this sphere. Gilbert published these findings in 1600 in a classic book entitled *De Magnete*.

In the three and a half centuries that have elapsed since Gilbert's work, no one has ever explained the earth's magnetism to everyone's satisfaction. For a long time scientists speculated that the earth might have a gigantic iron magnet as its core. Although the earth was indeed found to have an iron core, it is now certain that this core cannot be a magnet because iron, when heated, loses its strong magnetic properties ('ferromagnetism', the prefix coming from the Latin word for iron) at 760°C, and the temperature of the earth's core must be at least 1,000°C.

The temperature at which a substance loses its magnetism is called the 'Curie temperature', since it was first discovered by Pierre Curie in 1895. Cobalt and nickel, which resemble iron closely in many respects, are also ferromagnetic. The Curie temperature for nickel is 356°C; for cobalt it is 1,075°C. At low temperatures, certain other metals are ferromagnetic. Below −188°C, dysprosium is ferromagnetic, for instance.

In general, magnetism is a property of the atom itself, but in most materials the tiny atomic magnets are oriented in random directions, so that most of the effect is cancelled out. Even so, weak magnetic properties are often evidenced, and the result is 'paramagnetism'. The strength of magnetism is expressed in terms of 'permeability'. The permeability of a vacuum is 1·00 and that of paramagnetic substances is between 1·00 and 1·01.

Ferromagnetic substances have much higher permeabilities. Nickel has a permeability of 40, cobalt one of 55, and that of iron is in the thousands. In such substances, the existence of 'domains' was postulated in 1907 by the French physicist Pierre Weiss. These are tiny areas about 0·001 to 0·1 centimetres in diameter (which have actually been detected), in which the atomic magnets are so lined up as to reinforce one another, producing strong, over-all fields within the domain. In ordinary non-magnetized iron, the domains themselves are randomly oriented and cancel one another's effect. When the domains are brought into line by the action of another magnet, the iron is magnetized. The reorientation of domains during magnetism actually produces clicking and hissing noises that can be detected by suitable amplification, this being termed the 'Barkhausen effect' after its discoverer, the German physicist Heinrich Barkhausen.

In 'anti-ferromagnetic substances', such as manganese, the domains also line up, but in alternate directions, so that most of the magnetism is cancelled. Above a particular temperature, substances lose anti-ferromagnetism and become paramagnetic.

If the earth's iron core is not itself a permanent magnet because it is above the Curie temperature, then there must be some other way of explaining the earth's ability to affect a compass needle. What that might be grew out of the work of the English scientist Michael Faraday, who discovered the connection between magnetism and electricity.

In the 1820s, Faraday started with an experiment that had been first described by Peter Peregrinus (and which still amuses young students of physics). The experiment consists in sprinkling fine iron filings on a piece of paper above a magnet and gently tapping the paper. The shaken filings tend to line up along arcs from the north pole to the south pole of the magnet. Faraday decided that these marked actual 'magnetic lines of force', forming a magnetic 'field'.

Faraday, who had been attracted to the subject of magnetism by the Danish physicist Hans Christian Oersted's observation in 1820 that an electric current flowing in a wire deflected a nearby compass needle, came to the conclusion that the current must set up magnetic lines of force around the wire.

He felt this to be all the more so since the French physicist André Marie Ampère had gone on to study current-carrying wires immediately after Oersted's discovery. Ampère showed that two parallel wires with the current flowing in the same direction attracted each other; with currents flowing in opposite directions they repelled each other. This was very like the fashion in which two magnetic north poles (or two magnetic south poles) repelled each other while a magnetic north pole attracted a magnetic south pole. Better still, Ampère showed that a cylindrical coil of wire with an electric current flowing through it behaved like a bar magnet. In memory of his work, the unit of intensity of electric current was officially named the 'ampere' in 1881.

But if all this were so, thought Faraday (who had one of the most efficient intuitions in the history of science), and if electricity can set up a magnetic field so like the real thing that current-carrying wires can act like magnets, should not the reverse be true? Ought not a magnet produce a current of electricity that would be just like the current produced by chemical batteries?

In 1831, Faraday performed the experiment that was to change human history. He wound a coil of wire round one segment of an iron ring and a second coil of wire round another segment of the ring. Then he connected the first coil to a battery. His reasoning was that if he sent a current through the first coil, it would create magnetic lines of force which would be concentrated in the iron ring, and this induced magnetism in turn would produce a current in the second coil. To detect that current, he connected the second coil to a galvanometer – an instrument for measuring electrical currents, which had been devised by the German physicist Johann Salomo Christoph Schweigger in 1820.

The experiment did not work as Faraday had expected. The

flow of current in the first coil generated nothing in the second coil. But Faraday noticed that at the moment when he turned on the current, the galvanometer needle kicked over briefly, and it did the same thing, but in the opposite direction, when he turned the current off. He guessed at once that it was the movement of magnetic lines of force across a wire, not the magnetism itself, that set up the current. When a current began to flow in the first coil, it initiated a magnetic field that, as it spread, cut across the second

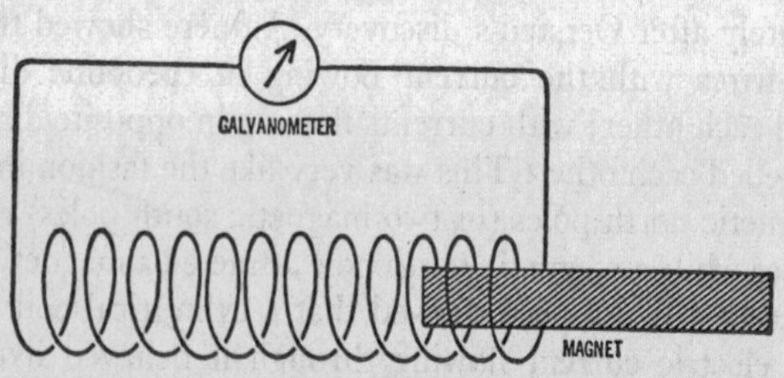

A Faraday experiment on the induction of electricity. When the magnet is moved in or out of the coil of wire, the cutting of its lines of force by the wire produces an electrical current in the coil.

coil, setting up a momentary electric current there. Conversely, when the current from the battery was cut off, the collapsing lines of magnetic force again cut across the wire of the second coil, causing a momentary surge of electricity in the direction opposite that of the first flow.

Thus Faraday discovered the principle of electrical induction and created the first 'transformer'. He proceeded to demonstrate the phenomenon more plainly by using a permanent magnet and moving it in and out of a coil of wire; although no source of electricity was involved, a current flowed in the coil whenever the magnet's lines of force cut across the wire.

Faraday's discoveries not only led directly to the creation of the dynamo for generating electricity but also laid the foundation for James Clerk Maxwell's 'electromagnetic' theory, which linked together light and other forms of radiation (such as radio) in a single family of 'electromagnetic radiations'.

Now the close connection between magnetism and electricity points to a possible explanation of the earth's magnetism. The compass needle has traced out its magnetic lines of force, which run from the 'north magnetic pole', located off northern Canada, to the 'south magnetic pole', located at the rim of Antarctica, each being about 15 degrees of latitude from the geographic poles. (The earth's magnetic field has been detected at great heights by rockets carrying 'magnetometers'.) The new suggestion is that the earth's magnetism may originate in the flow of electric currents deep in its interior.

The physicist Walter Maurice Elsasser has proposed that the

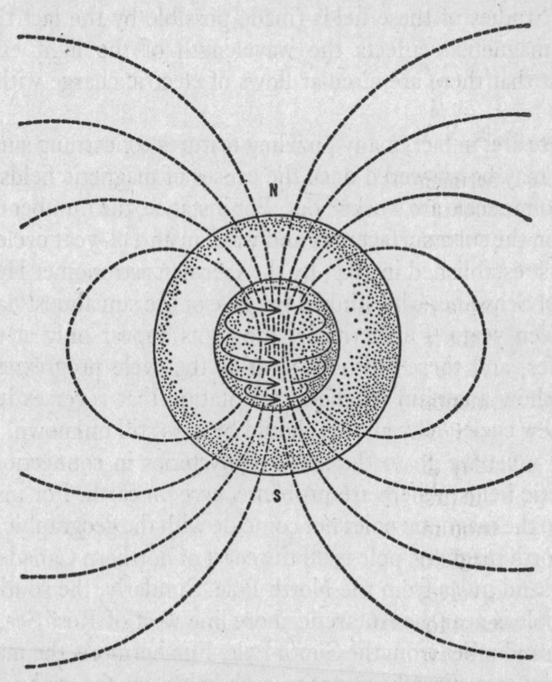

Elsasser's theory of the generation of the earth's magnetic field. Movements of material in the molten nickel–iron core set up electric currents which in turn generate magnetic lines of force. The dotted lines show the earth's magnetic field.

rotation of the earth sets up slow eddies in the molton iron core, circling west to east. These eddies have the effect of producing an electric current, likewise circling west to east. Just as Faraday's coil of wire produced magnetic lines of force within the coil, so the circling electric current does in the earth's core. It therefore creates the equivalent of an internal magnet extending north and south. This magnet in turn accounts for the earth's general magnetic field, oriented roughly along the axis of rotation, so that the magnetic poles are near the north and south geographic poles.

The sun also has a general magnetic field, which is two or three times as intense as that of the earth, and local fields apparently associated with the sunspots, which are thousands of times as intense. Studies of these fields (made possible by the fact that intense magnetism affects the wavelength of the light emitted) suggest that there are circular flows of electric charge within the sun.

There are, in fact, many puzzling features concerning sunspots, which may be answered once the causes of magnetic fields on an astronomic scale are worked out. For instance, the number of sunspots on the solar surface wax and wane in an 11½-year cycle. This was first established in 1843 by the German astronomer Heinrich Samuel Schwabe, who studied the face of the sun almost daily for seventeen years. Furthermore, the spots appear only at certain latitudes, and these latitudes shift as the cycle progresses. The spots show a certain magnetic orientation that reverses itself in each new cycle. Why all this should be so is still unknown.

Nor must we go to the sun for mysteries in connection with magnetic fields. There are problems here on earth. For instance, why do the magnetic poles not coincide with the geographic poles? The north magnetic pole is off the coast of northern Canada about a thousand miles from the North Pole. Similarly, the south magnetic pole is near the Antarctic shore line west of Ross Sea, about a thousand miles from the South Pole. Furthermore, the magnetic poles are not directly opposite each other on the globe. A line through the earth connecting them (the 'magnetic axis') does not pass through the centre of the earth.

Again, the deviation of the compass needle from 'true north' (i.e., the direction of the North Pole) varies irregularly as one travels east or west. In fact, the compass needle shifted on Columbus's first voyage, and Columbus hid this from his crew lest it excite terror that would force him to turn back.

This is one of the reasons why the use of a magnetic compass to determine direction is less than perfect. In 1911, a non-magnetic method for indicating direction was introduced by the American inventor Elmer Ambrose Sperry. It takes advantage of the tendency of a rapidly turning heavy-rimmed wheel (a 'gyroscope', first studied by the same Foucault who had demonstrated the rotation of the earth) to resist changes in its plan of rotation. This can be used to serve as a 'gyroscopic compass', which will maintain a fixed direction reference that will serve to guide ships or rockets.

But if the magnetic compass is less than perfect, it has been useful enough to serve mankind for centuries. The deviation of the magnetic needle from the true north can be allowed for. A century after Columbus, in 1581, the Englishman Robert Norman prepared the first map indicating the actual direction marked out by a compass needle ('magnetic declination') in various parts of the world. Lines connecting those points on the planet that show equal declinations ('isogonic lines') run crookedly from north magnetic pole to south magnetic pole.

Unfortunately, such maps must be periodically changed, for even at one spot the magnetic declination changes with time. For instance, the declination at London shifted 32 degrees of arc in two centuries; it was 8 degrees east of north in 1600 and steadily swung round counter-clockwise until it was 24 degrees west of north in 1800. Since then it has shifted back and in 1950 was only 8 degrees west of north.

Magnetic dip also changes slowly with time for any given spot on earth, and the map showing lines of equal dip ('isoclinic lines') must also be constantly revised. Moreover, the intensity of earth's magnetic field increases with latitude and is three times as strong near the magnetic poles as in the equatorial regions. This intensity also changes constantly, so that maps showing 'isodynamic lines' must also be periodically revised.

Like everything else about the magnetic field, the overall intensity of the field changes. For some time now, the intensity has been diminishing. The field has lost 15 per cent of its total strength since 1670; if this continues, it will reach zero by about the year 4000. What then? Will it continue decreasing, in the sense that it will reverse with the north magnetic pole in Antarctica and the south magnetic pole in the Arctic? In other words, does earth's magnetic field periodically diminish, reverse, intensify, diminish, reverse, and so on?

One way of telling if this can indeed happen is to study volcanic rocks. When lava cools, the crystals form in alignment with the magnetic field. As long ago as 1906, the French physicist Bernard Brunhes noted that some rocks were magnetized in the direction *opposite* to earth's present magnetic field. This finding was largely ignored at the time, but there is no denying it now. The telltale rocks inform us that not only has earth's magnetic field reversed, it has done so many times: nine times in the last 4 million years, at irregular intervals.

The most spectacular finding in this respect is on the ocean floor. If melted rock is indeed pushing up through the Global Rift and spreading out, then as one moves east or west from the Rift one comes across rock that has solidified a progressively longer time ago. By studying the magnetic alignment, one can indeed find reversals occurring in strips, progressively farther from the Rift, at intervals of anywhere from 50,000 to 20 million years. The only rational way of explaining this, so far, is to suppose that there *is* sea-floor spreading and there *are* magnetic-field reversals.

The fact of the reversals is easier to ascertain, however, than the reasons for it.

In addition to long-term drifts of the magnetic field, there are small changes during the course of the day. These suggest some connection with the sun. Furthermore, there are 'disturbed days' when the compass needle jumps about with unusual liveliness. The earth is then said to be experiencing a 'magnetic storm'. Magnetic storms are identical with electric storms and are usually accompanied by an increase in the intensity of auroral displays, an observation reported as long ago as 1759 by the English physicist John Canton.

The aurora borealis (a term introduced in 1621 by the French philosopher Pierre Gassendi, and Latin for 'northern dawn') is a beautiful display of moving coloured streamers or folds of light, giving an effect of unearthly splendour. Its counterpart in the Antarctic is called the aurora australis ('southern dawn'). In 1741, the Swedish astronomer Anders Celsius noted its connection with earth's magnetic field. The auroral streamers seem to follow the earth's magnetic lines of force and to concentrate, and become visible, at those points where the lines crowd most closely together – that is, at the magnetic poles. During magnetic storms the northern aurora can be seen as far south as Boston and New York.

Why the aurora should exist was not hard to understand. Once the ionosphere was discovered, it was understood that something (presumably solar radiation of one sort or another) was energizing the atoms in the upper atmosphere and converting them into electrically charged ions. At night, the ions would lose their charge and their energy, the latter making itself visible in the form of auroral light. It was a kind of specialized air glow, which followed the magnetic lines of force and concentrated near the magnetic poles because that would be expected of electrically charged ions. (The air glow itself involves uncharged atoms and therefore ignores the magnetic field.)

But what about the disturbed days and the magnetic storms? Again the finger of suspicion points to the sun.

Sunspot activity seems to generate magnetic storms. How such a disturbance 93 million miles away could affect the earth was a complete mystery until the spectrohelioscope, invented by the astronomer George Ellery Hale, brought forth a possible answer. This instrument allows the sun to be photographed in light of a particular colour – for instance, the red light of hydrogen. Furthermore, it shows the motions or changes taking place on the sun's surface. It gives good pictures of 'prominences' and 'solar flares' – great bursts of flaming hydrogen. These were first observed during the eclipse of the sun in 1842, which was visible on a line across the width of Europe and was the first solar eclipse to be systematically and scientifically observed.

Before the invention of the spectrohelioscope, only those flares shooting out at right angles to the direction of the earth could be seen. The spectroheliοscope, however, also showed those coming out in our direction from the centre of the sun's disk: in hydrogen light these hydrogen-rich bursts appear as light blotches against the darker background of the rest of the disk. It turned out that solar flares were followed by magnetic storms on the earth only when the flare was pointed towards the earth.

Apparently, then, magnetic storms were the result of bursts of charged particles, shot from the flares to the earth across 93 million miles of space. As long ago as 1896, something like this had been suggested by the Norwegian physicist Olaf Kristian Birkeland.

As a matter of fact, there was plenty of evidence that, wherever the particles might come from, the earth was bathed in an aura of them extending pretty far out in space. Radio waves generated by lightning had been found to travel along the earth's magnetic lines of force at great heights. (These waves, called 'whistlers' because they were picked up by receivers as odd whistling noises, had been discovered accidentally by the German physicist Heinrich Barkhausen during the First World War.) The radio waves could not follow the lines of force unless charged particles were present.

Yet it did not seem that these charged particles emerged from the sun only in bursts. In 1931, when Sydney Chapman was studying the sun's corona, he was increasingly impressed by its extent. What we can see during a total solar eclipse is only its innermost portion. The measurable concentrations of charged particles in the neighbourhood of the earth were, he felt, part of the corona. This meant then, in a sense, that the earth was revolving about the sun within that luminary's extremely attenuated outer atmosphere. Chapman drew the picture of the corona expanding outwards into space and being continually renewed at the sun's surface. There would be charged particles continuously streaming out of the sun in all directions, disturbing earth's magnetic field as it passed.

This suggestion became virtually inescapable in the 1950s, thanks to the work of the German astrophysicist Ludwig Franz Biermann. For half a century, it had been thought that the tails of comets, which always pointed generally away from the sun and which increased in length as the comet approached the sun, were formed by the pressure of light from the sun. Such light-pressure does exist, but Biermann showed that it wasn't nearly enough to produce cometary tails. Something stronger and with more of a push was required; this something could scarcely be anything but charged particles. The American physicist Eugene Norman Parker argued further in favour of a steady outflow of particles, with additional bursts at the time of solar flares and, in 1958, named the effect the 'solar wind'. The existence of this solar wind was finally demonstrated by the Soviet satellites Lunik I and Lunik II, which streaked outwards to the neighbourhood of the moon in 1959 and 1960, and by the American planetary probe Mariner II, which in 1962 passed near Venus.

The solar wind is no local phenomenon. There is reason to think it remains dense enough to be detectable at least as far out as the orbit of Saturn. Near the earth the velocity of solar-wind particles varies from 350 to 700 kilometres per second. Its existence represents a loss to the sun of a million tons of matter per second, but though this seems huge in human terms, it is utterly insignificant on the solar scale. In the entire lifetime of the sun less than a hundredth part of one per cent of its mass has been lost to the solar wind.

The solar wind may well affect man's everyday life. Beyond its effect on the magnetic field, the charged particles in the upper atmosphere may ultimately have an effect on the details of earth's weather. If so, the ebb and flow of the solar wind may yet become still another weapon in the armoury of the weather forecast.

An unforeseen effect of the solar wind was unexpectedly worked out as a result of satellite launchings. One of the prime jobs given to the man-made satellites was to measure the radiation in the upper atmosphere and nearby space, especially the intensity of the cosmic rays (charged particles of particularly high energy). How intense was this radiation up beyond the atmospheric shield? The satellites carried 'Geiger counters' (first devised by the German physicist Hans Geiger in 1907 and vastly improved in 1928), which measure particle radiation in the following way. The counter has a box containing gas under a voltage not quite strong enough to send a current through the gas. When a high-energy particle of radiation penetrates into the box, it converts an atom of the gas into an ion. This ion, hurtled forward by the energy of the blow, smashes neighbouring atoms to form more ions, which in turn smash their neighbours to form still more. The resulting shower of ions can carry an electric current, and for a fraction of a second a current pulses through the counter. The pulse is telemetered back to earth. Thus the instrument counts the particles, or flux of radiation, at the location where it happens to be.

When the first successful American satellite, Explorer I, went into orbit on 31 January 1958, its counter detected about the expected concentrations of particles at heights up to several hundred miles. But at higher altitudes (and Explorer I went as high as 1,575 miles) the count fell off; in fact, at times it dropped to zero! This might have been dismissed as due to some peculiar kind of accident to the counter, but Explorer III, launched on 26 March 1958 and reaching an apogee of 2,100 miles, had just the same experience. So did the Soviet Sputnik III, launched on 15 May 1958.

James A. Van Allen of the State University of Iowa, who was in charge of the radiation programme, and his aides came up with a possible explanation. The count fell virtually to zero, they decided, not because there was little or no radiation, but because there was too much. The instrument could not keep up with the particles entering it, and it blanked out in consequence. (This would be analogous to the blinding of our eyes by a flash of too-bright light.)

When Explorer IV went up on 26 July 1958, it carried special counters designed to handle heavy loads. One of them, for instance, was shielded with a thin layer of lead (analogous to dark sunglasses) that would keep out most of the radiation. And this time the counters did tell another story. They showed that the 'too-much-radiation' theory was correct. Explorer IV, reaching a height of 1,368 miles, sent down counts which, allowing for the shielding, disclosed that the radiation intensity up there was far higher than scientists had imagined. In fact, it was so intense that it raised a deadly danger to space flight by man.

It became apparent that the Explorer satellites had penetrated only the lower regions of this intense field of radiation. In the autumn of 1958 the two satellites shot by the United States in the direction of the moon (so-called 'moon probes') – Pioneer I, which went out 70,000 miles, and Pioneer III, which reached 65,000 miles – showed two main bands of radiation encircling the earth. They were named the 'Van Allen radiation belts', but were later named the 'magnetosphere' in line with the names given to other sections of space in the neighbourhood of the earth.

It was at first assumed that the magnetosphere was symmetrically placed about the earth, rather like a huge doughnut, and that the magnetic lines of force were themselves symmetrically arranged. This notion was upset when satellite data brought back other news. In 1963, in particular, the satellites Explorer XIV and Imp I were sent into highly elliptical orbits designed to carry them beyond the magnetosphere if possible.

It turned out that the magnetosphere had a sharp boundary, the 'magnetopause', which was driven back upon the earth on the side towards the sun by the solar wind, but which looped back around the earth and extended an enormous distance on the night side. The magnetopause was some 40,000 miles from the earth in the direction of the sun, but the tear-drop tail on the other side may extend outwards for a million miles or more. In 1966, the Soviet satellite Luna X, which circled the moon, detected a feeble magnetic field surrounding that world which may actually have been the tail of earth's magnetosphere sweeping past.

The entrapment of charged particles along the magnetic lines of force had been predicted in 1957 by an American-born Greek amateur scientist, Nicholas Christofilos, who made his living as a salesman for an American elevator firm. He had sent his calculations to scientists engaged in such research, but no one had paid much attention to them. (In science, as in other fields, professionals tend to disregard amateurs.) It was only when the professionals independently came up with the same results that Christofilos achieved recognition and was welcomed into the University of California, where he now works. His idea about particle entrapment is now called the 'Christofilos effect'.

To test whether the effect really occurs in space, the United

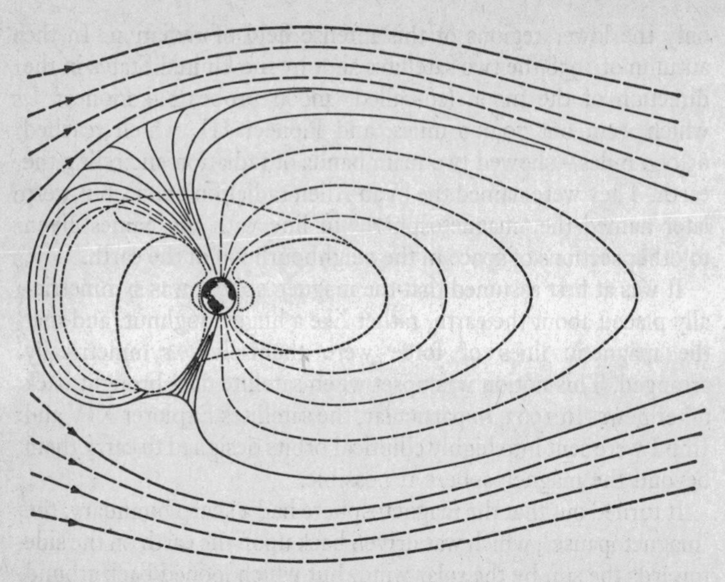

The Van Allen radiation belts, as traced by satellites. They appear to be made up of charged particles trapped in the earth's magnetic field.

States in August and September 1958 fired three rockets carrying nuclear bombs 300 miles up and there exploded the bombs – an experiment which was named 'Project Argus'. The flood of charged particles resulting from the nuclear explosions spread out along the lines of force and were indeed trapped there. The resulting band persisted for a considerable time; Explorer IV detected it during several hundred of its trips around the earth. The cloud of particles also gave rise to feeble auroral displays and disrupted radar for a while.

This was the prelude to other experiments that affected or even altered earth's near-space environment, and some of them met with opposition and vast indignation from sections of the scientific community. A nuclear bomb exploded in space on 9 July 1962 introduced marked changes in the Van Allen belts, changes that showed signs of persisting for a prolonged interval, as some disapproving scientists (such as Fred Hoyle) had predicted. The

Soviet Union carried out similar high-altit[illegible] Such tampering with the natural state of affair[illegible]th our understanding of the magnetosphere, a[illegible]hat this experiment will be soon repeated.

Then, too, attempts were made to s[illegible]in copper needles into orbit about the earth to te[illegible]eflect radio signals, in order to establish an unfa[illegible]ng-distance communication. (The ionosphere i[illegible]gnetic storms every once in a while and then ra[illegible]n may fail at a crucial moment.)

Despite the objection of rad[illegible]ho feared interference with the radio signals f[illegible] the project ('Project West Ford', after Westf[illegible]ts, where the preliminary work was done) w[illegible] on 9 May 1963. A satellite containing 400 mill[illegible]s, each three quarters of an inch long and finer t[illegible] – fifty pounds' weight altogether – was put into[illegible]s were ejected and then slowly spread into a wo[illegible] that was found to reflect radio waves just as the[illegible]cted to do. This band remained in orbit for th[illegible]ch thicker band would be required for useful p[illegible]er, and it is doubtful if the objections of the rad[illegible]an be overcome for that.

Naturally, scientists were [illegible]s to find out whether there were radiation belts about heavenly bodies other than the earth. One way of determining this was to send satellites upwards at velocities great enough to break them loose from the earth's grip altogether (7 miles per second – as compared with a satellite in orbit about earth, which travels a mere 5 miles per second). The first satellite to surpass escape velocity, break away from the earth altogether, and move into orbit about the sun as the first 'man-made planet' was the Soviet Union's Lunik I, launched on 2 January 1959. Their next moon probe, Lunik II, actually hit the moon in September 1959 (the first man-made object to land on a surface of a body other than the earth). They found no signs of radiation belts about the moon.

This was not surprising, for scientists had already surmised that the moon had no magnetic field of any consequence. The overall density of the moon has long been known to be but 3·3 grams per cubic centimetre (about three fifths that of the earth), and it could not have so low a density unless it were almost entirely silicate, with no iron core to speak of. The lack of a magnetic field would seem to follow, if present theories are correct.

But what of Venus? In size and mass it is almost the earth's twin, and there seems no doubt that it has an iron core. Does it also have a magnetosphere? Both the Soviet Union and the United States attempted to send out 'Venus probes' that would, in their orbits, pass close to Venus and send back useful data. The first such probe to be completely successful was Mariner II, launched by the United States on 27 August 1962. It passed within 21,600 miles of Venus on 14 December 1962, and found no signs of a magnetosphere.

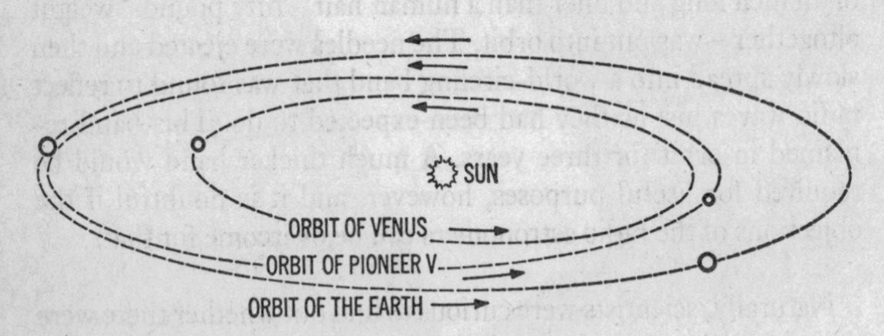

Orbit of the United States' 'artificial planet' Pioneer V, launched 11 March 1960, is shown in relation to the sun and the orbits of the earth and Venus. The dot on the rocket's orbit indicates roughly its position on 9 August 1960, when it was closest to the sun.

This does not necessarily deny the presence of an iron core in Venus, since an alternative explanation offers itself at once. The rotation of Venus is very slow, once in about eight months (the fact that the rotation is in the wrong direction has nothing to do with this particular case), and that is not enough to set up the kind of eddies in the core (if it exists) that would account for a magnetic field. Mercury, which rotates once in two months, has been reported to have a weak magnetic field about 1/60 the intensity of that of the earth.

What of Mars? Being a bit denser than the moon it may have a small iron core, and, since it rotates in twenty-four and a half hours, it may have a very weak magnetic field. The United States launched a 'Mars Probe' – Mariner IV – on 28 November 1964. In July 1965, it approached Mars closely, and, from the data it gathered, it would seem there was no magnetic field to speak of about that planet.

As for the solar system beyond Mars, good evidence turned up quickly to the effect that Jupiter and Saturn, at least, have radiation belts that are both more intense and more extensive than those of the earth. In fact, radio-wave radiation from Jupiter seems to indicate that it possesses a magnetic field at least twelve to sixteen times as strong as that of the earth. In 1965, radio-wave emission was detected from Uranus and Neptune.

One of the more dramatic reasons for intense curiosity in the magnetosphere is, of course, concern for the safety of human pioneers in outer space. In 1959, the United States selected seven men (popularly called 'astronauts') to take part in 'Project Mercury', which was to place men into orbit about the earth. The Soviet Union also initiated a training programme for what they called 'cosmonauts'.

The honour of reaching this goal first fell to the Soviet Union's cosmonaut Yuri Alexeyevich Gagarin, who was launched into orbit on 12 April 1961 (only three and a half years after the opening of the 'Space Age' with Sputnik I), and returned safely after circling the earth in an hour and a half. He was the first 'man in space'.

The Soviet Union sent other human beings in orbit over the next few years. For a while, the record for endurance was held by Valery F. Bykovsky, who, after being launched on 14 June 1963, circled the earth eighty-two times before coming down. Launched on 16 June 1963 was Valentina V. Tereshkova, the first woman in space. She completed forty-nine orbits.

The first American to be placed in orbit was John Herschel Glenn, who, after being launched on 20 February 1962, circled the earth three times. The American record for endurance, so far, is the trip of Leroy Gordon Cooper, who was launched on 15 May 1963 and successfully completed twenty-two orbits.

For years, manned space flights, both in the Soviet Union and the United States, have ended successfully and there were no casualties. There is some question as to whether or not the flights have had some long-term bad effects. Several astronauts, including Glenn, have suffered ailments of the middle ear, although this may not be directly connected with their experiences.

Flights in two-man and three-man capsules were carried through by the United States and the Soviet Union in 1964 and 1965. During the course of one two-man flight, on 18 March 1965, the Soviet cosmonaut Aleksei A. Leonov, encased in his spacesuit and holding on to a 'life-line', stepped out of his capsule and became the first human being to float freely in space.

Meteors

Even the Greeks knew that 'shooting stars' were not really stars, because no matter how many fell, the celestial population of stars remained the same. Aristotle reasoned that a shooting star, being a temporary phenomenon, had to be something within the atmosphere (and this time he was right). These objects were therefore called 'meteors', meaning 'things in the air'. Meteors that actually reach the earth's surface are called 'meteorites'.

The ancients even witnessed some falls of meteorites to the earth and found them to be lumps of iron. Hipparchus of Nicaea is said to have reported such a fall. The Kaaba, the sacred black stone in Mecca, is supposed to be a meteorite and to have gained its sanctity through its heavenly origin. The *Iliad* mentions a lump of rough iron being awarded as one of the prizes in the funeral games for Patroclus; this must have been meteoric in origin, because the time was the Bronze Age, before the metallurgy of iron ore had been developed. In fact, meteoric iron was probably in use as early as 3000 B.C.

During the eighteenth century, with the Age of Reason in full sway, science made a backward step in this particular respect. The scorners of superstition laughed at stories of 'stones from the sky'. Farmers who came to the Académie Française with samples of meteorites were politely, but impatiently, shown the door. When two Connecticut scholars in 1807 reported having witnessed a fall, President Thomas Jefferson (in one of his more unfortunate remarks) said that he would sooner believe that two Yankee professors would lie than that stones would fall from heaven.

However, on 13 November 1833, the United States was treated to a meteor shower of the type called 'Leonids' because they seem to radiate from a point in the constellation Leo. For some hours it turned the sky into a Roman-candle display more brilliant than any ever seen before or since. No meteorites reached the ground, as far as is known, but the spectacle stimulated the study of meteors, and astronomers turned to it for the first time in all seriousness.

The very next year, the Swedish chemist Jöns Jakob Berzelius began a programme for the chemical analyses of meteorites. Eventually such analyses gave astronomers valuable information on the general age of the solar system and even on the overall chemical make-up of the universe.

By noting the times of year when meteors came thickest, and the positions in the sky from which they seemed to come, the meteor watchers were able to work out orbits of various clouds of meteors. In this way they learned that a meteor shower occurred when the earth's orbit intersected the orbit of a meteor cloud.

Could it be that meteor clouds of this sort are actually the debris left over from disintegrated comets? That comets do disintegrate may be accepted, since one of them, Biela's Comet, virtually did just that before the eyes of astronomers in the nineteenth century and left a meteor cloud in its orbit.

Comets may be fragile because of their very structure. The American astronomer Fred Lawrence Whipple suggested in 1950 that comets consist of pebbles of rocky material cemented by 'ices' of such low-freezing gases as methane or ammonia. Some of the ices evaporate at each close approach to the sun, liberating dust and particles which are swept away from the sun by (as we now know) the solar wind. Eventually, the ices are all gone, and the comet either remains as a rocky core or disintegrates into the meteor cloud formed of its former pebbles. A comet may lose as much as 0·5 per cent of its mass at each approach; even a comet that approaches the sun not-too-closely may last not more than a million years. That comets still exist now that the solar system has lasted nearly 5,000 million years can only be because new comets are constantly entering the inner system from the vast cloud of comets that Oort has postulated to be far out in space.

Most of the meteorites found on the ground (about 1,700 are known altogether, of which 35 weigh over a ton each) were iron, and it seemed that iron meteorites must far outnumber the stony type. This proved to be wrong, however. A lump of iron lying half-buried in a stony field is very noticeable, whereas a stone among other stones is not. When astronomers made counts of meteorites found after they were actually seen to fall, they discovered that the stony meteorites outnumbered iron ones nine to one. (For a time, most stony meteorites were discovered in Kansas, which may seem odd until one realizes that in the stoneless, sedimentary soil of Kansas a stone is as noticeable as a lump of iron would be elsewhere.)

Meteorites seldom do damage. Although about 500 substantial meteorites strike the earth annually (with only some 20 recovered, unfortunately), the earth's surface is large and only small areas are thickly populated. No human being has ever been killed by a meteorite so far as is known, although a woman in Alabama reported being bruised by a glancing blow on 30 November 1955.

Yet meteorites have a devastating potentiality. In 1908, for instance, a strike in northern Siberia gouged out craters up to 150 feet in diameter and knocked down trees for 20 miles around. Fortunately, the meteorite fell in a wilderness; had it fallen from the same part of the sky five hours later in the earth's rotation, it might have hit St Petersburg, then the capital of Russia. If it had, the city would have been wiped out as thoroughly as by an H-bomb. One estimate is that the total weight of the meteorite was 40,000 tons. The largest strike since then, near Vladivostok (again in Siberia), was in 1947.

There are signs of even heavier strikes in prehistoric times. In Coconino County in Arizona there is a round crater about four fifths of a mile across and 600 feet deep, surrounded by a lip of earth 100 to 150 feet high. It looks like a miniature crater of the moon. It was long assumed to be an extinct volcano, but a mining engineer named Daniel Moreau Barringer insisted it was the result of a meteoric collision, and the hole now bears the name 'Barringer Crater'. The crater is surrounded by lumps of meteoric iron – thousands (perhaps millions) of tons of it altogether. Although only a small portion has been recovered so far, more meteoric iron has already been extracted from it and its surroundings than in all the rest of the world. The meteoric origin of the crater was also borne out by the discovery there in 1960 of forms of silica that could only have been produced by the momentary enormous pressures and temperatures accompanying meteoric impact.

Barringer Crater, formed in the desert an estimated 25,000 years ago, has been preserved fairly well. In most parts of the world similar craters would have been obliterated by water and plant overgrowth. Observations from aeroplanes, for instance, have sighted previously unnoticed circular formations, partly water-filled and partly overgrown, which are almost certainly meteoric. Several have been discovered in Canada, including Brent Crater in central Ontario and Chubb Crater in northern Quebec, each of which is two miles or more in diameter, and Ashanti Crater in Ghana, which is six miles in diameter. These are perhaps a million years old or more. Fourteen such 'fossil craters' are known, and subtle geological signs point to the existence of many more.

The craters of the moon visible to us with telescopes range from holes no larger than Barringer Crater to giants 150 miles across. The moon, lacking air, water, or life, is a nearly perfect museum for craters since they are subject to no wear except from the very slow action of temperature change resulting from the two-week alternation of lunar day and lunar night. Perhaps the earth would be pockmarked like the moon if it were not for the healing action of wind, water, and growing things.

It had been felt, at first, that the craters of the moon were volcanic in origin, but they do not really resemble earthly volcanic craters in structure. By the 1890s, the view that the craters had originated from meteoric strikes came into prominence and has gradually become accepted.

The large 'seas' or maria, which are vast, roughly circular stretches that are relatively crater-free, would in this view result from the impact of particularly large meteors. This view was bolstered in 1968 when satellites placed in orbit about the moon showed unexpected deviations in their circumlunar flights. The nature of these deviations forced the conclusion that parts of the lunar surface were denser than average and produced a slight increase in gravitational attraction, to which the satellite flying over those parts responded. These denser-than-average areas, which seemed to coincide with the maria, received the name 'mascons' (short for 'mass-concentrations'). The most obvious deduction was that the sizeable iron meteors that formed the seas were still buried beneath them and were considerably denser than the rocky material that generally made up the moon's crust. At least a dozen mascons were detected within a year of the initial discovery.

The view of the moon as a 'dead world' where no volcanic action is possible is, on the other hand, overdrawn. On 3 November 1958, the Russian astronomer N. A. Kozyrev observed a reddish spot in the crater Alphonsus. (William Herschel had reported seeing reddish spots on the moon as early as 1780.) Kozyrev's spectroscopic studies seemed to make it clear that gas and dust had been emitted. Since then, other red spots have been momentarily seen, and it seems quite certain that volcanic activity does occasionally take place on the moon. During the total lunar eclipse in December 1964, it was found that as many as 300 craters were hotter than the surrounding landscape, though of course they were not hot enough to glow.

Once the first satellite was put into orbit in 1957, it was only a matter of time before mankind began to investigate the moon at close quarters. The first successful 'moon probe' – that is, the first satellite to pass near the moon – was sent up by the Soviet Union on 2 January 1959. It was Lunik I, the first man-made object to take up an orbit about the sun. Within two months, the United States had duplicated the feat.

On 12 September 1959, the Soviets sent up Lunik II and aimed it to hit the moon. For the first time in history, a man-made object rested on the surface of another world. Then, a month later, the Soviet satellite Lunik III slipped beyond the moon and pointed a television camera at the side we never see from earth. Forty minutes of pictures of the other side were sent back from a distance of 40,000 miles. They were fuzzy and of poor quality, but they showed something interesting. The other side of the moon had scarcely any maria of the type that are so prominent a feature of our side. Why this asymmetry should exist is not entirely clear. Presumably the maria were formed comparatively late in the moon's history, when one side already faced the earth for ever and the large meteors that formed the seas were slanted towards the near face of the moon by earth's gravity.

But lunar exploration was only beginning. In 1964, the United States launched a moon probe, Ranger VII, which was designed to strike the moon's surface, taking photographs as it approached. On 31 July 1964, it completed its mission successfully, taking 4,316 pictures of an area now named 'Mare Cognitum' ('Known Sea'). In early 1965, Ranger VIII and Ranger IX had even greater success, if that were possible. These moon probes revealed the moon's surface to be hard (or crunchy, at worst) and not covered by the thick layer of dust some astronomers had suspected might exist. The probes showed those areas that seemed flat when seen through a telescope to be covered by craters too small to be seen from the earth.

The Soviet probe Luna IX succeeded in making a 'soft landing' (one not involving the destruction of the object making the landing) on the moon on 3 February 1966, and sent back photographs from ground levels. On 3 April 1966, the Soviets placed Luna X in a three-hour orbit about the moon; it measured radioactivity from the lunar surface, and the pattern indicated the rocks of the lunar surface were similar to the basalt that underlies earth's oceans.

American rocketmen followed this lead with even more elaborate rocketry. The first American soft landing on the moon was that of Surveyor I on 1 June 1966. By September 1967, Surveyor V was handling and analysing lunar soil under radio control from earth. It did indeed prove to be basalt-like and to contain iron particles that were probably meteoric in origin.

On 10 August 1966, the first of the American 'Lunar Orbiter' probes was sent circling round the moon. (These discovered the mascons.) The Lunar Orbiters took detailed photographs of every part of the moon, so that its surface features everywhere (including the part forever hidden from earth's surface) came to be known in fine detail. In addition, startling photographs were taken of earth as seen from the neighbourhood of the moon.

The lunar craters, by the way, have been named after astronomers and other great men of the past. Since most of the names were given by the Italian astronomer Giovanni Battista Riccioli about 1650, it is the older astronomers Copernicus, Tycho, and Kepler, as well as the Greek astronomers Aristotle, Archimedes, and Ptolemy, who are honoured by the larger craters.

The other side, first revealed by Lunik III, offered a new chance. The Russians, as was their right, pre-empted some of the more noticeable features. They named craters not only after Tsiolkovsky, the great prophet of space travel, but also after Lomonosov and Popov, two Russian chemists of the late eighteenth century. They have awarded craters to Western personalities, too, including Maxwell, Hertz, Edison, Pasteur, and the Curies, all of whom are mentioned in this book. One very fitting name placed on the other side of the moon is that of the French pioneer-writer of science fiction, Jules Verne.

In 1970, the other side of the moon was sufficiently well known to make it possible to name its features systematically. Under the leadership of the American astronomer Donald Howard Menzel, an international body assigned hundreds of names, honouring great men of the past who had contributed to the advance of science in one way or another. Very prominent craters were allotted to such Russians as Mendeleev (who first developed the periodic table that I will discuss in Chapter 5) and Gagarin, who was the first man to be placed in orbit about earth and who had since died in an aeroplane accident. Other prominent features were used to memorialize the Dutch astronomer Hertzsprung, the French mathematician Galois, the Italian physicist Fermi, the American mathematician Wiener, and the British physicist Cockcroft. In one restricted area, we can find Nernst, Roentgen, Lorentz, Moseley, Einstein, Bohr, and Dalton, all of great importance in the development of the atomic theory and sub-atomic structure.

Reflecting Menzel's interest in science writing and science fiction is his just decision to allot a few craters to those who helped rouse the enthusiasm of an entire generation for space flight when orthodox science dismissed it as a chimera. For that reason, there is a crater honouring Hugo Gernsback, who published the first magazines in the United States devoted entirely to science fiction, and another to Willy Ley, who, of all writers, most indefatigably and accurately portrayed the victories and potentialities of rocketry. (Ley died, tragically, six weeks before the first landing on the moon – a landing for which he had waited all his life.)

Yet all lunar exploration by instrument alone had to take a back seat, dramatically, to the greatest of all the rocket feats of the 1960s: manned exploration of space, something we will take up in Chapter 15.

Anyone who is inclined to be complacent about meteors or to think that colossal strikes were just a phenomenon of the solar system's early history might give some thought to the asteroids, or planetoids. Whatever their origin – whether they are surviving planetesimals or remnants of an exploded planet – there are some pretty big ones around and about. Most of them orbit the sun in a belt between Mars and Jupiter. But in 1898 a German astronomer G. Witt discovered one whose orbit, upon calculation, turned out to lie between Mars and the earth. He named it Eros, and ever since planetoids with unusual orbits have been given masculine names. (Those with ordinary orbits, between Mars and Jupiter, are given feminine names even when named after men, e.g., Rockefellia, Carnegia, Hooveria.)

The orbits of Eros and the earth approach to within 13 million miles of each other, which is half the minimum distance between the earth and Venus, our closest neighbour among the full-sized planets. In 1931 Eros reached a point only 17 million miles from the earth and it will make its next close approach in 1975. Several other 'earth grazers' have since been found. In 1932 two planetoids named Amor and Apollo were discovered with orbits approaching within 10 million and 7 million miles, respectively, of the earth's orbit. In 1936, there turned up a still-closer planetoid, named Adonis, which could approach to as close as 1·5 million miles from the earth. And in 1937 a planetoid given the name Hermes swam into sight in an orbit which might bring it within 200,000 miles of the earth, or actually closer than the moon. (The calculations of Hermes' orbit may not be entirely reliable, because the object did not stay in sight long and has not been sighted since.)

A particularly unusual earth grazer is Icarus, discovered in 1948 by Walter Baade. It approaches within 4 million miles of the earth, in an elongated comet-like orbit. At aphelion, it recedes as far as the orbit of Mars, and at perihelion it approaches to within 17 million miles of the sun. It is because of this that it has been named after the Greek mythological character who came to grief through flying too close to the sun on wings that were fixed in place with wax. Only certain comets ever approach the sun more closely than does Icarus. One of the large comets of the 1880s approached within less than a million miles of the sun.

Eros, the largest of the earth grazers, is a brick-shaped object perhaps fifteen miles long and five miles broad. Others, such as Hermes, are only about one mile in diameter. Still, even Hermes would gouge out a crater about one hundred miles across if it hit the earth or create tsunamis of unprecedented size if it struck the ocean. Fortunately the odds against an encounter are enormous.

Meteorites, as the only pieces of extraterrestrial matter we can examine, are exciting not only to astronomers, geologists, chemists, and metallurgists, but also to cosmologists, who are concerned with the origins of the universe and the solar system. Among the meteorites are puzzling glassy objects found in several places on earth. The first were found in 1787 in what is now western Czechoslovakia. Australian examples were detected in 1864. They received the name 'tektites', from a Greek word for 'molten', because they appear to have melted in their passage through the atmosphere.

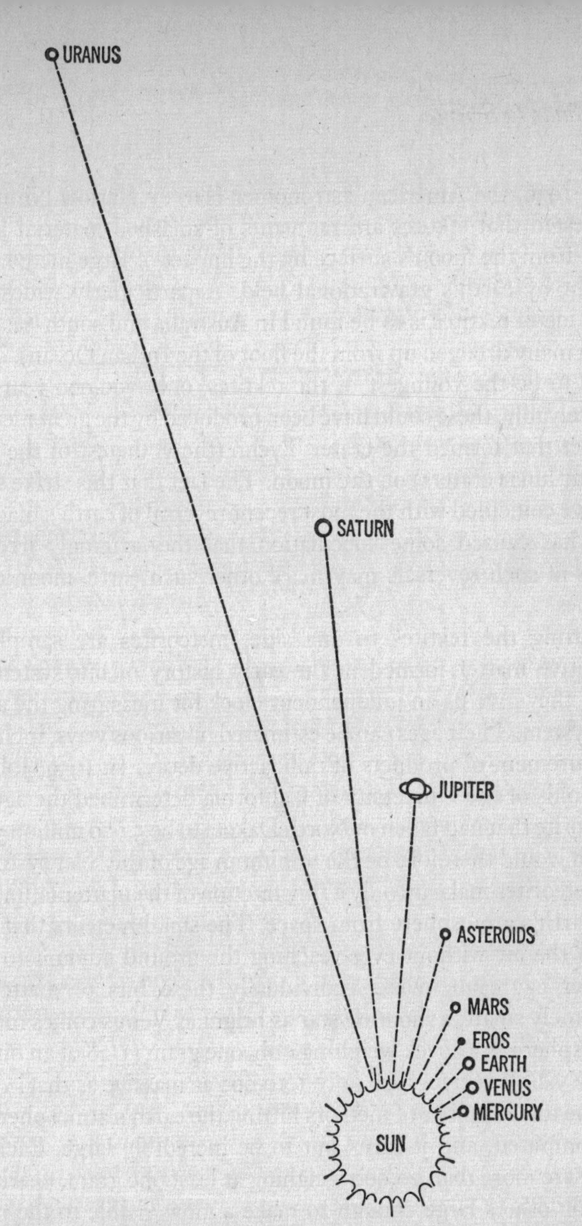

A schematic drawing of the radiuses of the orbits of most of the sun's planets, indicating their distances from the sun and the position of Eros and the asteroids. Roughly, each planet is about twice as far from the sun as the one before it.

In 1936, the American astronomer Harvey Harlow Ninninger suggested that tektites are remnants of splashed material forced away from the moon's surface by the impact of large meteors and caught by earth's gravitational field. A particularly widespread strewing of tektites is to be found in Australia and south-east Asia (with many dredged up from the floor of the Indian Ocean). These seem to be the youngest of the tektites, only 700,000 years old. Conceivably, these could have been produced by the great meteoric impact that formed the crater Tycho (the youngest of the spectacular lunar craters) on the moon. The fact that this strike seems to have coincided with the most recent reversal of earth's magnetic field has caused some speculation that the strikingly irregular series of such reversals may mark other such earth–moon catastrophes.

Putting the tektites to one side, meteorites are samples of primitive matter formed in the early history of our system. As such, they give us an independent clock for measuring the age of our system. Their ages can be estimated in various ways, including measurement of products of radioactive decay. In 1959, John H. Reynolds of the University of California determined the age of a meteorite that had fallen in North Dakota to be 5,000 million years, which would therefore be the minimum age of the solar system.

Meteorites make up only a tiny fraction of the matter falling into the earth's atmosphere from space. The small meteors that burn up in the air without ever reaching the ground amount to a far greater aggregate mass. Individually these bits of matter are extremely small; a shooting star as bright as Venus comes into the atmosphere as a speck weighing only one gram (1/28 of an ounce). Some visible meteors are only 1/10,000 as massive as that!

The total number of meteors hitting the earth's atmosphere can be computed, and it turns out to be incredibly large. Each day there are more than 20,000 weighing at least one gram, nearly 200 million others large enough to make a glow visible to the naked eye, and many thousands of millions more of still smaller sizes.

We know about these very small 'micro-meteors' because the air has been found to contain dust particles with unusual shapes and a high nickel content, quite unlike ordinary terrestrial dust.

Another evidence of the presence of micro-meteors in vast quantities is the faint glow in the heavens called 'zodiacal light' (first discovered about 1700 by G. D. Cassini) – so called because it is most noticeable in the neighbourhood of the plane of the earth's orbit, where the constellations of the zodiac occur. The zodiacal light is very dim and cannot be seen even on a moonless night unless conditions are favourable. It is brightest near the horizon where the sun has set or is about to rise and on the opposite side of the sky there is a secondary brightening called the 'Gegenschein' (German for 'opposite light'). The zodiacal light differs from the air glow: its spectrum has no lines of atomic oxygen or atomic sodium, but is just that of reflected sunlight and nothing more. The reflection agent presumably is dust concentrated in space in the plane of the planets' orbits – in short, micro-meteors. Their number and size can be estimated from the intensity of the zodiacal light.

Micro-meteors have now been counted with new precision by means of such satellites as Explorer XVI, launched in December 1962, and Pegasus I, launched on 16 February 1965. To detect them, some of the satellites are covered with patches of a sensitive material that signals each meteoric hit through a change in its electrical resistance. Others record the hits by means of a sensitive microphone behind the skin, picking up the 'pings'. The satellite counts have indicated that 3,000 tons of meteoric matter enter our atmosphere each day, five sixths of it consisting of micro-meteors too small to be detected as shooting stars. These micro-meteors may form a thin dust cloud about the earth, one that stretches out, in decreasing density, for 100,000 miles or so before fading out to the usual density of material in interplanetary space.

The Venus probe Mariner II, launched on 27 August 1962, showed the dust concentration in space generally to be only 1/10,000 the concentration near earth – which seems to be the centre of a dustball. The American astronomer Fred Lawrence Whipple suggests that the moon may be the source of the cloud, the dust being flung up from the moon's surface by the meteorite beating it has had to withstand. Venus, which has no moon (according to Mariner II), also has no dustball.

The geophysicist Hans Petterson, who has been particularly interested in this meteoric dust, took some samples of air in 1957 on a mountain top in Hawaii, which is as far from industrial dust-producing areas as one can get on the earth. His findings led him to believe that about 5 million tons of meteoric dust fall on the earth each year. (A similar measurement by James M. Rosen in 1964, making use of instruments borne aloft by balloons, set the figure at 4 million tons, though still others find reason to place the figure at merely 100,000 tons per year.) Hans Petterson tried to get a line on this fall in the past by analysing cores brought up from the ocean bottom for high-nickel dust. He found that, on the whole, there was more in the upper sediments than in the older ones below, which indicates – though the evidence is still scanty – that the rate of meteoric bombardment may have increased in recent ages. This meteoric dust may possibly be of direct importance to all of us, for, according to a theory advanced by the Australian physicist E. G. Bowen in 1953, this dust serves as nuclei for rain-drops. If this is so, then the earth's rainfall pattern reflects the rise and fall of the intensity with which micro-meteorites bombard us.

The Origin of Air

Perhaps we should wonder less about how the earth got its atmosphere than about how it managed to hang on to it through all the aeons the earth has been whirling and wheeling through space. The answer to the latter question involves something called 'escape velocity'.

If an object is thrown upwards from the earth, the pull of gravity gradually slows it until it comes to a momentary halt and then falls back. If the force of gravity were the same all the way up, the height reached by the object would be proportional to its initial velocity; that is, it would reach four times as high when launched with a speed of two miles an hour as it would when it started at one mile an hour (energy increases as the square of the velocity).

But of course the force of gravity does not remain constant: it weakens slowly with height. (To be exact, it weakens as the square of the distance from the earth's centre.) Let us say we shoot an object upwards with a velocity of one mile per second. It will reach a height of eighty miles before turning and falling (if we ignore air resistance). If we were to fire the same object upwards at two miles per second, it would climb higher than four times that distance. At the height of eighty miles, the pull of the earth's gravity is appreciably lower than at ground level, so that the object's further flight would be subject to a smaller gravitational drag. In fact, the projectile would rise to 350 miles, not 320.

Given an initial upwards velocity of 6·5 miles per second, an object will climb 25,800 miles. At that point the force of gravity is not more than one fortieth as strong as it is on the earth's surface. If we added just one tenth of a mile per second to the object's initial speed (i.e., launched it at 6·6 miles per second), it would go up to 34,300 miles.

It can be calculated that an object fired up at an initial speed of 6·98 miles per second will never fall back to the earth. Although the earth's gravity will gradually slow the object's velocity, its effect will steadily decline, so that it will never bring the object to a halt (zero velocity) with respect to the earth. (So much for the cliché that 'everything that goes up must come down'.) Lunik I and Pioneer IV, fired at better than 7 miles per second, will never come down.

The speed of 6·98 miles per second, then, is the earth's 'escape velocity'. The velocity of escape from any astronomical body can be calculated from its mass and size. From the moon, it is only 1·5 miles per second; from Mars, 3·2 miles per second; from Saturn, 23 miles per second; from Jupiter, the most massive planet in the solar system, it is 38 miles per second.

Now all this has a direct bearing on the earth's retention of its atmosphere. The atoms and molecules of the air are constantly flying about like tiny missiles. Their individual velocities may vary a great deal, and the only way they can be described is statistically: for example, giving the fraction of the molecules moving faster than a particular velocity, or giving the average velocity under given conditions. The formula for doing this was first worked out in 1860 by James Clerk Maxwell and the Austrian physicist Ludwig Boltzmann, and it is called the 'Maxwell–Boltzmann law'.

The mean velocity of oxygen molecules in air at room temperature turns out to be 0·3 mile per second. The hydrogen molecule, being only one sixteenth as heavy, moves on the average four times as fast, or 1·2 miles per second, because, according to the Maxwell–Boltzmann law, the velocity of a particular particle at a particular temperature is inversely proportional to the square root of its molecular weight.

It is important to remember that these are only average velocities. Half the molecules go faster than the average; a certain percentage go more than twice as fast as the average; a smaller percentage more than three times as fast, and so on. In fact, a tiny percentage of the oxygen and hydrogen molecules in the atmosphere go faster than 6·98 miles per second, the escape velocity.

In the lower atmosphere, these speedsters cannot actually escape, because collisions with their slower neighbours slow them down. But in the upper atmosphere, their chances are much better. First of all, the unimpeded radiation of the sun up there excites a large proportion of them to enormous energy and great speeds. In the second place, the probability of collisions is greatly reduced in the thinner air. Whereas a molecule at the earth's surface travels only four millionths of an inch (on the average) before colliding with a neighbour, at a height of 65 miles its average free path before colliding is 4 inches, and at 140 miles, it is 1,100 yards. There the average number of collisions encountered by an atom or molecule is only 1 per second, against 5,000 million per second at sea level. Thus a fast particle at a height of 100 miles or more stands a good chance of escaping from the earth. If it happens to be moving upwards, it is moving into regions of lesser and lesser density and experiences an ever smaller chance of collision, so that it may in the end depart into interplanetary space, never to return.

In other words, the earth's atmosphere leaks. But the leakage applies mainly to the lightest molecules. Oxygen and nitrogen are heavy enough so that only a tiny fraction of them achieves the escape velocity, and not much oxygen or nitrogen has been lost from the earth since their original formation. On the other hand, hydrogen and helium are easily raised to escape velocity. Consequently it is not surprising that no hydrogen or helium to speak of remains in the atmosphere of the earth today.

The more massive planets, such as Jupiter and Saturn, can hold even hydrogen and helium, so they may have large and deep atmospheres composed mostly of these five elements (which, after all, are the most common substances in the universe). The hydrogen present in vast quantities would react with other elements present, so that carbon, nitrogen, and oxygen would be present only in the form of hydrogen-containing compounds: methane (CH_4), ammonia (NH_3), and water (H_2O) respectively. The ammonia and methane in Jupiter's atmosphere, although present as relatively small-concentration impurities, were first discovered (in 1931, by the German-American astronomer Rupert Wildt) because these compounds produce noticeable absorption bands in the spectra, whereas hydrogen and helium do not. The presence of hydrogen and helium were detected by rather indirect methods in 1952.

On the basis of his findings, Wildt speculated about the structure of Jupiter and the other planets. He suggested that under the thick outermost shell of atmosphere, there was a layer of frozen water, and underneath that a rocky core. Similar structures were suggested for the major planets farther out. Saturn, which was distinctly less dense than Jupiter, would have a thicker atmosphere and a smaller core; Neptune, which was distinctly more dense, a thinner atmosphere and a larger core (for its size). However, all that can actually be seen of Jupiter is its upper atmosphere, and the radio-wave emissions are as yet insufficient to tell us much detail of what goes on below. For instance, it is possible to argue that Jupiter and the other 'gas giants' are hydrogen and helium, for the most part, all the way through to the centre, where pressures are so high that hydrogen assumes a metallic form.

Working in the other direction, a small planet like Mars is less able to hold even the comparatively heavy molecules and has an atmosphere only one tenth as dense as our own. The moon, with a smaller escape velocity, cannot hold any atmosphere to speak of and is airless.

Temperature is just as important a factor as gravity. The Maxwell–Boltzmann equation says that the average speed of particles is proportional to the square root of the absolute temperature. If the earth were at the temperature of the sun's surface, all the atoms and molecules in its atmosphere would be speeded up by four to five times, and the earth could no more hold on to its oxygen and nitrogen than it could to hydrogen or helium.

On the other hand, if temperatures were lower, the chance of holding molecules of a particular kind is increased. In 1943, for instance, Kuiper managed to detect an atmosphere of methane on Titan, the largest satellite of Saturn. Titan is not very much larger than the moon, and if it were at the moon's distance from the sun, it would have no atmosphere. At the frigid temperatures of the outer solar system, it manages. It is possible that the other large outer satellites – Neptune's satellite, Triton; and Jupiter's four satellites, Io, Europa, Ganymede, and Callisto – may have thin atmospheres of some sort, but these have not yet been detected. For the time being, Titan remains unique among the satellites of the planetary system.

The earth's possession of an atmosphere is a strong point against the theory that it and the other planets of the solar system originated from some catastrophic accident, such as near-collision between another sun and ours. It argues, rather, in favour of the dust-cloud and planetesimal theory. As the dust and gas of the cloud condensed into planetesimals and these in turn collected to form a planetary body, gas might have been trapped within a spongy mass, like air in a snowbank. The subsequent gravity contraction of the mass might then have squeezed out the gases towards the surface. Whether a particular gas would be held in the earth would depend in part on its chemical reactivity. Helium and neon, though they must have been among the most common gases in the original cloud, are so inert chemically that they form no compounds and would have escaped as gases in short order. Therefore the concentrations of helium and neon on the earth are insignificant fractions of their concentrations in the universe generally. It has been calculated, for instance, that the earth has retained only one out of every 50,000 million neon atoms present in the original cloud of gas, and our atmosphere has even fewer, if any, of the original helium atoms. I say 'if any' because, while there is a little helium in the atmosphere today, all of it may come from the breakdown of radioactive elements and leakage of helium trapped in cavities underground.

On the other hand, hydrogen, though lighter than helium or neon, has been captured with greater efficiency because it has combined with other substances, notably with oxygen to form water. It is estimated that the earth still has one out of every 5 million hydrogen atoms that were in the original cloud.

Nitrogen and oxygen illustrate the chemical aspect even more neatly. Although the nitrogen molecule and the oxygen molecule are about equal in mass, the earth has held on to 1 out of 6 of the original atoms of highly reactive oxygen but only 1 out of every 800,000 of inert nitrogen.

When we speak of gases of the atmosphere, we have to include water vapour, and here we get into the interesting question of how the oceans originated. In the early stages of the earth's history, even if it was only moderately hot, all the water must have been in the form of vapour. Some geologists believe that the water was then concentrated in the atmosphere as a dense cloud of vapour, and, after the earth cooled, it fell in torrents to form the ocean. On the other hand, some geologists maintain that our oceans have been built up mainly by water seeping up from the earth's interior. Volcanoes show that there is a great deal of water in the crust, for the gas they discharge is mostly water vapour. If that is so, the oceans may still be growing, albeit slowly.

But was the earth's atmosphere always what it is today, at least since its formation in the first place? It seems very unlikely. For one thing, molecular oxygen, which makes up one fifth of the volume of the atmosphere, is so active a substance that its presence in free form is extremely unlikely, unless it were continuously being produced. Furthermore, no other planet has an atmosphere anything like our own, so that one is strongly tempted to conclude that earth's atmosphere is the result of unique events (as, for instance, the presence of life on this planet, but not on the others).

Harold Urey has presented detailed arguments in favour of the idea that the original atmosphere was composed of ammonia and methane. Hydrogen, helium, carbon, nitrogen, and oxygen are the predominant elements in the universe, with hydrogen far and away the most common. In the presence of such a preponderance of hydrogen, carbon would be likely to combine with hydrogen to form methane (CH_4), nitrogen with hydrogen to form ammonia (NH_3), and oxygen with hydrogen to form water (H_2O). Helium and excess hydrogen would, of course, escape; the water would form the oceans; the methane and ammonia, as comparatively heavy gases, would be held by the earth's gravity and so constitute the major portion of the atmosphere.

If all the planets with sufficient gravity to hold an atmosphere at all began with atmospheres of this type, they would nevertheless not all keep such an atmosphere. Ultra-violet radiation from the sun would introduce changes. These changes would be minimal for the outer planets, which in the first place received comparatively little radiation from the distant sun, and in the second place had vast atmospheres capable of absorbing considerable radiation without being perceptibly changed. The outer planets, therefore, would keep the hydrogen/helium/ammonia/methane atmospheres to the present day.

Not so the five inner worlds of Mars, Earth, Moon, Venus, and Mercury. Of these, the Moon and Mercury are too small, too hot, or both to retain any perceptible atmosphere. This leaves Mars, Earth, and Venus, with thin atmospheres of chiefly ammonia, methane, and water to begin with. What would happen?

Ultra-violet radiation striking water molecules in the upper primordial atmosphere of the earth would break them apart to hydrogen and oxygen ('photo-dissociation'). The hydrogen would escape, leaving oxygen behind. Being reactive, however, the molecules would react with almost any other molecule in the neighbourhood. They would react with methane (CH_4) to form carbon dioxide (CO_2) and water (H_2O). They would react with ammonia (NH_3) to form free nitrogen (N_2) and water.

Very slowly, but steadily, the atmosphere would be converted from methane and ammonia to nitrogen and carbon dioxide. The nitrogen would tend to react slowly with the minerals of the crust to form nitrates, leaving carbon dioxide as the major portion of the atmosphere.

Will water continue to photo-dissociate, however? Will hydrogen continue to escape into space, and will oxygen continue to collect in the atmosphere? And if oxygen does collect and finds nothing to react with (it cannot react further with carbon dioxide), then will it not add a proportion of molecular oxygen to the carbon dioxide present (thus accounting for earth's atmospheric oxygen)? The answer is, No!

Once carbon dioxide becomes the major component of the atmosphere, ultra-violet radiation does not bring about further changes through dissociation of the water molecule. When the oxygen begins to collect in free form, a thin ozone layer is formed in the upper atmosphere. This absorbs the ultra-violet, blocking it from the lower atmosphere and preventing further photo-dissociation. A carbon-dioxide atmosphere is stable.

But carbon dioxide introduces the greenhouse effect (see p. 163). If the carbon-dioxide atmosphere is thin and if it is relatively far from the sun, the effect is small. This is the case with Mars, for instance. Its atmosphere, chiefly carbon dioxide, is thinner than that of the earth – how much thinner was not realized till the American Mars-probe Mariner IV passed close to Mars in July 1965. We now know that the Martian atmosphere is not more than 1/100 as dense as earth's.

Suppose, though, that a planet's atmosphere is more like that of earth, and it is as close to the sun (or closer). The greenhouse effect will then be enormous: temperatures will rise, vaporizing the oceans to a greater and greater extent. The water vapour will add to the greenhouse effect, accelerating the change, forcing more and more carbon dioxide into the air as well through temperature effects on the crust. In the end, the planet will be enormously hot, will have all its water in the atmosphere in the form of a vapour which will forever hide its surface under eternal clouds, and will have a very thick atmosphere of carbon dioxide.

This is precisely the case with Venus. The American Venus-probe, skimming past Venus in December 1962, corroborated earlier reports, based on radio-wave emission from Venus's atmosphere, that Venus was considerably hotter than would be expected from its position *vis-à-vis* the sun. The Soviet Union has sent a series of probes actually into Venus's atmosphere beginning in 1967; in December 1970, they managed to place one on the surface itself while the probe's instruments were still working. (The temperature and pressures quickly destroy the instruments.) Venus's surface turns out to be at a temperature of 900°F (500°C), which is nearly at the red-hot stage, and its atmosphere, largely carbon dioxide, is about 100 times as dense as that of earth.

Earth did not move in the direction of either Mars of Venus. The nitrogen content of its atmosphere did not soak into the crust, leaving a thin, cold carbon-dioxide wind. Nor did the greenhouse effect turn it into a choking desert world of great heat. Something happened, and that something was the development of life, even while the atmosphere was still in its ammonia/methane stage (Chapter 12).

Life-induced reactions in earth's oceans broke down nitrogen compounds to liberate molecular nitrogen and thus kept that gas in the atmosphere in large quantities. Furthermore, cells developed the capacity to break down the water molecules to hydrogen and oxygen by using the energy of *visible* light, which is not blocked by ozone. The hydrogen was combined with carbon dioxide to form the complicated molecules that made up the cell, while the oxygen was liberated into the atmosphere. In this way, thanks to life, earth's atmosphere altered from nitrogen-and-carbon-dioxide to nitrogen-and-oxygen. The greenhouse effect was reduced to very little; the earth remained cool, capable of retaining its unique possession of an ocean of liquid water and an atmosphere containing large quantities of free oxygen.

In fact, our oxygenated atmosphere may be a characteristic only of the last 10 per cent of earth's existence, and even as recently as 600 million years ago, our atmosphere may have had only a tenth as much oxygen as it has now.

But we do have it now, and we may be thankful for the life that made the free atmospheric oxygen possible, and for the life that such oxygen in turn makes possible.

5 The Elements

The Periodic Table

The early Greek philosophers, whose approach to most problems was theoretical and speculative, decided that the earth was made of a very few 'elements', or basic substances. Empedocles of Akragas, about 430 B.C., set the number at four – earth, air, water, and fire. Aristotle, a century later, supposed the heavens to consist of a fifth element, 'aether'. The successors of the Greeks in the study of matter, the medieval alchemists, got mired in magic and quackery, but they came to shrewder and more reasonable conclusions than the Greeks because they at least handled the materials they speculated about.

Seeking to explain the various properties of substances, the alchemists attached these properties to certain controlling elements that they added to the list. They identified mercury as the element that imparted metallic properties to substances and sulphur as the element that imparted the property of flammability. One of the last and best of the alchemists, the sixteenth-century Swiss physician Theophrastus Bombastus von Hohenheim, better known as Paracelsus, added salt as the element that imparted resistance to heat.

The alchemists reasoned that one substance could be changed into another by merely adding and subtracting elements in the proper proportions. A metal such as lead, for instance, might be changed into gold by adding the right amount of mercury to the lead. The search for the precise technique of converting 'base metal' to gold went on for centuries. In the process, the alchemists discovered substances vastly more important than gold – such as the mineral acids and phosphorus.

The mineral acids – nitric acid, hydrochloric acid, and, particularly, sulphuric acid – introduced a virtual revolution in alchemical experiments. These substances were much stronger acids than the strongest previously known (the acetic acid of vinegar), and with them substances could be decomposed without the use of high temperatures and long waits. Even today, the mineral acids, particularly sulphuric acid, are of vital use in industry. It is even said that the extent of the industrialization of a nation can be judged by its annual consumption of sulphuric acid.

Nevertheless, few alchemists allowed themselves to be diverted by these important side issues from what they considered to be the main quest. Unscrupulous members of the craft indulged in outright fakery, producing gold by sleight-of-hand, to win what we would call today 'research grants' from rich patrons. This brought the profession into such disrepute that the very word 'alchemist' had to be abandoned. By the seventeenth century, 'alchemist' had become 'chemist' and 'alchemy' had graduated to a science called 'chemistry'.

In the bright birth of science, one of the first of the new chemists was Robert Boyle, the author of Boyle's law of gases (pp. 169–70). In his *The Sceptical Chymist*, published in 1661, Boyle first laid down the specific modern criterion of an element: a basic substance that can be combined with other elements to form 'compounds' and that, conversely, cannot be broken down to any simpler substance after it is isolated from a compound.

Boyle retained a medieval view about what the actual elements were, however. For instance, he believed that gold was not an element and could be formed in some way from other metals. So, in fact, did his contemporary, Isaac Newton, who devoted a great deal of time to alchemy. (Indeed, Emperor Francis Joseph of Austria-Hungary subsidized experiments for making gold as late as 1867.)

In the century after Boyle, practical chemical work began to make clear which substances could be broken down into simpler substances and which could not. Henry Cavendish showed that hydrogen would combine with oxygen to form water, so water could not be an element. Later Lavoisier resolved the supposed element air into oxygen and nitrogen. It became plain that none of the Greek 'elements' was an element by Boyle's criterion.

As for the elements of the alchemists, mercury and sulphur did indeed turn out to be elements 'according to Boyle'. But so did iron, tin, lead, copper, silver, gold, and such non-metals as phosphorus, carbon, and arsenic. And Paracelsus's 'element' salt eventually was broken down into two simpler substances.

Of course, the definition of elements depended on the chemistry of the time. As long as a substance could not be broken down by the chemical techniques of the day, it could still be considered an element. For instance, Lavoisier's list of thirty-three elements included such items as lime and magnesia. But fourteen years after Lavoisier's death on the guillotine in the French Revolution, the English chemist Humphry Davy, using an electric current to split the substances, divided lime into oxygen and a new element he called 'calcium' and similarly split magnesia into oxygen and another new element he named 'magnesium'.

On the other hand, Davy was able to show that a green gas that the Swedish chemist Carl Wilhelm Scheele had made from hydrochloric acid was not a compound of hydrochloric acid and oxygen, as had been thought, but a true element, and he named it 'chlorine' (from the Greek word for 'green').

At the beginning of the nineteenth century, the English chemist John Dalton came up with a radically new way of looking at elements. Oddly enough, this view harked back to some of the Greeks, who had, after all, contributed what turned out to be perhaps the most important single concept in the understanding of matter.

The Greeks argued about whether matter was continuous or discrete: that is, whether it could be divided and subdivided indefinitely into ever finer dust or would be found in the end to consist of indivisible particles. Leucippus of Miletus and his pupil

Democritus of Abdera insisted about 450 B.C. that the latter was the case. Democritus, in fact, gave the particles a name: he called them 'atoms' (meaning 'non-divisible'). He even suggested that different substances were composed of different atoms or combinations of atoms and that one substance could be converted into another by rearranging the atoms. Considering that all this was only an intelligent guess, one is thunderstruck by the correctness of his intuition. Although this idea may seem obvious today, it was so far from obvious at the time that Plato and Aristotle rejected it out of hand.

It survived, however, in the teachings of Epicurus of Samos, who wrote about 300 B.C., and in the philosophic school ('Epicureanism') to which he gave rise. An important Epicurean was the Roman philosopher Lucretius, who, about 60 B.C., embodied atomic notions in a long poem *On the Nature of Things*. Lucretius's poem survived through the Middle Ages and was one of the earlier works to be printed once that technique had been invented.

The notion of atoms never entirely passed out of the consciousness of Western scholarship. Prominent among the atomists in the dawn of modern science were the Italian philosopher Giordano Bruno and the French philosopher Pierre Gassendi. Bruno had many unorthodox scientific views, such as a belief in an infinite universe with the stars distant suns about which planets revolved, and expressed himself over-boldly. He was burned as a heretic in 1600 – the outstanding martyr to science of the Scientific Revolution. The Russians have named a crater on the other side of the moon in his honour.

Gassendi's views impressed Boyle. Boyle's own experiments showing that gases could easily be compressed and expanded seemed to show that these gases must be composed of widely spaced particles. Both Boyle and Newton were therefore among the convinced atomists of the seventeenth century.

Dalton showed that the various rules governing the behaviour of gases could indeed be explained on the basis of the atomic nature of matter. (He recognized the priority of Democritus by using the word 'atoms'.) According to Dalton, each element represented a particular kind of atom, and any quantity of the element was made up of identical atoms of this kind. What distinguished one element from another was the nature of its atoms. And the basic physical difference between atoms was in their weight. Thus sulphur atoms were heavier than oxygen atoms, which in turn were heavier than nitrogen atoms, they in turn heavier than carbon atoms, and these in turn heavier than hydrogen atoms.

The Italian chemist Amedeo Avogadro applied the atomic theory to gases in such a way as to show that it made sense to suppose that equal volumes of gas (of whatever nature) were made up of equal numbers of particles. This is 'Avogadro's hypothesis'. These particles were at first assumed to be atoms, but eventually were shown to be composed, in most cases, of small groups of atoms called 'molecules'. If a molecule contains atoms of different kinds (as the water molecule, which consists of an oxygen atom and two hydrogen atoms), it is a molecule of a 'chemical compound'.

Naturally it became important to measure the relative weights of different atoms – to find the 'atomic weights' of the elements, so to speak. The tiny atoms themselves were hopelessly beyond the reach of nineteenth-century weighing techniques. But by weighing the quantity of each element separated from a compound, and making deductions from the elements' chemical behaviour, it was possible to work out the relative weights of the atoms. The first to go about this systematically was the Swedish chemist Jöns Jacob Berzelius. In 1828, he published a list of atomic weights based on two standards – one giving the atomic weight of oxygen the arbitrary value of 100, the other taking the atomic weight of hydrogen as equal to 1.

Berzelius's system did not catch on at once, but in 1860, at the first International Chemical Congress in Karlsruhe, Germany, the Italian chemist Stanislao Cannizzaro presented new methods for determining atomic weights, making use of Avogadro's hypothesis, which had hitherto been neglected. He described his views so forcefully that the world of chemistry was won over.

The weight of oxygen rather than hydrogen was adopted as the standard, because oxygen could more easily be brought into combination with various elements (and combination with other elements was the key step in the usual method of determining atomic weights). Oxygen's atomic weight was arbitrarily taken by the Belgian chemist Jean Servais Stas, in 1850, as exactly 16, so that the atomic weight of hydrogen, the lightest known element, would be just about one – 1·008, to be exact.

Ever since Cannizzaro's time, chemists have sought to work out atomic weights with greater and greater accuracy. This reached a climax, as far as purely chemical methods were concerned, in the work of the American chemist Theodore William Richards, who, in 1904 and thereafter, determined the atomic weights with an unapproached accuracy. For this he received the Nobel Prize in chemistry in 1914. On the basis of later discoveries about the physical constitution of atoms, Richards's figures have since been corrected to still more refined values.

Throughout the nineteenth century, although much work was done on atoms and molecules and scientists generally were convinced of their reality, there existed no direct evidence that they were anything more than convenient abstractions. Some quite prominent scientists, such as the German chemist Wilhelm Ostwald, refused to accept them in any other way. To him, they were useful but not 'real'.

The reality of molecules was made clear by 'Brownian motion'. This was first observed in 1827 by the Scottish botanist Robert Brown, who noted that pollen grains suspended in water jiggled erratically. At first it was thought that this was due to the life in the pollen grains, but equally small particles of completely inanimate dyes also showed the motion.

In 1863, it was first suggested that the movement was due to unequal bombardment of the particles by surrounding water molecules. For large objects, a slight inequality in the number of molecules striking from left and from right would not matter. For microscopic objects, bombarded by perhaps only a few hundred molecules per second, a few in excess – this side or that – can induce a perceptible jiggle. The random movement of the tiny particles is almost visible proof of the 'graininess' of water, and of matter generally.

Einstein worked out a theoretical analysis of this view of Brownian motion and showed how one could work out the size of the water molecules from the extent of the little jiggling movements of the dye particles. In 1908, the French physicist Jean Perrin studied the manner in which particles settled downwards through water under the influence of gravity. The settling was opposed by molecular collisions from below, so that a Brownian movement was opposing gravitational pull. Perrin used this finding to calculate the size of the water molecules by means of the equation Einstein had worked out, and even Ostwald had to give in. For his investigations Perrin received the Nobel Prize for physics in 1926.

So atoms have steadily been translated from semi-mystical abstractions into almost tangible objects. Indeed, today we can say that man has at last 'seen' the atom. This is accomplished with the so-called 'field ion microscope', invented in 1955 by Erwin W. Mueller of Pennsylvania State University. His device strips positively charged ions off an extremely fine needle tip and shoots them to a fluorescent screen in such a way as to produce a 5 million-fold magnified image of the needle tip. This image actually makes the individual atoms composing the tip visible as bright little dots. The technique was improved to the point where images of single atoms could be obtained. The American physicist Albert Victor Crewe reported the detection of individual atoms of uranium and thorium by means of a scanning electron-microscope in 1970.

As the list of elements grew in the nineteenth century, chemists began to feel as if they were becoming entangled in a thickening jungle. Every element had different properties, and they could see no underlying order in the list. Since the essence of science is to try to find order in apparent disorder, scientists hunted for some sort of pattern in the properties of the elements.

In 1862, after Cannizzaro had established atomic weight as one of the important working tools of chemistry, a French geologist, Alexandre Émile Beguyer de Chancourtois, found that he could arrange the elements in the order of increasing atomic weight in a tabular form, such that elements with similar properties fell in the same vertical column. Two years later, a British chemist,

John Alexander Reina Newlands, independently arrived at the same arrangement. But both were ignored or ridiculed. Neither could get his suggestions properly published at the time. Many

1 Hydrogen (H) 1.008								
3 Lithium (Li) 6.939	4 Beryllium (Be) 9.012							
11 Sodium (Na) 22.990	12 Magnesium (Mg) 24.312							
19 Potassium (K) 39.102	20 Calcium (Ca) 40.08	21 Scandium (Sc) 44.956	22 Titanium (Ti) 47.90	23 Vanadium (V) 50.942	24 Chromium (Cr) 51.996	25 Manganese (Mn) 54.938	26 Iron (Fe) 55.847	27 Cobalt (Co) 58.933
37 Rubidium (Rb) 85.47	38 Strontium (Sr) 87.62	39 Yttrium (Y) 88.905	40 Zirconium (Zr) 91.22	41 Niobium (Nb) 92.906	42 Molybdenum (Mo) 95.94	43* Technetium (Tc) 98.91	44 Ruthenium (Ru) 101.07	45 Rhodium (Rh) 102.905
55 Cesium (Cs) 132.905	56 Barium (Ba) 137.34	57 Lanthanum (La) 138.91	58 Cerium (Ce) 140.12	59 Prasodymium (Pr) 140.907	60 Neodymium (Nd) 144.24	61* Promethium (Pm) 145	62 Samarium (Sm) 150.35	63 Europium (Eu) 151.96
			72 Hafnium (Hf) 178.49	73 Tantalum (Ta) 180.948	74 Tungsten (W) 183.85	75 Rhenium (Re) 186.2	76 Osmium (Os) 190.2	77 Iridium (Ir) 192.2
87* Francium (Fr) 223	88* Radium (Ra) 226.05	89* Actinium (Ac) 227	90* Thorium (Th) 232.038	91* Protactinium (Pa) 231	92* Uranium (U) 238.03	93* Neptunium (Np) 237	94* Plutonium (Pu) 242	95* Americium (Am) 243
			104* Rutherfordium (Rf) 259	105* Hahnium (Ha) 260				

The periodic table of the elements. The black areas of the table represent the two rare-earth series: the lanthanides and the actinides, named after their respective first members. The number in the lower right-hand corner of each box indicates the atomic weight of the element. An asterisk marks elements that are radioactive. Each element's atomic number appears at top centre of its box.

years later, after the importance of the periodic table had become universally recognized, their papers were published at last. Newlands even got a medal.

								2 Helium (He) 4.003
			5 Boron (B) 10.811	6 Carbon (C) 12.011	7 Nitrogen (N) 14.007	8 Oxygen (O) 15.999	9 Fluorine (F) 18.998	10 Neon (Ne) 20.183
			13 Aluminium (Al) 26.982	14 Silicon (Si) 28.086	15 Phosphorus (P) 30.974	16 Sulphur (S) 32.064	17 Chlorine (Cl) 35.453	18 Argon (A) 39.948
28 Nickel (Ni) 58.71	29 Copper (Cu) 63.54	30 Zinc (Zn) 65.37	31 Gallium (Ga) 69.72	32 Germanium (Ge) 72.59	33 Arsenic (As) 74.922	34 Selenium (Se) 78.96	35 Bromine (Br) 79.909	36 Krypton (Kr) 83.80
46 Palladium (Pd) 106.4	47 Silver (Ag) 107.870	48 Cadmium (Cd) 112.40	49 Indium (In) 114.82	50 Tin (Sn) 118.69	51 Antimony (Sb) 121.75	52 Tellurium (Te) 127.60	53 Iodine (I) 126.904	54 Xenon (Xe) 131.30
64 Gadolinium (Gd) 157.25	65 Terbium (Tb) 158.924	66 Dysprosium (Dy) 162.50	67 Holmium (Ho) 164.930	68 Erbium (Er) 167.26	69 Thulium (Tm) 168.934	70 Ytterbium (Yb) 173.04	71 Lutetium (Lu) 174.97	
78 Platinum (Pt) 195.09	79 Gold (Au) 196.967	80 Mercury (Hg) 200.59	81 Thallium (Tl) 204.37	82 Lead (Pb) 207.19	83 Bismuth (Bi) 208.98	84* Polonium (Po) 210	85* Astatine (At) 210	86* Radon (Rn) 222
96* Curium (Cm) 244	97* Berkelium (Bk) 245	98* Californium (Cf) 246	99* Einsteinium (Es) 253	100* Fermium (Fm) 255	101* Mendelevium (Md) 256	102* Nobelium (No) 255	103* Lawrencium (Lw) 257	

It was the Russian chemist Dmitri Ivanovich Mendeleev who got the credit for finally bringing order into the jungle of the elements. In 1869, he and the German chemist Julius Lothar Meyer proposed tables of the elements, making essentially the same point that de Chancourtois and Newlands had already made. But Mendeleev received the recognition because he had the courage and confidence to push the idea further than the others.

In the first place, Mendeleev's 'periodic table' (so-called because it showed the periodic occurrence of similar chemical properties) was more complicated than that of Newlands and nearer what we now believe to be correct. Second, where the properties of an element placed it out of order according to its atomic weight, he boldly switched the order, on the ground that the properties were more important than the atomic weight. He was eventually proved correct in this. For instance, tellurium, with an atomic weight of 127·60, should, on the weight basis, come after iodine, whose atomic weight is 126·91. But in the columnar table, putting tellurium ahead of iodine places it under selenium, which it closely resembles, and similarly puts iodine under its cousin bromine.

Finally, and most important, where Mendeleev could find no other way to make his arrangement work, he did not hesitate to leave holes in the table and to announce with what seemed infinite gall, that elements must be discovered that belonged in those holes. He went further. For three of the holes, he described the elements that would fit each, utilizing as his guide the properties of the elements above and below the holes in the table. And here Mendeleev had a stroke of luck. Each of his three predicted elements was found in his own lifetime, so that he witnessed the triumph of his system. In 1875, the French chemist Lecoq de Boisbaudran discovered the first of these missing elements and named it 'gallium' (after the Latin name for France). In 1879, the Swedish chemist Lars Fredrik Nilson found the second and named it 'scandium' (after Scandinavia). And in 1886, the German chemist Clemens Alexander Winkler isolated the third and named it 'germanium' (after Germany, of course). All three elements had almost precisely the properties predicted by Mendeleev!

With the discovery of X-rays, a new era opened in the history of the periodic table. In 1911, the British physicist Charles Glover Barkla discovered that when X-rays were scattered by a metal, the scattered rays had a sharply defined penetrating power, depending on the metal; in other words, each element produced its own 'characteristic X-rays'. For this discovery Barkla was awarded the Nobel Prize in physics for 1917.

There was some question as to whether X-rays were streams of tiny particles or consisted of wave-like radiations after the manner of light. One way of checking this was to see if X-rays could be diffracted (that is, forced to change direction) by a 'diffraction grating' consisting of a series of fine scratches. However, for proper diffraction, the distance between the scratches must be roughly equal to the size of the waves in the radiation. The most finely spaced scratches that could be prepared sufficed for ordinary light, but the penetrating power of X-rays made it likely that, if X-rays were wave-like, the waves would have to be much smaller than those of light. Therefore, no ordinary diffraction gratings would suffice to diffract X-rays.

However, it occurred to the German physicist Max Theodore Felix von Laue that crystals were a natural diffraction grating far finer than any man-made one. A crystal is a solid with a neat geometric shape, with its plane faces meeting at characteristic angles, and with a characteristic symmetry. This visible regularity is the result of an orderly array of atoms making up its structure. There were reasons for thinking that the space between one layer of atoms and the next was about the size of an X-ray wavelength. If so, crystals would diffract X-rays.

Laue experimented and found that X-rays passing through a crystal were indeed diffracted and formed a pattern on a photographic plate that showed that they had the properties of waves. Within the same year, the English physicist William Lawrence Bragg and his equally distinguished father William Henry Bragg developed an accurate method of calculating the wavelength of a particular type of X-ray from its diffraction pattern. Conversely, X-ray diffraction patterns were eventually used to determine the exact orientation of the atom layers that did the diffracting. In this way, X-rays opened the door to a new understanding of the atomic structure of crystals. For their work on X-rays, Laue received the Nobel Prize for physics in 1914, while the Braggs shared the Nobel Prize for physics in 1915.

Then, in 1914, the young English physicist Henry Gwyn-Jeffreys Moseley determined the wavelengths of the characteristic X-rays produced by various metals and made the important discovery that the wavelength decreased in a very regular manner as one went up the periodic table.

This pinned the elements into definite position in the table. If two elements, supposedly adjacent in the table, yielded X-rays that differed in wavelength by twice the expected amount, then there must be a gap between them belonging to an unknown element. If they differed by three times the expected amount, there must be two missing elements. If, on the other hand, the two elements' characteristic X-rays differed by only the expected amount, one could be certain that there was no missing element between the two.

It was now possible to give the elements definite numbers. Until then there had always been the possibility that some new discovery might break into the sequence and throw any adopted numbering system out of kilter. Now there could no longer be unsuspected gaps.

Chemists proceeded to number the elements from 1 (hydrogen) to 92 (uranium). These 'atomic numbers' were found to be significant in connection with the internal structures of the atoms (see Chapter 6) and to be more fundamental than the atomic weight. For instance, the X-ray data proved that Mendeleev had been right in placing tellurium (atomic number 52) before iodine (53), in spite of tellurium's higher atomic weight.

Moseley's new system proved its worth almost at once. The French chemist Georges Urbain, after discovering 'lutetium' (named after the old Latin name of Paris), had later announced that he had discovered another element which he called 'celtium'. According to Moseley's system, lutetium was element 71 and 'celtium' should be 72. But when Moseley analysed 'celtium's' characteristic X-rays, it turned out to be lutetium all over again.

Element 72 was not actually discovered until 1923, when the Danish physicist Dirk Coster and the Hungarian chemist Georg von Hevesy detected it in a Copenhagen laboratory and named it 'hafnium', from the Latinized name of Copenhagen.

Moseley was not present for this verification of the accuracy of his method; he had been killed at Gallipoli in 1915 at the age of twenty-eight – certainly one of the most valuable lives lost in the First World War. Moseley probably lost a Nobel Prize through his early death. The Swedish physicist Karl Manne George Siegbahn extended Moseley's work, discovering new series of X-rays and accurately determining X-ray spectra for the various elements. He was awarded the Nobel Prize for physics in 1924.

In 1925, Walter Noddack, Ida Tacke, and Otto Berg, of Germany filled another hole in the periodic table. After a three-year search through ores containing elements related to the one they were hunting for, they turned up element 75 and named it 'rhenium', in honour of the Rhine River. This left only four holes: elements 43, 61, 85, and 87.

It was to take two decades to track those four down. Although chemists did not realize it at the time, they had found the last of the stable elements. The missing ones were unstable species so rare on the earth today that all but one of them would have to be created in the laboratory to be identified. And thereby hangs a tale.

Radioactive Elements

After the discovery of X-rays, many scientists were impelled to investigate these new and dramatically penetrating radiations. One of them was the French physicist Antoine-Henri Becquerel. Henri's father, Alexandre Edmond (the physicist who had first photographed the solar spectrum), had been particularly interested in 'fluorescence', which is visible radiation given off by substances after exposure to the ultra-violet rays in sunlight.

The elder Becquerel had, in particular, studied a fluorescent substance called potassium uranyl sulphate (a compound made up of molecules each containing an atom of uranium). Henri wondered whether the fluorescent radiations of the potassium uranyl sulphate contained X-rays. The way to check this was to expose the sulphate to sunlight (whose ultra-violet light would excite the fluorescence) while the compound lay on a photographic plate wrapped in black paper. Since the sunlight could not penetrate the black paper, it would not itself affect the plate, but, if the fluorescence it excited contained X-rays, they *would* penetrate the paper and darken the plate. Becquerel tried the experiment in 1896, and it worked. Apparently there were X-rays in the fluorescence. Becquerel even got the supposed X-rays to pass through thin sheets of aluminium and copper, and that seemed to clinch the matter, for no radiation except X-rays was known to do this.

But then, by a great stroke of good fortune, a period of cloudy weather intervened. Waiting for the return of sunlight, Becquerel put away his photographic plates, with pinches of sulphate lying on them, in a drawer. After several days, he grew impatient and decided to develop his plates anyway, with the thought that even without direct sunlight some trace of X-rays might have been produced. When he saw the developed pictures, Becquerel experienced one of those moments of deep astonishment and delight that are the dream of all scientists. The photographic plate was deeply darkened by strong radiation! Something other than fluorescence or sunlight was responsible for it. Becquerel decided (and experiments quickly proved) that this something was the uranium in the potassium uranyl sulphate.

This discovery further electrified scientists, already greatly excited by the recent discovery of the X-rays. One of the scientists who at once set out to investigate the strange radiation from uranium was a young Polish-born chemist named Marie Sklodowska, who just the year before had married Pierre Curie, the discoverer of the Curie temperature (see p. 189).

Pierre Curie, in collaboration with his brother Jacques, had discovered that certain crystals, when put under pressure, developed a positive electric charge on one side and a negative charge on the other. This phenomenon is called 'piezoelectricity' (from a Greek word meaning 'to press'). Marie Curie decided to measure the radiation given off from uranium by means of piezoelectricity. She set up an arrangement whereby this radiation would ionize the air between two electrodes, a current would then flow, and the strength of this small current would be measured by the amount of pressure that had to be placed on a crystal to produce a balancing counter-current. This method worked so well that Pierre Curie dropped his own work at once and, for the rest of his life, joined Marie as an eager second.

It was Marie Curie who suggested the term 'radioactivity' to describe the ability of uranium to give off radiations and who went on to demonstrate the phenomenon in a second radioactive substance – thorium. In fast succession, enormously important discoveries were made by other scientists as well. The penetrating radiations from radioactive substances proved to be even more penetrating and more energetic than X-rays; they are now called 'gamma rays'. Radioactive elements were found to give off other types of radiation also, which led to discoveries about the internal structure of the atom, but this is a story for another chapter (see Chapter 6). What has the greatest bearing on our discussion of the elements is the discovery that the radioactive elements, in giving off the radiation, changed to other elements – a modern version of transmutation.

Marie Curie was the first to come on the implications of this phenomenon, and she did so accidentally. In testing pitchblende for its uranium content, to see if samples of the ore had enough uranium to be worth the refining effort, she and her husband found to their surprise that some of the pieces had more radioactivity than they ought to have even if they were made of pure uranium. This meant, of course, that there had to be other radioactive elements in the pitchblende. These unknown elements could be present only in small quantities, because ordinary chemical analysis did not detect them, so they must be very radioactive indeed.

In great excitement, the Curies obtained tons of pitchblende, set up shop in a small shed, and under primitive conditions and with only their unbeatable enthusiasm to drive them on they proceeded to struggle through the heavy, black ore for the trace quantities of new elements. By July of 1898, they had isolated a trace of black powder 400 times as intensely radioactive as the same quantity of uranium.

This contained a new element with chemical properties like those of tellurium, and it therefore probably belonged beneath it in the periodic table. (It was later given the atomic number 84.) The Curies named it 'polonium', after Marie's native land.

But polonium accounted for only part of the radioactivity. More work followed, and, by December 1898, the Curies had a preparation that was even more intensely radioactive than polonium. It contained still another element, which had properties like those of barium (and was eventually placed beneath barium with the atomic number 88). The Curies called it 'radium', because of its intense radioactivity.

They worked on for four more years to collect enough pure radium so that they could see it. Then Marie Curie presented a summary of her work as her Ph.D. dissertation in 1903. It was probably the greatest doctoral dissertation in scientific history. It earned her not one but two Nobel Prizes. Marie and her husband, along with Becquerel, received the Nobel Prize for physics in 1903 for their studies of radioactivity, and, in 1911, Marie alone (her husband having died in a traffic accident in 1906) was awarded the Nobel Prize for chemistry for the discovery of polonium and radium.

Polonium and radium are far more unstable than uranium or thorium, which is another way of saying that they are far more radioactive. More of their atoms break down each second. Their lifetimes are so short that practically all the polonium and radium in the universe should have disappeared within a matter of a million years or so. Why do we still find them in the thousands-of-millions-of-years-old earth? The answer is that radium and polonium are continually being formed in the course of the breakdown of uranium and thorium to lead. Wherever uranium and thorium are found, small traces of polonium and radium are likewise to be found. They are intermediate products on the way to lead as the end product.

Three other unstable elements on the path from uranium and

thorium to lead were discovered by means of the careful analysis of pitchblende or by researches into radioactive substances. In 1899, André Louis Debierne, on the advice of the Curies, searched pitchblende for other elements and came up with one he called 'actinium' (from the Greek word for 'ray'), which eventually received the atomic number 89. The following year, the German physicist Friedrich Ernst Dorn demonstrated that radium, when it broke down, formed a gaseous element. A radioactive gas was something new! Eventually the element was named 'radon' (from radium and argon, its chemical cousin) and was given the atomic number 86. Finally, in 1917, two different groups – Otto Hahn and Lise Meitner in Germany and Frederick Soddy and John A. Cranston in England – isolated from pitchblende element 91, named protactinium.

By 1925, then, the score stood at eighty-eight identified elements – eighty-one stable and seven unstable. The search for the missing four – numbers 43, 61, 85, 87 – became avid indeed.

Since all the known elements from number 84 to 92 were radioactive, it was confidently expected that 85 and 87 would be radioactive as well. On the other hand, 43 and 61 were surrounded by stable elements, and there seemed no reason to suspect that they were not themselves stable as well. Consequently, they should be found in nature.

Element 43, lying just above rhenium in the periodic table, was expected to have similar properties and to be found in the same ores. In fact, the team of Noddack, Tacke, and Berg, which had discovered rhenium, felt certain that it had also detected X-rays of a wavelength that went along with element 43. So they announced its discovery, too, and named it 'masurium', after a region in East Prussia. However, their identification was not confirmed, and in science a discovery is not a discovery unless and until it has been confirmed by at least one independent researcher.

In 1926, two University of Illinois chemists announced that they had found element 61 in ores containing its neighbouring elements (60 and 62), and they named their discovery 'illinium'. The same year, a pair of Italian chemists at the University of Florence thought that they had isolated the same element and named it 'florentium'. But other chemists could not confirm the work of either group.

A few years later, an Alabama Polytechnic Institute physicist, using a new analytical method of his own devising, reported that he had found small traces of element 87 and of element 85; he called them 'virginium' and 'alabamine', after his native and adopted states, respectively. But these discoveries could not be confirmed, either.

Events were to show that the 'discoveries' of elements 43, 61, 85, and 87 had been mistaken.

The first of the four to be identified beyond doubt was element 43. The American physicist Ernest Orlando Lawrence, who was to receive the Nobel Prize in physics for his invention of the cyclotron (see p. 321), made the element in his accelerator by bombarding molybdenum (element 42) with high-speed particles. His bombarded material developed radioactivity, and Lawrence sent it for analysis to the Italian chemist Emilio Gino Segrè, who was interested in the element-43 problem. Segrè and his colleague C. Perrier, after separating the radioactive part from the molybdenum, found that it resembled rhenium in its properties, but was not rhenium. They decided that it could only be element number 43 and that element number 43, unlike its neighbours in the periodic table, was radioactive. Because it is not being produced as a breakdown product of a higher element, virtually none of it is left in the earth's crust, and so Noddack and company were undoubtedly mistaken in thinking they had found it. Segrè and Perrier eventually were given the privilege of naming element 43; they called it 'technetium', from a Greek word meaning 'artificial', because it was the first man-made element. By 1960, enough technetium had been accumulated to determine its melting point – close to 2,200°C. (Segrè was later to receive a Nobel Prize for quite another discovery, having to do with another man-made bit of matter – see p. 325.)

In 1939, element number 87 was finally discovered in nature. The French chemist Marguerite Perey isolated it from among the breakdown products of uranium. It was present in extremely small amounts, and only improvements in technique enabled it to be found where earlier it had been missed. She later named the new element 'francium', after her native land.

Element 85, like technetium, was produced in the cyclotron, by bombardment of bismuth (element 83). In 1940, Segrè, Dale Raymond Corson, and K. R. MacKenzie isolated element 85 at the University of California, Segrè having by then emigrated from Italy to the United States. The Second World War interrupted their work on the element, but after the war they returned to it and in 1947 proposed the name 'astatine' for the element, from a Greek word meaning 'unstable'. (By that time, tiny traces of astatine had, like francium, been found in nature among the breakdown products of uranium.)

Meanwhile, the fourth and final missing element, number 61, had been discovered among the products of the fission of uranium, a process that is explained in Chapter 9. (Technetium, too, turned up among these products.) Three chemists at the Oak Ridge National Laboratory – J. A. Marinsky, L. E. Glendenin, and Charles DuBois Coryell – isolated element 61 in 1945. They named it 'promethium', after the Greek demigod Prometheus, who had stolen fire for mankind from the sun. Element number 61, after all, had been stolen from the sunlike fires of the atomic furnace.

So the list of elements, from 1 to 92, was at last complete. And yet, in a sense, the strangest part of the adventure had only begun. For scientists had broken through the bounds of the periodic table; uranium was not the end.

A search for elements beyond uranium – 'transuranium elements' – had actually begun as early as 1934. Enrico Fermi in Italy had found that when he bombarded an element with a newly discovered sub-atomic particle called the 'neutron' (see p. 308), this often transformed the element into the one of the next higher atomic number. Could uranium be built up to element 93 – a totally synthetic element that did not exist in nature? Fermi's group proceeded to attack uranium with neutrons, and they got a product that they thought was indeed element 93. They called it 'uranium X'.

In 1938, Fermi received the Nobel Prize in physics for his studies in neutron bombardment. At the time, the real nature of his discovery, or its consequences for mankind, was not even suspected. Like that other Italian, Columbus, he had found, not what he was looking for, but something far more important of which he was not aware.

Suffice it to say here that, after a series of chases up a number of false trails, it was finally discovered that what Fermi had done was, not to create a new element, but to split the uranium atom into two nearly equal parts. When physicists turned in 1940 to studies of this process, element 93 cropped up as an almost casual result of their experiments. In the mélange of elements that came out of the bombardment of uranium by neutrons, there was one that at first defied identification. Then it dawned on Edwin McMillan of the University of California that perhaps the neutrons released by fission had converted some of the uranium atoms to a higher element, as Fermi had hoped would happen. McMillan and Philip Abelson, a physical chemist, were able to prove that the unidentified element was in fact number 93. The proof of its existence lay in the nature of its radioactivity, as was to be the case in all subsequent discoveries.

McMillan suspected that another transuranium element might be mixed with number 93. The chemist Glenn Theodore Seaborg, together with his co-workers Arthur Charles Wahl and J. W. Kennedy, soon showed that this was indeed so and that the element was number 94.

Since uranium, the supposed end of the periodic table, had been named, at the time of its discovery, after the then newly discovered planet, Uranus, elements 93 and 94 were now named after Neptune and Pluto, planets discovered after Uranus. They were called 'neptunium' and 'plutonium', respectively. It turned out that they existed in nature, for small traces of neptunium and plutonium were later found in uranium ores. So uranium was not the heaviest natural element after all.

Seaborg and a group at the University of California, in which Albert Ghiorso was prominent, went on to build more transuranium elements, one after the other. By bombarding plutonium with sub-atomic particles, in 1944 they created elements 95 and 96, named respectively 'americium' (after America) and 'curium' (after the Curies). When they had manufactured a sufficient quantity of americium and curium to work with, they bombarded those elements and successfully produced number 97 in 1949 and number 98 in 1950. These they named 'berkelium' and 'californium', after Berkeley and California. In 1951, Seaborg and McMillan shared the Nobel Prize in chemistry for this train of achievements.

The next elements were discovered in more catastrophic fashion. Elements 99 and 100 emerged in the first hydrogen bomb explosion, detonated in the Pacific in November 1952. Although their existence was detected in the explosion debris, the elements were not confirmed and named until after the University of California group made small quantities of both in the laboratory in 1955. The names given them were 'einsteinium' and 'fermium', after Albert Einstein and Enrico Fermi, both of whom had died some months before. Then the group bombarded a small quantity of einsteinium and formed element 101, which they called 'mendelevium', after Mendeleev.

The next step came through a collaboration between California and the Nobel Institute in Sweden. The Institute carried out a particularly complicated type of bombardment that apparently produced a small quantity of element 102. It was named 'nobelium', in honour of the Institute, but the experiment has not been confirmed. The element has been formed by methods other than those described by the first group of workers, so that there was a delay before nobelium was officially accepted as the name of the element.

In 1961, a few atoms of element 103 were detected at the University of California, and it was given the name 'lawrencium', after E. O. Lawrence, who had recently died. In 1964, a group of Soviet scientists under Georgii Nikolaevich Flerov reported the formation of element 104, and, in 1967, the formation of element 105. In both cases, the methods used to form the elements could not be confirmed, and American teams under Albert Ghiorso formed them in other ways. There is a dispute raging over priorities; both groups claim the right to name the elements. The Soviet group has named 104 'kurchatovium', after Igor Vasilievich Kurchatov, who had led the Soviet team that developed their atomic bomb, and who had died in 1960. The American group named 104 'rutherfordium' and 105 'hahnium', after Ernest Rutherford and Otto Hahn, both of whom made key discoveries in sub-atomic structure.

Each step in this climb up the transuranium scale was harder than the one before. At each successive stage, the element became harder to accumulate and more unstable. When mendelevium was reached, identification had to be made on the basis of seventeen atoms, no more. Fortunately, radiation-detecting techniques were marvellously refined by 1955. The Berkeley scientists actually hooked up their instruments to a fire alarm, so that every time a mendelevium atom was formed, the characteristic radiation it emitted on breaking down announced the event by a loud and triumphant ring of the bell. (The fire department soon put a stop to this.)

Electrons

When Mendeleev and his contemporaries found that they could arrange the elements in a periodic table composed of families of substances showing similar properties, they had no notion as to why the elements fell into such groups or why the properties were related. Eventually a clear and rather simple answer emerged, but it came only after a long series of discoveries that at first seemed to have nothing to do with chemistry.

It all began with studies of electricity. Faraday performed every experiment with electricity he could think of, and one of the things he tried to do was to send an electric discharge through a vacuum. He was not able to get a vacuum good enough for the purpose. But, by 1854, a German glass blower named Heinrich Geissler had invented an adequate vacuum pump and produced a glass tube enclosing metal electrodes in an unprecedentedly good vacuum. When experimenters succeeded in producing electric discharges in the 'Geissler tube', they noticed that a green glow appeared on the tube wall opposite the negative electrode. The German physicist Eugen Goldstein suggested in 1876 that this green glow was caused by the impact on the glass of some sort of radiation originating at the negative electrode, which Faraday had named the 'cathode'. Goldstein called the radiation 'cathode rays'.

Were the cathode rays a form of electromagnetic radiation? Goldstein thought so, but the English physicist William Crookes and some others said no: they were a stream of particles of some kind. Crookes designed improved versions of the Geissler tube (called 'Crookes tubes'), and with these he was able to show that the rays were deflected by a magnet. This meant that they were probably made up of electrically charged particles.

In 1897, the physicist Joseph John Thomson settled the question beyond doubt by demonstrating that the cathode rays could also be deflected by electric charges. What, then, were these cathode 'particles'? The only negatively charged particles known at the time were the negative ions of atoms. Experiments showed that the cathode-ray particles could not possibly be such ions, for they were so strongly deflected by an electromagnetic field that they must have an unthinkably high electric charge or else must be extremely light particles with less than 1/1,000 the mass of a hydrogen atom. The latter interpretation turned out to fit the evidence best. Physicists had already guessed that the electric current was carried by particles, and so these cathode-ray particles were accepted as the ultimate particles of electricity. They were called 'electrons' – a name that had been suggested in 1891 by the Irish physicist George Johnstone Stoney. The electron was finally determined to have 1/1,837 the mass of a hydrogen atom. (For establishing its existence, Thomson was awarded the Nobel Prize in physics in 1906.)

The discovery of the electron at once suggested that it might be a sub-particle of the atom – in other words, that atoms were not the ultimate, indivisible units of matter that Democritus and John Dalton had pictured them to be.

This was a hard pill to swallow, but the lines of evidence converged inexorably. One of the most convincing items was Thomson's showing that negatively charged particles that came out of a metal plate when it was struck by ultra-violet radiation (the 'photoelectric effect') were identical with the electrons of the cathode rays. The photoelectric electrons must have been knocked out of the atoms of the metal.

Since electrons could easily be removed from atoms (by other means as well as by the photoelectric effect), it was natural to conclude that they were located in the outer regions of the atom. If this was so, there must be a positively charged region within the atom balancing the electrons' negative charges, because the atom as a whole was normally neutral. It was at this point that investigators began to close in on the solution of the mystery of the periodic table.

To remove an electron from an atom takes a little energy. Conversely, when an electron falls into the vacated place in the atom, it must *give up* an equal amount of energy. (Nature is usually symmetrical, especially when it comes to considerations of energy.) This energy is released in the form of electromagnetic radiation. Now since the energy of radiation is measured in terms of wavelength, the wavelength of the radiation emitted by an electron falling into a particular atom will indicate the force with which the electron is held by that atom. The energy of radiation increases with shortening wavelength: the greater the energy, the shorter the wavelength.

We arrive, then, at Moseley's discovery that metals (i.e., the heavier elements) produced X-rays, each at a characteristic wavelength, which decreased in regular fashion as one went up the periodic table. Each successive element, it seemed, held its electrons more strongly than the one before, which is another way of saying that each had a successively stronger positive charge in its internal region.

Assuming that each unit of positive charge corresponded to the negative charge on an electron, it followed that the atom of each successive element must have one more electron. The simplest way of picturing the periodic table, then, was to suppose that the first element, hydrogen, had 1 unit of positive charge and 1 electron; the second element, helium, 2 positive charges and 2 electrons; the third, lithium, 3 positive charges and 3 electrons; and so on all the way up to uranium, with 92 electrons. So the atomic numbers of the elements turned out to represent the number of electrons in their atoms.

One more major clue and the atomic scientists had the answer to the periodicity of the periodic table. It developed that the electronic radiation of a given element was not necessarily restricted to a single wavelength; it might emit radiations at two, three, four, or even more different wavelengths. These sets of radiations were named the K-series, the L-series, the M-series, and so on. The investigators interpreted this to mean that the electrons were arrayed in 'shells' around the positively charged core of the atom. The electrons of the innermost shell were most strongly held, and their removal took the most energy. An electron falling into this shell would emit the most energetic radiation, that is, of the shortest wavelengths, or the K-series. The electrons of the next innermost shell were responsible for the L-series of radiations; the next shell produced the M-series; and so on. Consequently, the shells were called the K-shell, the L-shell, the M-shell, and so on.

By 1925, the Austrian physicist Wolfgang Pauli advanced his 'exclusion principle', which explained just how electrons were distributed within each shell, since no two electrons could possess, according to this principle, exactly the same energy and spin. For this, Pauli received the Nobel Prize for physics in 1945.

In 1916, the American chemist Gilbert Newton Lewis worked out the kinship of properties and the chemical behaviour of some of the simpler elements on the basis of their shell structure. There was ample evidence, to begin with, that the innermost shell was limited to two electrons. Hydrogen has only one electron; therefore the shell is unfilled. The atom's tendency is to fill this K-shell, and it can do so in a number of ways. For instance, two hydrogen atoms can pool their single electrons and, by sharing the two electrons, mutually fill their K-shells. This is why hydrogen gas almost always exists in the form of a pair of atoms – the hydrogen molecule. To separate the two atoms and free them as 'atomic hydrogen' takes a good deal of energy. Irving Langmuir of the General Electric Company, who independently worked out a similar scheme involving electrons and chemical behaviour, presented a practical demonstration of the strong tendency of the hydrogen atom to keep its electron shell filled. He made an 'atomic hydrogen torch' by blowing hydrogen gas through an electric arc, which split the molecules' atoms apart; when the atoms recombined after passing the arc, they liberated the energy they had absorbed in splitting apart, and this was sufficient to yield temperatures up to 3,400°C!

In helium, element number 2, the K-shell is filled with two electrons. Helium atoms therefore are stable and do not combine with other atoms. When we come to lithium, element 3, we find that two of its electrons fill the K-shell and the third starts the L-shell. The succeeding elements add electrons to this shell one by one: beryllium has 2 electrons in the L-shell, boron has 3, carbon 4, nitrogen 5, oxygen 6, fluorine 7, and neon 8. Eight is the limit for the L-shell, and therefore neon corresponds to helium in having its outermost electron shell filled. And sure enough, it, too, is an inert gas with properties like helium's.

Every atom with an unsatisfied outer shell has a tendency to enter into combination with other atoms in a manner that leaves it with a filled outer shell. For instance, the lithium atom readily surrenders its one L-shell electron so that its outer shell is the filled K, while fluorine tends to seize an electron to add to its seven and complete the L-shell. Therefore lithium and fluorine have an affinity for each other; when they combine, lithium donates its L-electron to fluorine to fill the latter's L-shell. Since the atoms' interior positive charges do not change, lithium, with one electron subtracted, now carries a net positive charge, while fluorine, with one extra electron, carries a net negative charge. The mutual attraction of the opposite charges holds the two ions together. The compound is called lithium fluoride.

L-shell electrons can be shared as well as transferred. For instance, each of two fluorine atoms can share one of its electrons with the other, so that each atom has a total of eight in its L-shell, counting the two shared electrons. Similarly, two oxygen atoms will pool a total of four electrons to complete their L-shells; and two nitrogen atoms will share a total of six. Thus fluorine, oxygen, and nitrogen all form two-atom molecules.

The carbon atom, with only four electrons in its L-shell, will share each of them with a different hydrogen atom, thereby filling the K-shells of the four hydrogen atoms and in turn filling its own L-shell by sharing *their* electrons. This stable arrangement is the methane molecule, CH_4.

In the same way, a nitrogen atom will share electrons with three

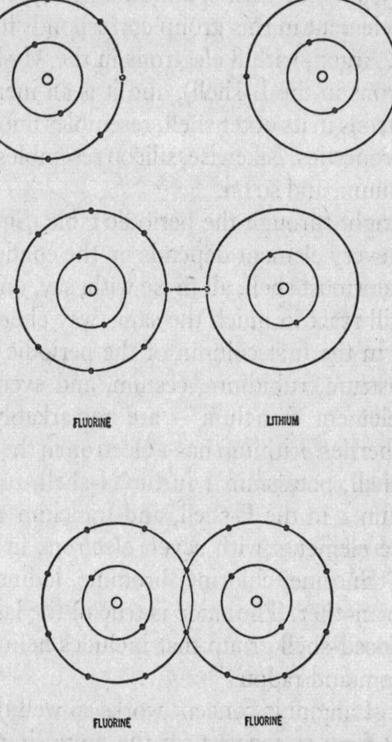

Transfer and sharing of electrons. Lithium transfers the electron in its outer shell to fluorine in the combination of lithium fluoride; each atom then has a full outer shell. In the fluorine molecule (Fl_2), two electrons are shared, filling both atoms' outer shells.

hydrogen atoms to form ammonia; an oxygen atom will share electrons with two hydrogen atoms to form water; a carbon atom will share electrons with two oxygen atoms to form carbon dioxide; and so on. Almost all the components formed by the elements in the first part of the periodic table can be accounted for on the basis of this tendency to complete the outermost shell by giving up electrons, accepting electrons, or sharing electrons.

The element after neon, sodium, has 11 electrons, and the eleventh must start a third shell. Then follow magnesium, with 2 electrons in the M-shell, aluminium with 3, silicon with 4, phosphorus with 5, sulphur with 6, chlorine with 7, and argon with 8.

Now each element in this group corresponds to one in the preceding series. Argon, with 8 electrons in the M-shell, is like neon (with 8 electrons in the L-shell), and it is an inert gas. Chlorine, having 7 electrons in its outer shell, resembles fluorine very closely in chemical properties. Likewise, silicon resembles carbon, sodium resembles lithium, and so on.

So it goes right through the periodic table. Since the chemical behaviour of every element depends on the configuration of electrons in its outermost shell, all those with, say, one electron in the outer shell will react in much the same way chemically. Thus all the elements in the first column of the periodic table – lithium, sodium, potassium, rubidium, cesium, and even the man-made radioactive element francium – are remarkably alike in their chemical properties. Lithium has 1 electron in the L-shell, sodium 1 in the M-shell, potassium 1 in the N-shell, rubidium 1 in the O-shell, cesium 1 in the P-shell, and francium 1 in the Q-shell. Again, all the elements with seven electrons in their respective outer shells – fluorine, chlorine, bromine, iodine, and astatine – resemble one another. The same is true of the last column in the table – the closed-shell group that includes helium, neon, argon, krypton, xenon, and radon.

The Lewis–Langmuir concept works so well that it still serves in its original form to account for the more simple and straightforward varieties of behaviour among the elements. However, not all the behaviour was quite as simple and straightforward as might be thought.

For instance, each of the inert gases – helium, neon, argon, krypton, xenon, and radon – has eight electrons in the outermost shell (except for helium, which has two electrons in its only shell), and this is the most stable possible situation. Atoms of these elements have a minimum tendency to lose or gain electrons and therefore a minimum tendency to engage in chemical reactions. The gases would be 'inert', as their name proclaims.

However, a 'minimum tendency' is not really the same as 'no tendency', but most chemists forgot this and acted as though it was ultimately impossible for the inert gases to form compounds. This was not true of all of them, of course. As long ago as 1932, the American chemist Linus Pauling considered the ease with which electrons could be removed from different elements and noted that all elements without exception, even the inert gases, can be deprived of electrons. It was just that for this to happen requires more energy in the case of the inert gases than in that of other elements near them in the periodic table.

The amount of energy required to remove electrons among the elements in any particular family decreases with increasing atomic weight, and the heaviest inert gases, xenon and radon, do not have unusually high requirements. It is no more difficult to remove an electron from a xenon atom, for instance, than from an oxygen atom.

Pauling therefore predicted that the heavier inert gases might form chemical compounds with elements that were particularly prone to accept electrons. The element most eager to accept electrons is fluorine, and that seemed to be the natural target.

Now radon, the heaviest inert gas, is radioactive and is unavailable in any but trace quantities. Xenon, however, the next heaviest, is stable and occurs in small quantities in the atmosphere. The best chance, therefore, would be to attempt to form a compound between xenon and fluorine. However, for thirty years nothing was done in this respect, chiefly because xenon was expensive and fluorine very hard to handle, and chemists felt they had better things to do than chase this particular will-o'-the-wisp.

In 1962, however, the British-Canadian chemist Neil Bartlett, working with a new compound, platinum hexafluoride (PtF_6), found that it was remarkably avid for electrons, almost as much so as fluorine itself. This compound would take electrons away from oxygen, an element that is normally avid to gain electrons rather than lose them. If PtF_6 could take electrons from oxygen, it ought to be able to take them from xenon too. The experiment was tried, and xenon fluoroplatinate ($XePtF_6$), the first compound of an inert gas, was reported.

Other chemists at once sprang into the fray, and a number of xenon compounds with fluorine, with oxygen, or with both were formed, the most stable being xenon difluoride (XeF_2). A compound of krypton and fluorine, krypton tetrafluoride (KrF_4), has also been formed, as well as a radon fluoride. Compounds with oxygen were also formed. There were, for instance, xenon oxytetrafluoride ($XeOF_4$), xenic acid (H_2XeO_4) and sodium perxenate (Na_4XeO_6). Most interesting, perhaps, was xenon trioxide (Xe_2O_3), which explodes easily and is dangerous. The smaller inert gases – argon, neon, and helium – are more resistant to sharing their electrons than the larger ones, and they remain inert (for all anything chemists can do even yet).

Chemists quickly recovered from the initial shock of finding that the inert gases could form compounds. Such compounds fit into the general picture after all. Consequently, there is now a general reluctance to speak of the gases as 'inert gases'. The alternative name of 'noble gases' is now preferred, and one speaks of 'noble gas compounds' and 'noble gas chemistry'. (I think this is a change for the worse. After all the gases are still inert, even if not completely inert. The concept 'noble', in this context, implies 'standoffish' or 'disinclined to mix with the common herd', and this is just as inappropriate as 'inert' and, moreover, does not suit a democratic society.)

In addition to the fact that the Lewis–Langmuir scheme was applied too rigidly to the inert gases, it can scarcely be applied at all to many of the elements with atomic numbers higher than 20. In particular, refinements had to be added to deal with a very puzzling aspect of the periodic table having to do with the so-called 'rare earths' – elements 57 to 71, inclusive.

To go back a bit, the early chemists considered any substance that was insoluble in water and unchanged by heat to be an 'earth' (a hangover of the Greek view of 'earth' as an element). Such substances included what we would today call calcium oxide, magnesium oxide, silicon dioxide, ferric oxide, aluminium oxide, and so on – compounds which actually constitute about 90 per cent of the earth's crust. Calcium oxide and magnesium oxide are slightly soluble and in solution display 'alkaline' properties (that is, opposite to those of acids), and so they were called the 'alkaline earths'; when Humphry Davy isolated the metals calcium and magnesium from these earths, they were named alkaline earth metals. The same name was eventually applied to all the elements that fall into the column of the periodic table containing magnesium and calcium: that is, to beryllium, strontium, barium, and radium.

The puzzle to which I have referred began in 1794, when a Finnish chemist, Johan Gadolin, examined an odd rock that had been found near the Swedish hamlet Ytterby and decided that it was a new 'earth'. Gadolin gave this 'rare earth' the name 'yttria', after Ytterby. Later the German chemist Martin Heinrich Klaproth found that yttria could be divided into two 'earths', for one of which he kept the name yttria, while he named the other 'ceria' (after the newly discovered planetoid Ceres). But the Swedish chemist Carl Gustav Mosander subsequently broke these down further into a series of different earths. All eventually proved to be oxides of new elements named the 'rare-earth metals'. By 1907, fourteen such elements had been identified. In order of increasing atomic weight they are:

lanthanum (from a Greek word meaning 'hidden')
cerium (from Ceres)
praseodymium (from the Greek for 'green twin', after a green line in its spectrum)
neodymium ('new twin')
samarium (from 'samarskite', the mineral in which it was found)
europium (from Europe)

gadolinium (in honour of Johan Gadolin)
terbium (from Ytterby)
dysprosium (from a Greek word meaning 'hard to get at')
holmium (from Stockholm)
erbium (from Ytterby)
thulium (from Thule, an old name for Scandinavia)
ytterbium (from Ytterby)
lutetium (from Lutetia, an old name for Paris).

On the basis of their X-ray properties, these elements were assigned the atomic numbers from 57 (lanthanum) to 71 (lutetium). As I related earlier, there was a gap at 61 until the missing element, promethium, emerged from the fission of uranium. It made the fifteenth in the list.

Now the trouble with the rare-earth elements is that they apparently cannot be made to fit into the periodic table. It is fortunate that only four of them were definitely known when Mendeleev proposed the table; if they had all been on hand, the table might have been altogether too confusing to be accepted. There are times, even in science, when ignorance is bliss.

The first of the rare-earth metals, lanthanum, matches up all right with yttrium, number 39, the element above it in the table. (Yttrium, though found in the same ores as the rare earths and similar to them in properties, is not a rare-earth metal. It is, however, named after Ytterby. Four elements honour that hamlet – which is overdoing it.) The confusion begins with the rare earth after lanthanum, namely, cerium, which ought to resemble the element following yttrium, that is, zirconium. But it does nothing of the sort; instead, it resembles yttrium again. And the same is true of all fifteen of the rare-earth elements: they strongly resemble yttrium and one another (in fact, they are so alike chemically that at first they could not be separated except by the most tedious procedures), but they are not related to any other elements preceding them in the table. We have to skip the whole rare-earth group and go on to hafnium, element 72, to find the element related to zirconium, the one after yttrium.

Baffled by this state of affairs, chemists could do no better than to group all the rare-earth elements into one box beneath yttrium and list them individually in a kind of footnote to the table.

The answer to the puzzle finally came as a result of details added to the Lewis–Langmuir picture of the electron-shell structure of the elements.

In 1921, C. R. Bury suggested that the shells were not necessarily limited to eight electrons apiece. Eight always sufficed to satisfy the outer shell. But a shell might have a greater capacity when it was not on the outside. As one shell built on another, the inner shells might absorb more electrons, and each succeeding shell might hold more than the one before. Thus the K-shell's total capacity would be 2 electrons, the L-shell's 8, the M-shell's 18, the N-shell's 32, and so on – the step-ups going according to a pattern of successive squares multiplied by two (i.e., 2×1, 2×4, 2×9, 2×16, etc.).

This view was backed up by a detailed study of the spectra of the elements. The Danish physicist Niels Henrik David Bohr showed that each electron shell was made up of sub-shells at slightly different energy levels. In each succeeding shell, the spread of the sub-shells was greater so that soon the shells overlapped. As a result, the outermost sub-shell of an interior shell (say, the M-shell) might actually be farther from the centre, so to speak, than the innermost sub-shell of the next shell beyond it (i.e., the N-shell). This being so, the N-shell's inner sub-shell might fill with electrons while the M-shell's outer sub-shell was still empty.

An example will make this clearer. The M-shell, according to the theory, is divided into three sub-shells, whose capacities are 2, 6, and 10 electrons respectively, making a total of 18. Now argon, with 8 electrons in its M-shell, has filled only two inner sub-shells. And, in fact, the M-shell's third, or outermost, sub-shell will not get the next electron in the element-building process, because it lies beyond the innermost sub-shell of the N-shell. That is, in potassium, the element after argon, the nineteenth electron goes, not into the outermost sub-shell of M, but into the innermost sub-shell of N. Potassium, with 1 electron in its N-shell, resembles sodium, which has 1 electron in its M-shell. Calcium, the next element (20), has 2 electrons in the N-shell and resembles magnesium, which has 2 in the M-shell. But now the innermost sub-shell of the N-shell, having room for only 2 electrons, is full. The next electrons to be added can start filling the outermost sub-shell of the M-shell which so far has not been touched. Scandium (21) begins the process, and zinc (30) completes it. In zinc, the outermost sub-shell of the M-shell has at last acquired its complement of 10 electrons. The 30 electrons of zinc are distributed as follows: 2 in the K-shell, 8 in the L-shell, 18 in the M-shell, and 2 in the N-shell. At this point, electrons can resume the filling of the N-shell. The next electron gives the N-shell 3 electrons and forms gallium (31), which resembles aluminium, with 3 in the M-shell.

The point of all this is that elements 21 to 30, formed on the road to filling a sub-shell which had been skipped temporarily, are 'transitional' elements. Note that calcium resembles magnesium and gallium resembles aluminium. Now magnesium and aluminium are adjacent members of the periodic table (numbers 12 and 13). But calcium (20) and gallium (31) are not. Between them lie the transitional elements, and these introduce a complication in the periodic table.

The N-shell is larger than the M-shell and is divided into four sub-shells instead of three: they can hold 2, 6, 10, and 14 electrons, respectively. Krypton, element 36, fills the two innermost sub-shells of the N-shell, but here the innermost sub-shell of the overlapping O-shell intervenes, and, before electrons can go on to N's two outer sub-shells, they must fill that one. The element after krypton, rubidium (37), has its thirty-seventh electron in the O-shell. Strontium (38) completes the filling of the two-electron O sub-shell. Thereupon a new series of transitional elements proceeds to fill the skipped third sub-shell of the N-shell. With cadmium (48) this is completed; now N's fourth and outermost sub-shell is skipped while electrons fill O's second innermost sub-shell, ending with xenon (54).

But even now N's fourth sub-shell must bide its turn, for by this stage the overlapping has become so extreme that even the P-shell interposes a sub-shell that must be filled before N's last. After xenon come cesium (55) and barium (56), with one and two electrons, respectively, in the P-shell. It is still not N's turn: the fifty-seventh electron, surprisingly, goes into the third sub-shell of the O-shell, creating the element lanthanum. Then, and only then, an electron at long last enters the outermost sub-shell of the N-shell. One by one the rare-earth elements add electrons to the N-shell until element 71, lutetium, finally fills it. Lutetium's

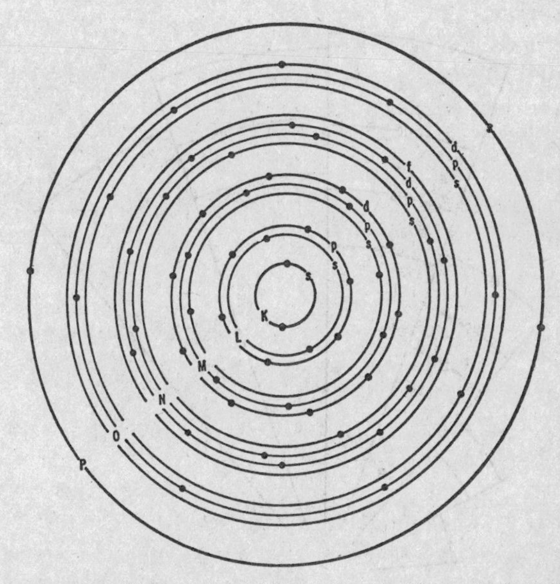

The electron shells of lanthanum. Note that the fourth sub-shell of the N-shell has been skipped and is empty.

electrons are arranged thus: 2 in the K-shell, 8 in the L-shell, 18 in the M-shell, 32 in the N-shell, 9 in the O-shell (two sub-shells full plus one electron in the next sub-shell), and 2 in the P-shell (inner-most sub-shell full).

Now at last we begin to see why the rare-earth elements, and some other groups of transitional elements, are so alike. The decisive thing that differentiates elements, as far as their chemical properties are concerned, is the configuration of electrons in their outermost shell. For instance, carbon, with 4 electrons in its outermost shell, and nitrogen, with 5, are completely different in their properties. On the other hand, in sequences where electrons are busy filling inner sub-shells while the outermost shell remains unchanged, the properties vary less. Thus iron, cobalt, and nickel (elements 26, 27, and 28), all of which have the same outer-shell

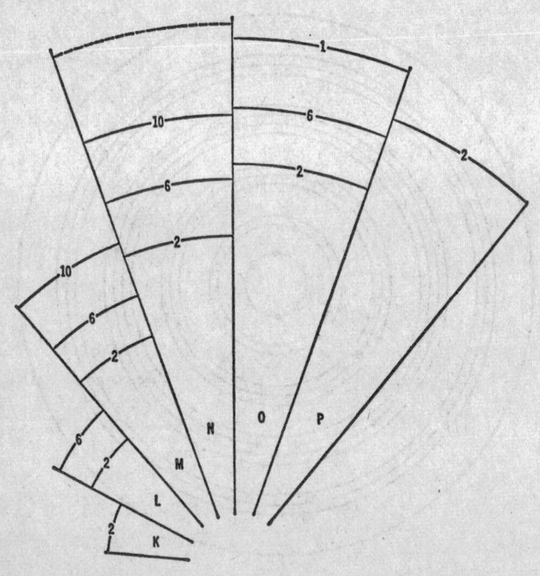

Schematic representation of the overlapping of electron shells and sub-shells in lanthanum. The outermost sub-shell of the N-shell has yet to be filled.

electronic configuration – an N sub-shell filled with 2 electrons – are a good deal alike in chemical behaviour. Their internal electronic differences (in an M sub-shell) are largely masked by their surface electronic similarity. And this goes double for the rare-earth elements. Their differences (in the N-shell) are buried under, not one, but two outer electronic configurations (in the O-shell and the P-shell), which in all these elements are identical. Small wonder that the elements are chemically as alike as peas in a pod.

Because the rare-earth metals have so few uses, and are so difficult to separate, chemists made little effort to do so – until the uranium atom was fissioned. Then it became an urgent matter indeed, because radioactive varieties of some of these elements were among the main products of fission, and in the atomic bomb project it was necessary to separate and identify them quickly and cleanly.

The problem was solved in short order by use of a chemical technique first devised in 1906 by a Russian botanist named Mikhail Semeyonovich Tswett. He named it 'chromatography' ('writing in colour'). Tswett had found that he could separate plant pigments, chemically very much alike, by washing them down a column of powdered limestone with a solvent. He dissolved his mixture of plant pigments in petroleum ether and poured this on the limestone. Then he proceeded to pour in clear solvent. As the pigments were slowly washed down through the limestone powder, each pigment moved down at a different rate, because each differed in strength of adhesion to the powder. The result was that they separated into a series of bands, each of a different colour. With continued washing, the separated substances trickled out separately at the bottom of the column, one after the other.

The world of science for many years ignored Tswett's discovery, possibly because he was only a botanist and only a Russian, while the leaders of research on separating difficult-to-separate substances at the time were German biochemists. But, in 1931, a German biochemist, Richard Willstätter, rediscovered the process, whereupon it came into general use. (Willstätter had received the 1915 Nobel Prize in chemistry for his excellent work on plant pigments. Tswett, so far as I know, has gone unhonoured.)

Chromatography through columns of powder was found to work on almost all sorts of mixtures – colourless as well as coloured. Aluminium oxide and starch proved to be better than limestone for separating ordinary molecules. Where ions are separated, the process is called ion exchange, and compounds known as zeolites were the first efficient agents applied for this purpose. Calcium and magnesium ions could be removed from 'hard' water, for instance, by pouring the water through a zeolite column. The calcium and magnesium ions adhere to the zeolite and are replaced in solution by the sodium ions originally present on the zeolite, so 'soft' water drips out of the bottom of the column. The sodium ions of zeolite have to be replenished from time to time by pouring in a concentrated solution of salt (sodium chloride). In 1935, a refinement came with the development of 'ion-exchange resins'. These synthetic substances can be designed for the job to be done. For instance, certain resins will substitute hydrogen ions for positive ions, while others substitute hydroxyl ions for negative ions; a combination of both types will remove most of the salts from sea water. Kits containing such resins were part of the survival equipment on life rafts during the Second World War.

It was the American chemist Frank Harold Spedding who adapted ion-exchange chromatography to the separation of the rare earths. He found that these elements came out of an ion-exchange column in the reverse order of their atomic number, so that they were not only quickly separated but also identified. In fact, the discovery of promethium, the missing element 61, was confirmed in this way from the tiny quantities found among the fission products.

Thanks to chromatography, purified rare-earth elements can now be prepared by the pound or even by the ton. It turns out that the rare earths are not particularly rare: the rarest of them (excepting promethium) are more common than gold or silver, and the most abundant – lanthanum, cerium, and neodymium – are more plentiful than lead. Together the rare-earth metals make up a larger percentage of the earth's crust than copper and tin combined. So scientists have pretty well dropped the term 'rare earths' and now call this series of elements the 'lanthanides', after its lead-off member. To be sure, the individual rare earths have not had many uses in the past, but in 1965 certain europium–yttrium compounds turned out to be particularly useful as red-sensitive 'phosphors' in colour television. Obviously, big things may come of this.

As if to reward the chemists and physicists for their decipherment of the rare-earth mystery, the new knowledge provided a key to the chemistry of the elements at the end of the periodic table, including the man-made ones.

The series of heavy elements in question begins with actinium, number 89. In the table it falls under lanthanum. Actinium has two electrons in the Q-shell, just as lanthanum has two electrons in the P-shell. Actinium's eighty-ninth and last electron entered the P-shell, just as lanthanum's fifty-seventh and last entered the O-shell. Now the question is: Do the elements after actinium continue to add electrons to the P-shell and remain ordinary transition elements? Or do they, perchance, follow the pattern of the elements after lanthanum, where the electrons dive down to fill the skipped sub-shell below? If the latter is true, then actinium may start a new series of 'rare-earth metals'.

The natural elements in this series are actinium, thorium, protactinium, and uranium. They were not much studied before 1940. What little was known about their chemistry suggested that they were ordinary transition elements. But when the man-made elements neptunium and plutonium were added to the list and studied intensively, these two showed a strong chemical resemblance to uranium. This prompted Glenn Seaborg to propose that the heavy elements were in fact following the rare-earth pattern and filling the buried, unfilled sub-shell. As more transuranium elements were added to the list, studies of their chemistry bore out this view, and it is now generally accepted.

The shell being filled is the fourth sub-shell of the O-shell. With lawrencium, element number 103, that sub-shell is filled. All the elements from actinium to lawrencium share much the same chemical properties and resemble lanthanum and the lanthanides. With element 104, the 104th electron will have to be added to the P-shell, and its properties should be like those of hafnium. It will be the final touch that clinches the existence of the second rare-earth series, and that is why chemists look forward so eagerly to obtaining and studying element 104.

They already have one independent confirmation. Ion-exchange chromatography separates the transuranium elements beautifully and in an exactly analogous manner to the separation of the lanthanides.

In token of the parallelism, the heavier 'rare-earth metals' are now called 'actinides', just as the members of the first series are called lanthanides.

Gases

From the dawn of chemistry, it was recognized that many substances could exist in the form of a gas, liquid, or solid, depending on the temperature. Water is the most common example: sufficiently cooled, it becomes solid ice, and sufficiently heated, it becomes gaseous steam. Van Helmont, who first used the word 'gas', differentiated between substances that were gases at ordinary temperatures, such as carbon dioxide, and those that, like steam, were gases only at elevated temperatures. He called the latter 'vapours', and we still speak of 'water vapour' rather than 'water gas'.

The study of gases, or vapours, continued to fascinate chemists, partly because they lent themselves to quantitative studies. The rules governing their behaviour were simpler and more easily worked out than those governing the behaviour of liquids and solids.

In 1787, the French physicist Jacques Alexandre César Charles discovered that, when a gas was cooled, each degree of cooling caused its volume to contract by about 1/273 of the volume it had at 0°C, and, conversely, each degree of warming caused it to expand by the same 1/273. The expansion with warmth raised no logical difficulties, but, if shrinkage with cold were to continue according to Charles' law (as it is called to this day), at −273°C a gas should have shrunk to nothing! This paradox did not particularly bother chemists, for they realized that Charles' law could not hold all the way down, and they had no way of getting to very low temperatures to see what happened.

The development of the atomic theory, picturing gases as collections of molecules, presented the situation in new terms. The volume was now seen to depend on the velocity of the molecules. The higher the temperature, the faster they moved, the more 'elbow room' they required, and the greater the volume. Conversely, the lower the temperature, the more slowly they moved, the less room they required, and the smaller the volume. In the 1860s, the British physicist William Thomson, who had just been raised to the peerage as Lord Kelvin, suggested that it was the molecules' average energy content that declined by 1/273 for every degree of cooling. Whereas volume could not be expected to disappear completely, energy could. Thomson maintained that at $-273°C$ the energy of molecules would sink to zero. Therefore $-273°C$ must represent the lowest possible temperature. So this temperature (now put at $-273{\cdot}16°C$ according to refined modern instruments) would be 'absolute zero', or, as it is often stated, 'zero Kelvin'. On this absolute scale the melting point of ice is 273°K.

Naturally, among physicists there would be great interest in trying to reach absolute zero. There is something about any distant horizon that calls for conquest. Men had been exploring extremes of coldness even before Thomson defined the ultimate goal. This exploration involved attempts to liquefy gases. Michael Faraday had found that even at ordinary temperatures some gases could be liquefied by putting them under pressure; he had liquefied chlorine, sulphur dioxide, and ammonia in this way in the 1820s. Now, once liquefied, a gas could act as a cooling agent. When the pressure above the liquid was slowly reduced, the gas evaporated, and the evaporation absorbed heat from the remaining liquid. (When you blow on a moistened finger, the coolness you feel is the effect of the water evaporation drawing heat from the finger.) The general principle is well known today as the basis of modern refrigeration.

As early as 1755, the Scottish chemist William Cullen had produced ice mechanically by forming a vacuum over quantities of water, enforcing rapid evaporation of the water, and, of course, cooling to the freezing point. Nowadays, an appropriate gas is liquefied by a compressor and then circulated in coils or pipe where, as the liquid evaporates, it withdraws heat from the surrounding space.

Water itself is inappropriate for the purpose, as the ice that forms would clog the pipes. In 1834, an American inventor, Jacob Perkins, patented (in Great Britain) the use of ether as a refrigerant. Other gases such as ammonia and sulphur dioxide also came into use. All these refrigerants had the disadvantage of being poisonous or flammable. In 1930, however, the American chemist Thomas Midgley discovered dichlorodifluoromethane (CF_2Cl_2), better known under the trade-name of 'Freon'. This is non-toxic and non-flammable and suits the purpose perfectly. With Freon, home refrigeration became widespread and commonplace.

Refrigeration applied, in moderation, to large volumes is 'air conditioning', so called because the air is also conditioned, i.e., filtered and dehumidified. The first practical air-conditioning unit was designed in 1902 by the American inventor Willis H. Carrier; since the Second World War air conditioning has become so common in the major American cities as to be nearly universal.

But the refrigeration principle can be carried to extremes, too. If a liquefied gas is enclosed in a well-insulated container, so that its evaporation draws heat only from the liquid itself, very low temperatures can be attained. By 1835, physicists had reached temperatures as low as −110°C (163°K).

Hydrogen, oxygen, nitrogen, carbon monoxide, and some other common gases, however, defied liquefaction at this temperature even with the use of high pressures. For a time, their liquefaction was despaired of, and they were called 'the permanent gases'.

In 1869, however, the Irish physicist Thomas Andrews deduced from his experiments that every gas had a 'critical temperature' above which it could not be liquefied even under pressure. This was later put on a firm theoretical basis by the Dutch physicist, Johannes Diderik Van der Waals, who, in this fashion, earned the 1910 Nobel Prize for physics.

To liquefy any gas one had to be certain, therefore, that one was working at a temperature below the critical value, or it was labour thrown out. Efforts were made to reach still lower temperatures to conquer the stubborn gases. A 'cascade' method, lowering temperatures by steps, turned the trick. First, liquefied sulphur dioxide, cooling through evaporation, was used to liquefy carbon dioxide, then the liquid carbon dioxide to liquefy a more resistant gas, and so on. In 1877, the Swiss physicist Raoul Pictet finally managed to liquefy oxygen, at a temperature of −140°C (133°K) and under a pressure of 500 atmospheres (7,500 pounds per square inch). The French physicist Louis Paul Cailletet at about the same time liquefied, not only oxygen, but also nitrogen and carbon monoxide. Naturally these liquids made it possible to go on at once to still lower temperatures. The liquefaction point of oxygen at ordinary air pressure was eventually found to be −183°C (90°K), that of carbon monoxide −190°C (83°K), and that of nitrogen −195°C (78°K).

Hydrogen resisted all efforts at liquefaction until 1900. The Scottish chemist James Dewar then accomplished the feat by bringing a new stratagem into play. Lord Kelvin (William Thomson) and the English physicist James Prescott Joule had shown that even in the gaseous state a gas could be cooled simply by letting it expand and preventing heat from leaking into the gas from outside, provided the temperature was low enough to begin with. Dewar therefore cooled compressed hydrogen to a temperature of −200°C in a vessel surrounded by liquid nitrogen, let this super-frigid hydrogen expand and cool further, and repeated the cycle again and again by conducting the ever-cooling hydrogen back through pipes. The compressed hydrogen, subjected to this 'Joule–Thomson effect', finally became liquid at a temperature of about −240°C (33°K). At still lower temperatures he managed to obtain solid hydrogen.

To preserve his super-frigid liquids, Dewar devised special silver-coated glass flasks. These were double-walled with a vacuum between. Heat could be lost (or gained) through a vacuum only by the comparatively slow process of radiation, and the silver coating reflected the incoming (or, for that matter, outgoing) radiation. Such 'Dewar flasks' are the direct ancestor of the household Thermos flask.

By 1895, the British inventor William Hampson and the German physicist Carl Lindé had developed methods of liquefying air on a commercial scale. Pure liquid oxygen, separated from the nitrogen, became a highly useful article. Its main use, in terms of quantity, was in blow-torches, principally for welding. But more dramatic were its services in medicine (e.g., oxygen tents), in aviation, in submarines, and so on.

With the coming of rocketry, liquefied gases suddenly rose to new heights of glamour. Rockets require an extremely rapid chemical reaction, yielding large quantities of energy. The most convenient type of fuel is a combination of a liquid combustible, such as alcohol or kerosene, and liquid oxygen. Oxygen, or some alternative oxidizing agent, must be carried by the rocket in any case, because it runs out of any natural supply of oxygen when it leaves the atmosphere. And the oxygen must be in liquid form, since liquids are denser than gases and more oxygen can be squeezed into the fuel tanks in liquid form than in gaseous. Consequently, liquid oxygen has come into high demand in rocketry.

The efficiency of a mixture of fuel and oxidizer is measured by a quantity known as the 'specific impulse'. This represents the number of pounds of thrust produced by the combination of one pound of the fuel–oxidizer mixture in one second. For a mixture of kerosene and oxygen, the specific impulse is equal to 242. Since the payload a rocket can carry depends on the specific impulse, there has been an avid search for more efficient combinations. The best chemical fuel, from this point of view, is liquid hydrogen. Combined with liquid oxygen, it can yield a specific impulse equal to 350 or so. If liquid ozone or liquid fluorine could be used in place of oxygen, the specific impulse could be raised to something like 370.

Research to find even better fuels for rockets is being pursued in several directions. Certain light metals, such as lithium, boron, magnesium, aluminium, and, particularly, beryllium deliver more energy on combining with oxygen than even hydrogen does. Some of these are rare, however, and all involve technical difficulties in the burning, difficulties arising from smokiness, oxide deposits, and so on.

Attempts are also being made to work out new solid fuels that serve as their own oxidizers (like gunpowder, which was the first rocket propellant, but much more efficient). Such fuels are called 'monopropellants', since they need no separate supply of oxidizer and make up the one propellant required. Fuels that also require oxidizers are 'bipropellants' (two propellants). Monopropellants, it is hoped, would be easy to store and handle and would burn in a rapid but controlled fashion. The principal difficuty is probably that of developing a monopropellant with a specific impulse approaching those of the bipropellants.

Another possiblity is atomic hydrogen, which Langmuir put to use in his blow-torch. It had been calculated that a rocket engine operating on the recombination of hydrogen atoms into molecules could develop a specific impulse of more than 1,300. The main problem is how to store the atomic hydrogen. So far the best hope seems to be to cool the free atoms very quickly and very drastically immediately after they are formed. Researches at the National Bureau of Standards seem to show that free hydrogen atoms are best preserved if trapped in a solid material at extremely low temperatures – say frozen oxygen or argon. If it could be arranged to push a button, so to speak, to let the frozen gases start warming up and evaporating, the hydrogen atoms would be freed and allowed to recombine. If such a solid could hold even as much as 10 per cent of its weight in free hydrogen atoms, the result would be a better fuel than any we now possess. But, of course, the temperature would have to be very low indeed – considerably below that of liquefied hydrogen. These solids would have to be kept at about −272°C, or just one degree above absolute zero.

In another direction altogether lies the possibility of driving ions backwards (rather than the exhaust gases of burnt fuel). The individual ions, of tiny mass, would produce tiny impulses, but this could be continued over long periods. A ship placed in orbit by the high but short-lived force of chemical fuel could then, in the virtually frictionless medium of space, slowly accelerate under the long-lived lash of ions to near light velocity. The material best suited to such an ionic drive is cesium, the substance that can most easily be made to lose electrons and form cesium ion. An electric field can then be made to accelerate the cesium ion and shoot it out of the rocket opening.

But to return to the world of low temperature. Even the liquefaction and solidification of hydrogen did not represent the final victory. By the time hydrogen yielded, the inert gases had been discovered; of these the lightest, helium, remained a stubborn hold-out against liquefaction at the lowest temperatures attainable. Then, in 1908, the Dutch physicist Heike Kammerlingh Onnes finally subdued helium. He carried the Dewar system one step further. Using liquid hydrogen, he cooled helium gas under pressure to about −255°C (18°K) and then let the gas expand to cool itself further. By this method he liquefied the gas. Thereafter, by letting the liquid helium evaporate, he got down to the temperature at which helium could be liquefied under normal atmospheric pressure (4·2°K) and even to temperatures as low as 0·7°K. For his low-temperature work, Onnes received the Nobel Prize in physics in 1913. (Nowadays the liquefaction of helium is a simpler matter. In 1947, the American chemist Samuel Cornette Collins invented the 'cryostat', which, by alternate compressions and expansions, can produce as much as two gallons of liquid helium an hour.) Onnes, however, did more than reach new depths of temperature. He was the first to show that unique properties of matter existed at those depths.

One of these properties is the strange phenomenon called 'super-conductivity'. In 1911, Onnes was testing the electrical resistance of mercury at low temperatures. It was expected that resistance to an electric current would steadily decrease as the removal of heat reduced the normal vibration of the atoms in the metal. But at 4·12°K the mercury's electrical resistance suddenly disappeared altogether! An electric current coursed through it without any loss of strength whatever. It was soon found that other metals also could be made super-conductive. Lead, for instance, became super-conductive at 7·22°K. An electric current of several hundred amperes set up in a lead ring kept at that temperature by liquid helium went on circling through the ring for two and a half years with absolutely no detectable decrease in quantity.

As temperatures were pushed lower and lower, more metals were added to the list of super-conductive materials. Tin became super-conductive at 3·73°K, aluminium at 1·20°K, uranium at 0·8°K, titanium at 0·53°K, hafnium at 0·35°K. (Some 1,400 different elements and alloys are now known to display super-conductivity.) But iron, nickel, copper, gold, sodium, and potassium must have still lower transition points – if they can be made super-conductive at all – because they have not been reduced to this state at the lowest temperatures reached. The highest transition point found for a metal is that of technetium, which becomes super-conductive at temperatures under 11·2°K.

A low-boiling liquid can easily maintain substances immersed in it at the temperature of its boiling point. To attain lower temperatures, the aid of a still-lower-boiling liquid must be called upon. Liquid hydrogen boils at 20·4°K, and it would be most useful to find a super-conducting substance with a transition temperature at least this high. Only then can super-conductivity be studied in systems cooled by liquid hydrogen. Failing that, only the one lower-boiling liquid, liquid helium – much rarer, more expensive, and harder to handle – must be used. A few alloys, particularly those involving the metal niobium, have transition temperatures higher than those of any pure metal. Finally, in 1968, an alloy of niobium, aluminium, and germanium was found that remained super-conductive at 21°K. Super-conductivity at liquid-hydrogen temperatures became feasible – but just barely.

A useful application of super-conductivity suggests itself at once in connection with magnetism. A current of electricity through a coil of wire around an iron core can produce a strong magnetic field – the greater the current, the stronger the field. Unfortunately, the greater the current, the greater the heat produced under ordinary circumstances, and this puts a limit to what can be done. In super-conductive wires, however, electricity flows without producing heat and, it would seem, more and more electric current could be squeezed into the wires to produce unprecedentedly strong 'electromagnets' at only a fraction of the power that must be expended under ordinary conditions. There is, however, a catch.

Along with super-conductivity goes another property involving magnetism. At the moment that a substance becomes super-conductive, it also becomes perfectly 'diamagnetic'; that is, it excludes the lines of force of a magnetic field. This was discovered by W. Meissner in 1933 and is therefore called the 'Meissner effect'. By making a magnetic field strong enough, however, one can destroy the substance's super-conductivity and the hope for super-magnetism, even at temperatures well below its transition point. It is as if, once enough lines of force have been concentrated in the surroundings, some at last manage to penetrate the substance, and when that happens, gone is the super-conductivity as well.

Attempts have been made to find super-conductive substances that will tolerate high magnetic fields. There is, for instance, a tin-niobium alloy with the high transition temperature of 18°K. It can support a magnetic field of some 250,000 gauss, which is high indeed. This fact was discovered in 1954, but it was only in 1960 that techniques were developed for forming wires of this ordinarily brittle alloy. A compound of vanadium and gallium may do even better, and super-conductive electromagnets reaching field intensities of 500,000 gauss have been constructed.

Another startling phenomenon at low temperatures was discovered in helium itself. It is called 'super-fluidity'.

Helium is the only known substance that cannot be frozen solid at ordinary pressures, even at absolute zero. There is a small irreducible energy content, even at absolute zero, which cannot possibly be removed (so that the energy content is 'zero' in a practical sense), but which is enough to keep the extremely 'non-sticky' atoms of helium free of each other, and therefore liquid. Actually, the German physicist Hermann Walther Nernst showed in 1905 that it is not the energy of a substance that becomes zero at absolute zero, but a closely related property: the 'entropy'. For this he received the 1920 Nobel Prize in chemistry. This does not mean, however, that solid helium doesn't exist under any conditions. It can be produced at temperatures below 1°K, by a pressure of about twenty-five atmospheres.

In 1935, Willem Hendrik Keesom and his sister, A. P. Keesom, working at the Onnes laboratory in Leyden, found that liquid helium at a temperature below 2·2°K conducted heat almost perfectly. It conducted heat so quickly, at the speed of sound, in fact, that all parts of the helium were always at the same temperature. It would not boil – as any ordinary liquid will by reason of localized hot spots forming bubbles of vapour – because there were no localized hot spots in the liquid helium (if you can speak of hot spots in connection with a liquid below 2°K). When it evaporated, the top of the liquid simply slipped off quietly – peeling off, so to speak, in sheets.

The Russian physicist Peter Leonidovich Kapitza went on to investigate this property and found that the reason helium conducted heat so well was that it flowed with remarkable ease, carrying the heat from one part of itself to another almost instan-

1. Stellar Populations I and II. On the left a spiral arm of the Andromeda galaxy, photographed in blue light, shows giant and super-giant stars of Population I; on the right the galaxy NGC 205, a companion of the Andromeda galaxy, photographed in yellow light, shows stars of Population II, the brightest of which are red stars one-hundredth as bright as the blue giants of Population I. (The large stars in both pictures are foreground stars belonging to our own Milky Way.)

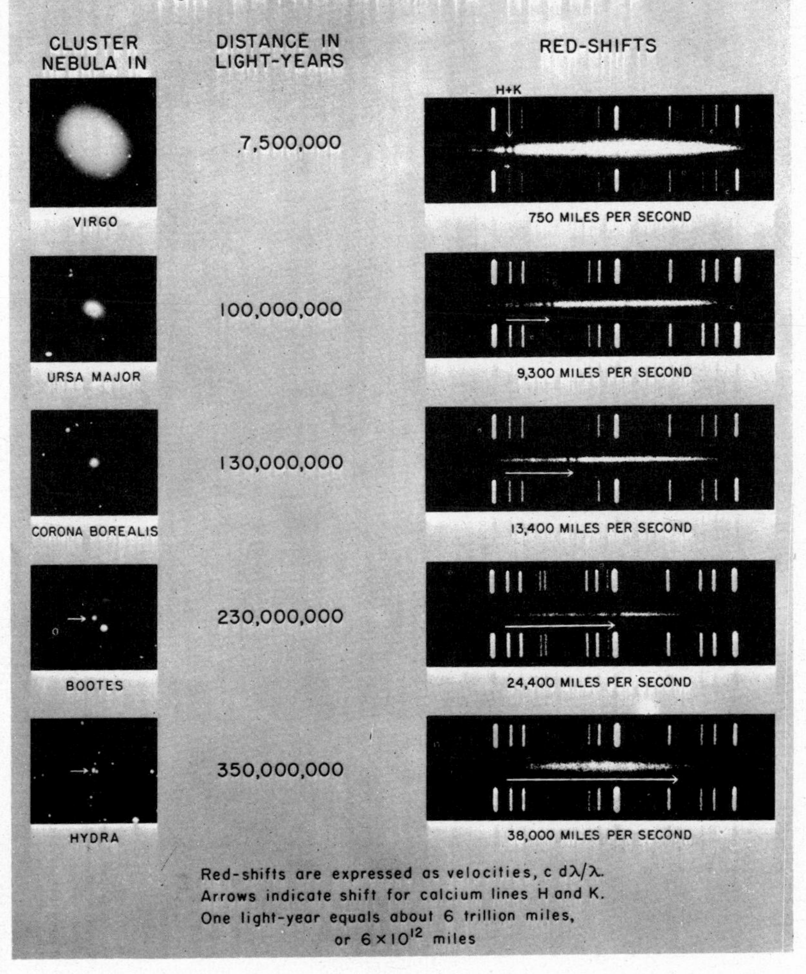

2a. (*above left*) A spiral galaxy in broadside view – the 'whirlpool nebula' in Canes Venatici.

2b. (*below left*) Two colliding galaxies – NGC 4038 and 4039.

3. (*above*) Red shifts of distant galaxies.

4. (*above*) A globular cluster in Canes Venatici.
5a. (*above right*) Cornell University's new radio telescope. The reflector of this radio-radar telescope at Arecibo, Puerto Rico, is 1,000 feet in diameter and is suspended in a natural bowl.
5b. (*below right*) A solar flare, stretching 140,000 miles from the sun, photographed in the light of calcium. The white circle represents the earth on a similar scale.

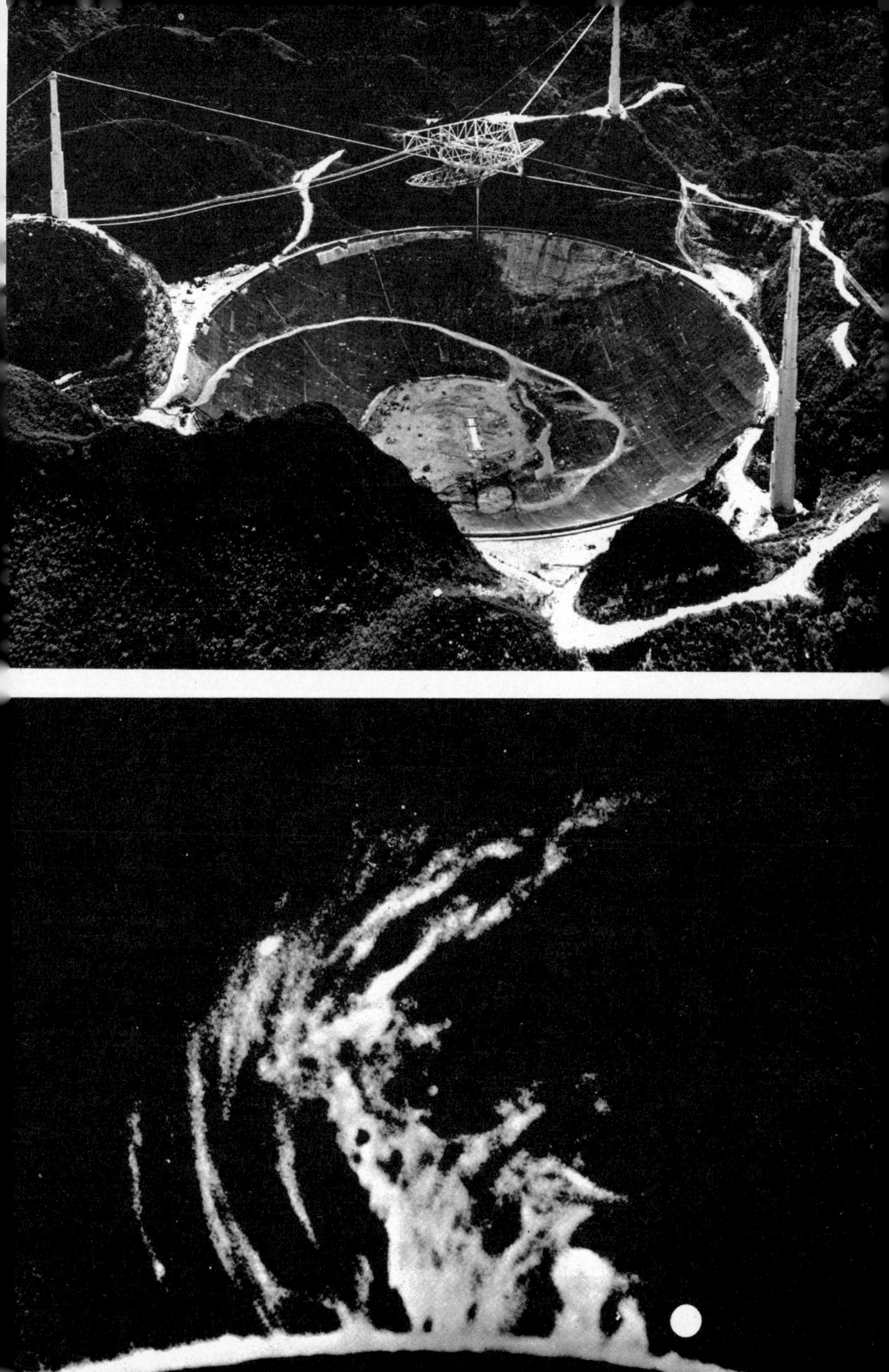

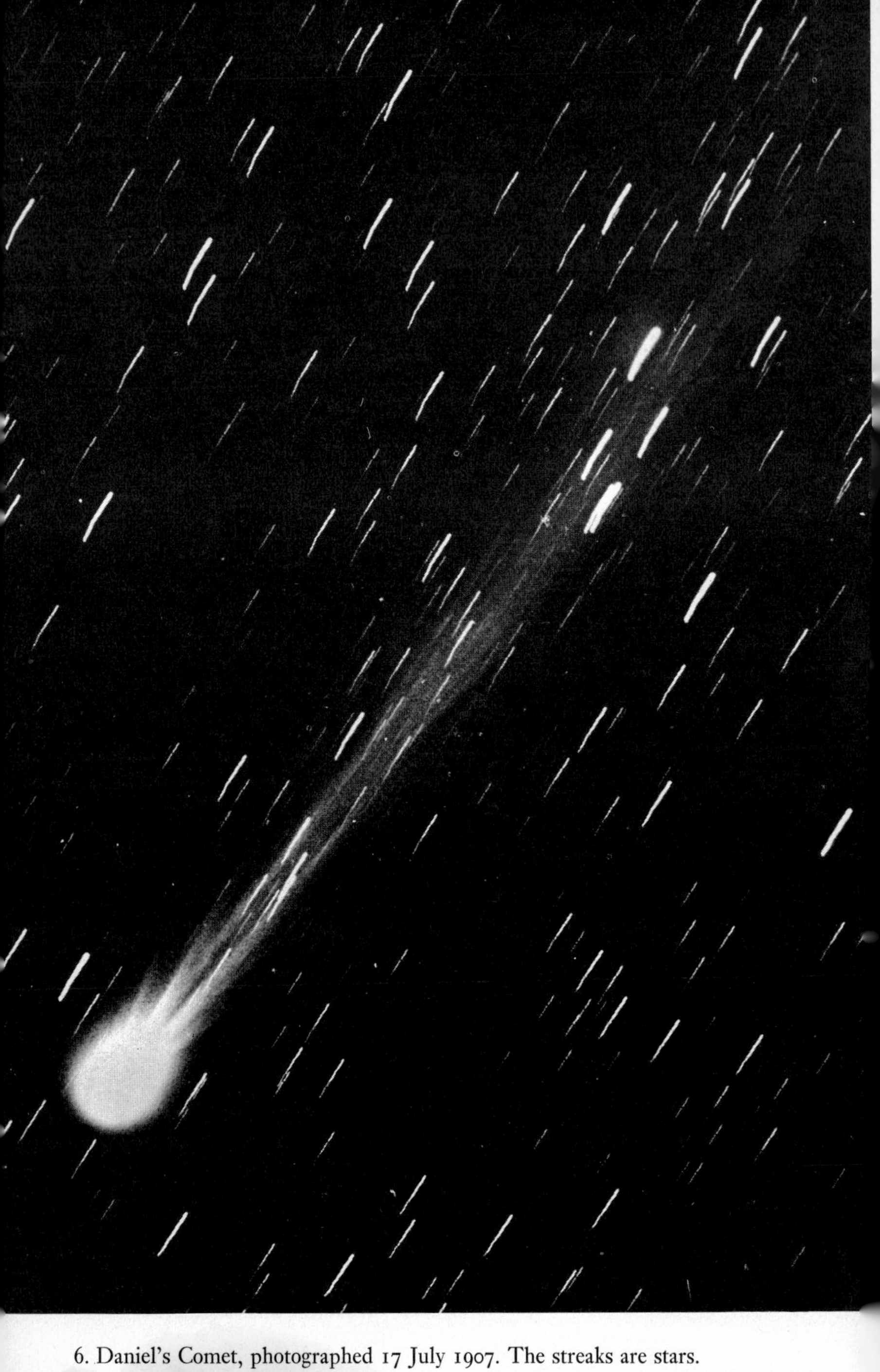

6. Daniel's Comet, photographed 17 July 1907. The streaks are stars.

7. The Coal Sack, a huge cloud of dust and gas in the Southern Cross.

8a. (*above*) A photograph of the crater Copernicus taken from 28·4 miles above the surface of the moon by Lunar Orbiter II.
8b. (*below*) This view of the rising earth greeted the Apollo 8 astronauts as they came from behind the moon after the lunar orbit insertion burn. On the earth 240,000 statute miles away, the sunset terminator bisects Africa.

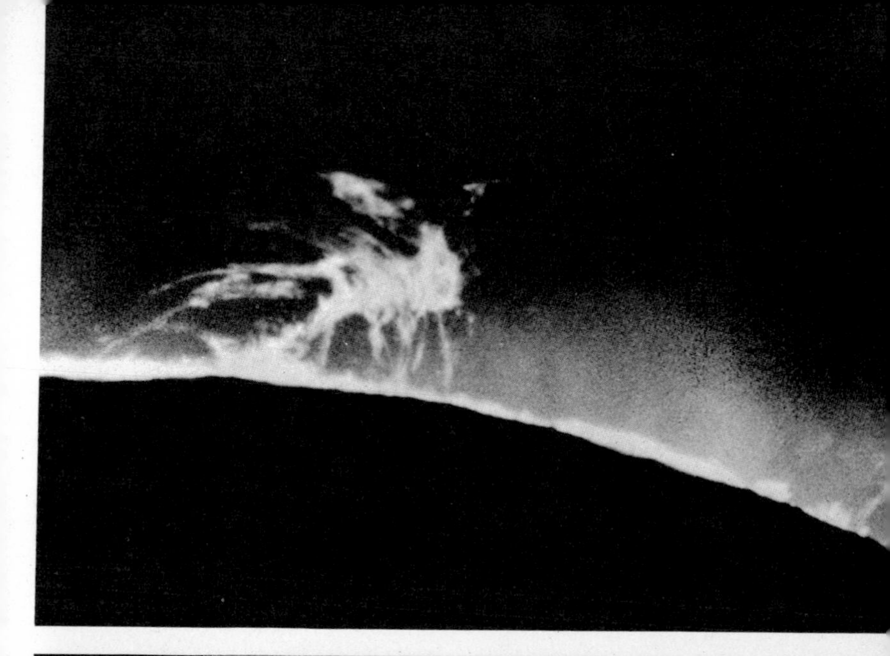

9a. (*above*) Solar prominences.

9b. (*below*) Aurora photographed in Alaska during the IGY.

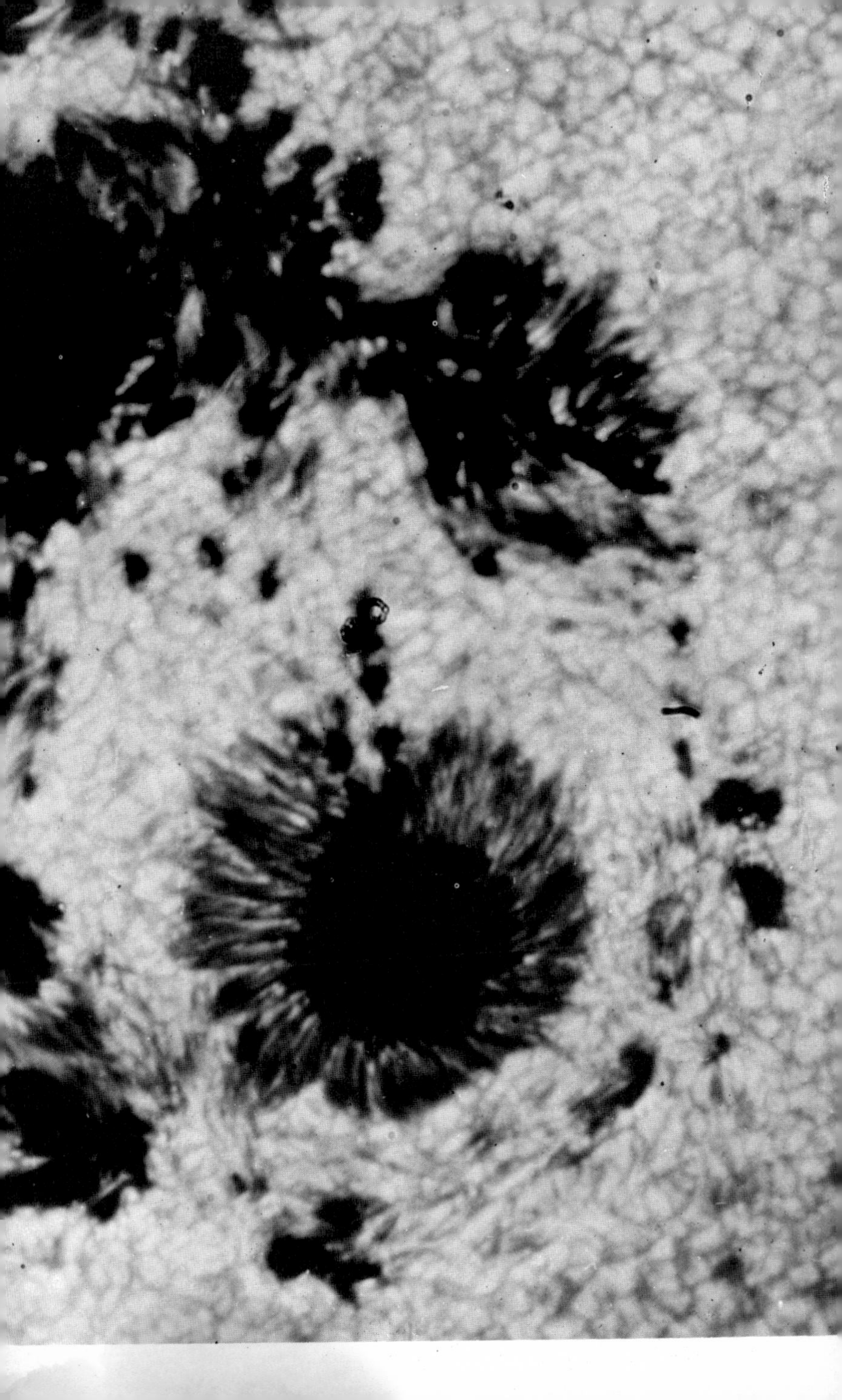

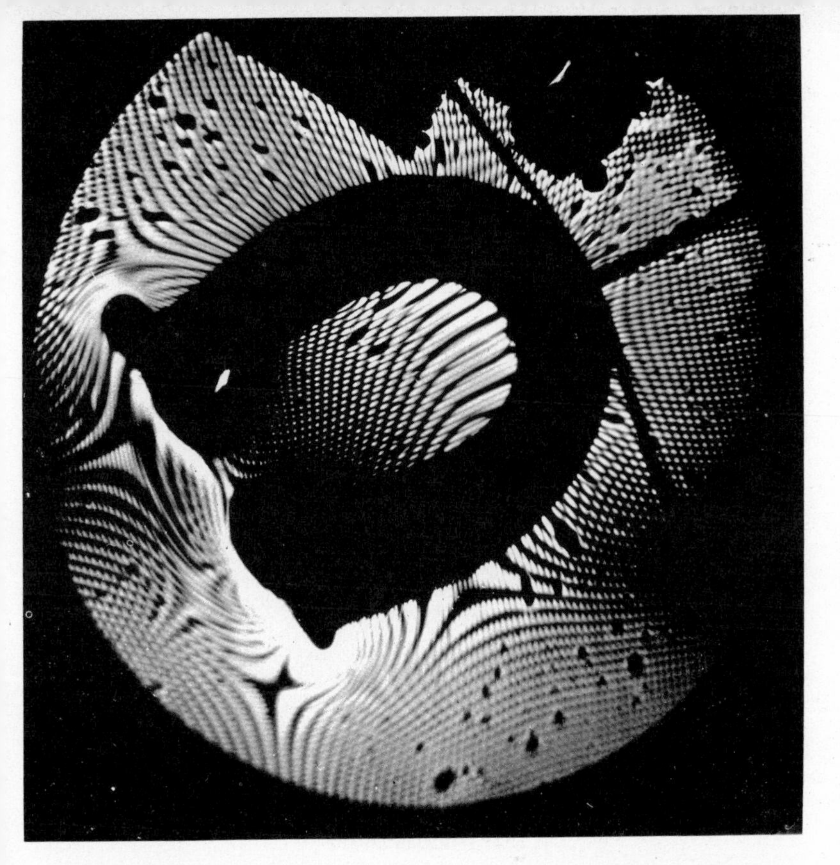

10. (*left*) Sunspots photographed with unprecedented sharpness by the U.S. Stratoscope project from a balloon at 80,000 feet. The spots consist of a dark core of relatively cool gases embedded in a strong magnetic field. This group of particularly active spots produced a brilliant aurora and a vigorous magnetic storm on the earth.

11. (*above*) Magnetic field, photographed with an electron microscope by means of a new shadow technique developed by the U.S. National Bureau of Standards. The small horseshoe magnet used here is only about a quarter of an inch wide.

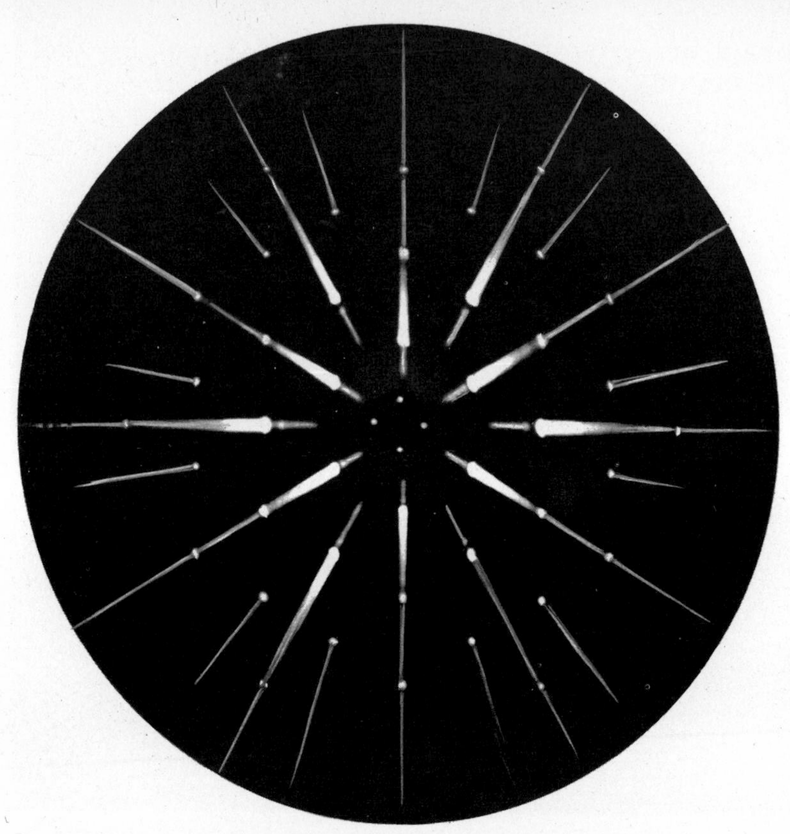

12. A single ice crystal photographed by X-ray diffraction, showing the symmetry and balance of the physical forces holding the structure together.

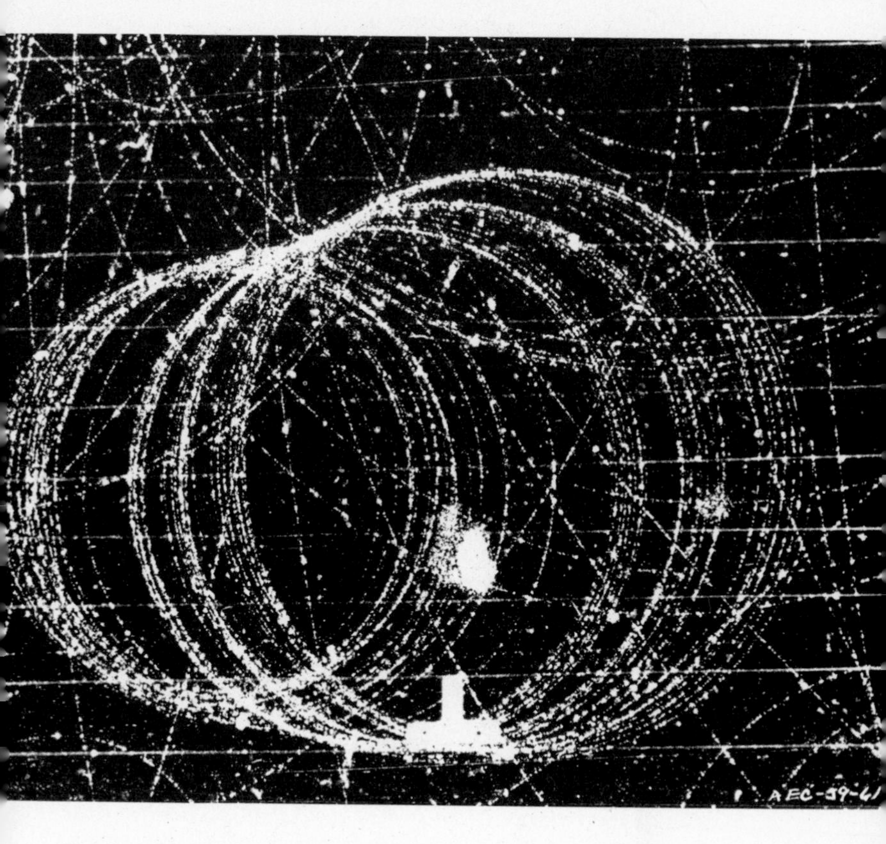

13. Tracks of electrons and positrons formed in a bubble chamber by high-energy gamma rays. The circular pattern was made by an electron revolving in the magnetic field.

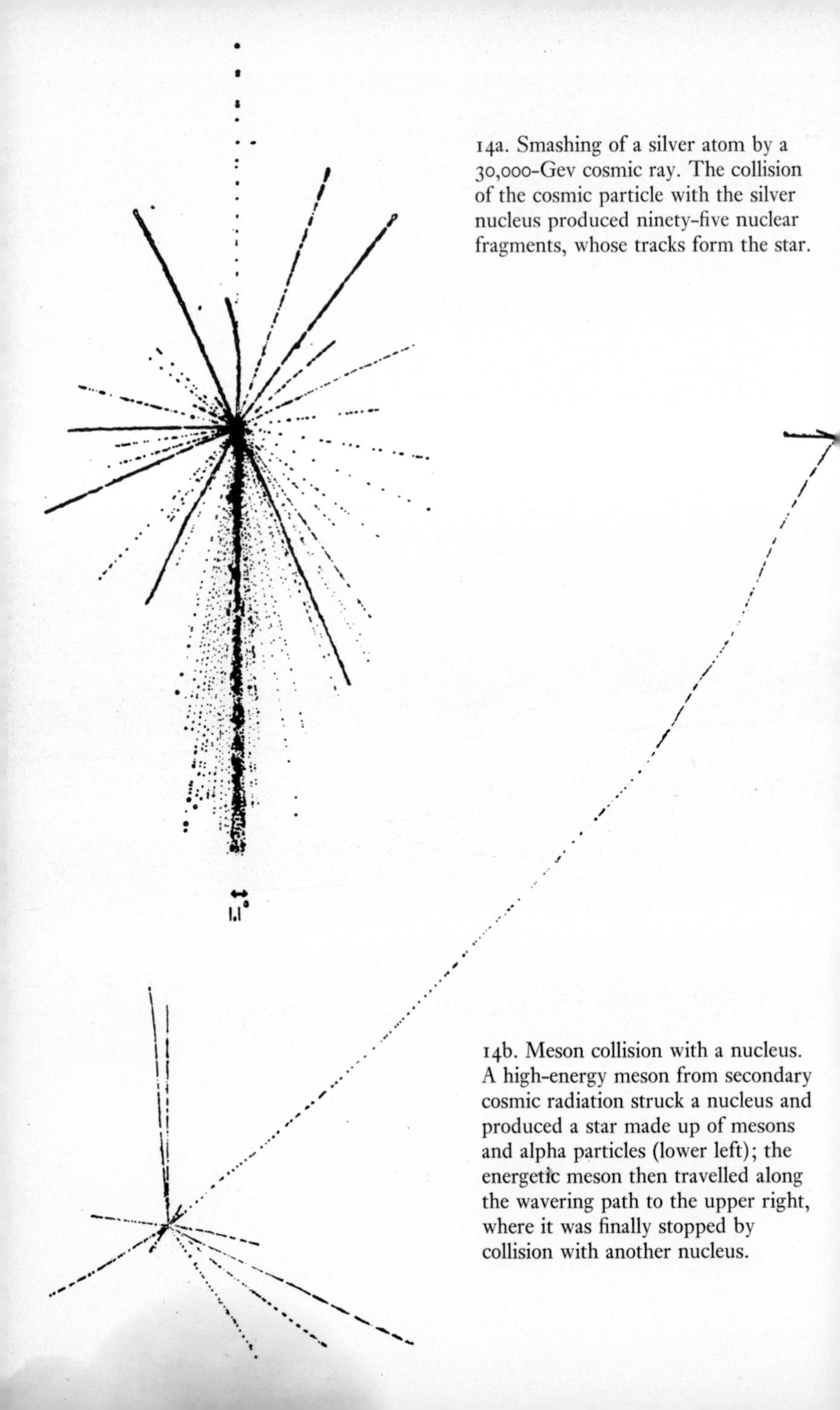

14a. Smashing of a silver atom by a 30,000-Gev cosmic ray. The collision of the cosmic particle with the silver nucleus produced ninety-five nuclear fragments, whose tracks form the star.

14b. Meson collision with a nucleus. A high-energy meson from secondary cosmic radiation struck a nucleus and produced a star made up of mesons and alpha particles (lower left); the energetic meson then travelled along the wavering path to the upper right, where it was finally stopped by collision with another nucleus.

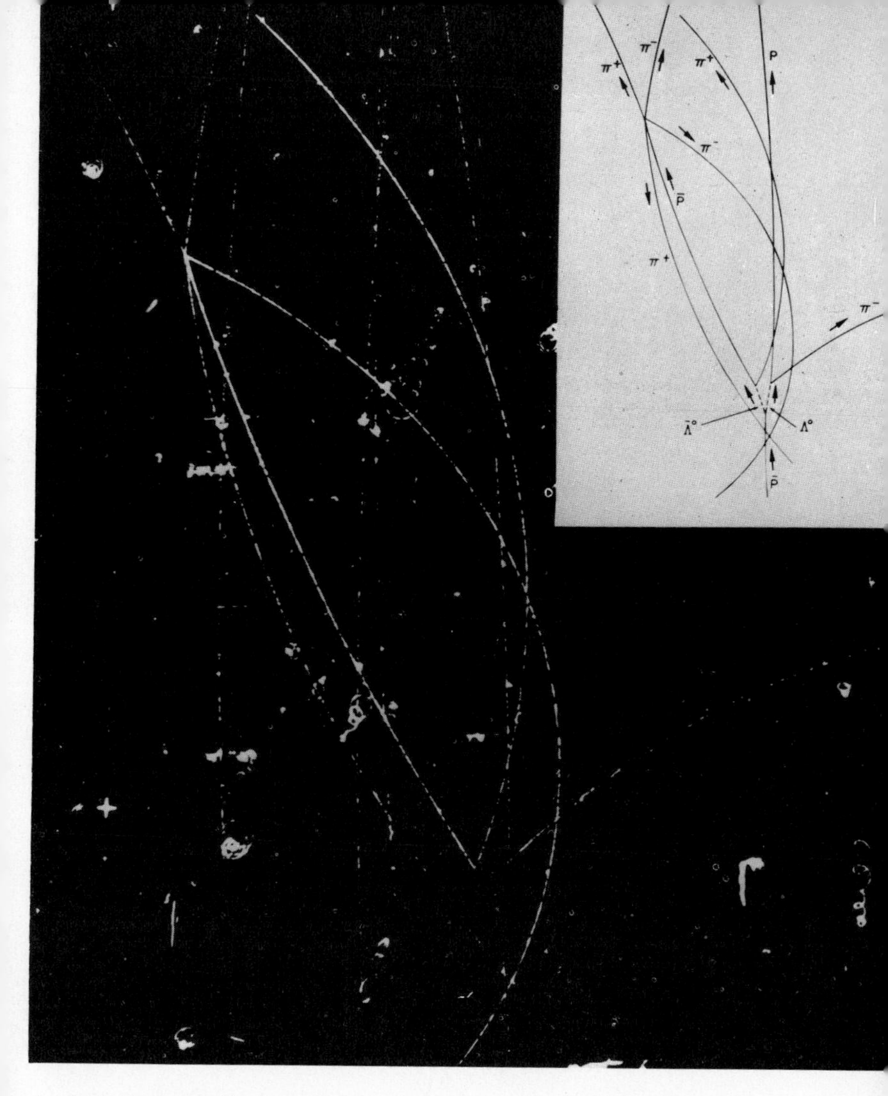

15. Anti-protons and lambda particles produced this picture in the large bubble chamber at the University of California's Radiation Laboratory. As the diagram shows, an anti-proton (p) from the bevatron entered at the bottom; where its track ends there is a gap (dashed lines) that represents the travel of the undetectable neutral lambda and anti-lambda particles. The anti-lambda then decayed into a positive pi meson and an anti-proton, which went on to produce four pi mesons (upper left); the lambda decayed into a proton and a negative pi meson (right side of the fork).

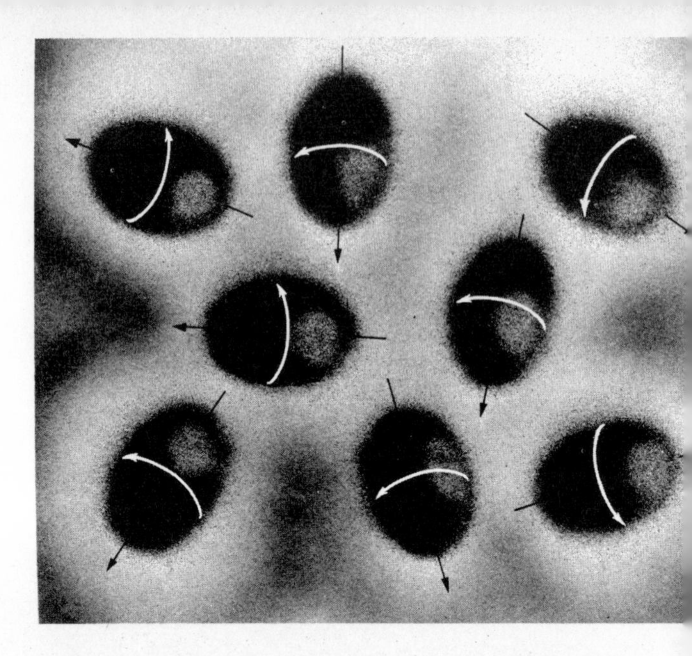

16a. (*above*) Spinning protons in this schematic drawing are oriented in random directions. The white arrow shows the direction of the spin.

16b. (*below*) Protons lined up by a steady magnetic field. Those oriented in the opposite-to-normal direction (arrows pointing downwards) are in the excited spin state.

taneously, at least 200 times as rapidly as copper, the next best heat conductor. It flowed even more easily than a gas, having a viscosity only 1/1,000 that of gaseous hydrogen, and it would leak through apertures so tiny that they stopped a gas. Furthermore, the super-fluid liquid would form a film on glass and flow along it as quickly as it would pour through a hole. If an open container of the liquid was placed in a larger container filled to a lower level, the fluid would creep up the side of the glass and over the rim into the outer container, until the levels in both were equalized.

Helium is the only substance that exhibits this phenomenon of super-fluidity. In fact, the super-fluid behaves so differently from the way helium itself does above 2·2°K that it has been given a separate name, helium II, to distinguish it from liquid helium above that temperature, called helium I.

Only helium permits investigation of temperatures close to absolute zero, and, consequently, it has become a very important element in both pure and applied science. The atmospheric supply is negligible, and the most important sources are natural gas wells into which helium, formed from uranium and thorium breakdown in the earth's crust, sometimes seeps. The gas produced by the richest known well (in New Mexico) is 7·5 per cent helium.

Spurred by the odd phenomena discovered in the neighbourhood of absolute zero, physicists have naturally made every effort to get down as close to absolute zero as possible and expand their knowledge of what is now known as 'cryogenics'. The evaporation of liquid helium can, under special conditions, produce temperatures as low as 0·5°K. (Temperatures at such a level, by the way, are measured by special methods involving electricity – e.g., by the size of the current generated in a thermocouple, by the resistance of a wire made of some non-superconductive metal, by changes in magnetic properties, or even by the speed of sound in helium. The measurement of extremely low temperatures is scarcely easier than their attainment.) Temperatures substantially lower than 0·5°K have been reached by a technique first suggested in 1925 by the Dutch physicist Peter Joseph Wilhelm Debye. A 'paramagnetic' substance (i.e., a substance that concentrates lines of magnetic force) is placed almost in contact with liquid helium, separated from it by helium gas, and the temperature of the whole system is reduced to about 1°K. The system is then placed within a magnetic field. The molecules of the paramagnetic substance line up parallel to the field's lines of force and in doing so give off heat. This heat is removed by further slight evaporation of the surrounding helium. Now the magnetic field is removed. The paramagnetic molecules immediately fall into a random orientation. In going from an ordered to a random orientation, the molecules must absorb heat, and the only thing they can absorb it from is the liquid helium. The temperature of the liquid helium therefore drops.

This can be repeated and repeated, each time lowering the temperature of the liquid helium, the technique being perfected by the American chemist William Francis Giauque, who received the Nobel Prize for chemistry in 1949 in consequence. In this way, a temperature of 0·00002°K was reached in 1957.

In 1962, the German-British physicist Heinz London and his co-workers suggested the possibility of using a new device to attain still lower temperatures. Helium occurs in two varieties, helium 4 and helium 3. Ordinarily they mix perfectly, but at temperatures below about 0·8°K, they separate, with the helium 3 in a top layer. Some of the helium 3 is in the bottom layer with the helium 4, and it is possible to cause helium 3 to shift back and forth across the boundary, lowering the temperature each time in a fashion analogous to the shift between liquid and vapour in the case of an ordinary refrigerant such as Freon. Cooling devices making use of this principle were first constructed in the Soviet Union in 1965.

The Russian physicist Isaak Yakovievich Pomeranchuk suggested in 1950 a method of deep cooling using other properties of helium 3, while as long ago as 1934, the Hungarian-British physicist Nicholas Kurti suggested the use of magnetic properties similar to those taken advantage of by Giauque, but involving the atomic nucleus – the innermost structure of the atom – rather than entire atoms and molecules.

As a result of the use of these new techniques, temperatures as low as 0·000001°K have been attained. And as long as physicists find themselves within a millionth of a degree of absolute zero, might they not just get rid of what little entropy is left and finally reach the mark itself?

No! Absolute zero is unattainable, something Nernst demonstrated in his Nobel-Prize-winning treatment of the subject (sometimes referred to as 'the Third Law of Thermodynamics'). In any lowering of temperature, only part of the entropy can be removed. In general, removing half of the entropy of a system is equally difficult regardless of what the total is. Thus it is just as hard to go from 300°K (about room temperature) to 150°K (colder than any temperature Antarctica attains) as to go from 20° to 10°K. It is then just as hard to go from 10° to 5°K and from 5° to 2·5°K and so on. Having attained a millionth of a degree above absolute zero, the task of going from that to half-a-millionth of a degree is as hard as going from 300° to 150°K, and if that is attained it is an equally difficult task to go from half-a-millionth to a quarter-of-a-millionth, and so on forever. Absolute zero lies at an infinite distance no matter how closely it seems to be approached.

One of the new scientific horizons opened up by the work on liquefaction of gases was the development of an interest in producing high pressures. It seemed that putting various kinds of matter (not only gases) under great pressure might bring out some fundamental information about the nature of matter and also about the interior of the earth. At a depth of 7 miles, for instance, the pressure is 1,000 atmospheres; at 400 miles, 200,000 atmospheres; at 2,000 miles, 1,400,000 atmospheres; and at the centre of the earth, 4,000 miles down, it reaches 3,000,000 atmospheres. (Of course, earth is a rather small planet. The central pressures within Saturn are estimated to be over 50 million atmospheres; within the even larger Jupiter, 100 million.)

The best that nineteenth-century laboratories could do was about 3,000 atmospheres, attained by E. H. Amagat in the 1880s. But, in 1905, the American physicist Percy Williams Bridgman began to devise new methods that soon reached pressures of 20,000 atmospheres and burst the tiny metal chambers he used for his experiments. He went to stronger materials and eventually succeeded in producing pressures of over a million atmospheres. For his work on high pressure he received the Nobel Prize in physics in 1946.

Under his ultra-high pressures, Bridgman was able to force the atoms and molecules of a substance into more compact arrangements, which were sometimes retained after the pressure was released. For instance, he converted ordinary yellow phosphorus, a non-conductor of electricity, into a black, conducting form of phosphorus. He brought about startling changes even in water. Ordinary ice is less dense than liquid water. Using high pressure, Bridgman produced a series of ices ('ice-II', 'ice-III', etc.) that were not only denser than the liquid, but were ice only at temperatures well above the normal freezing point of water. Ice-VII is a solid at temperatures higher than the boiling point of water.

The word 'diamond' brings up the most glamorous of all the high-pressure feats. Diamond, of course, is crystallized carbon, as is also graphite. (When an element appears in two different forms, these forms are 'allotropes'. Diamond and graphite are the most dramatic examples of the phenomenon. Ozone and ordinary oxygen are another example.) The chemical nature of diamond was first proved in 1772 by Lavoisier and some French fellow chemists. They pooled their funds to buy a diamond and proceeded to heat it to a temperature high enough to burn it up. The gas that resulted was found to be carbon dioxide. Later the British chemist Smithson Tennant showed that the amount of carbon dioxide measured could be produced only if diamond was pure carbon, and, in 1799, the French chemist Guyton de Morveau clinched the case by converting a diamond into a lump of graphite.

That was an unprofitable manoeuvre, but now why could not matters be reversed? Diamond is 55 per cent denser than graphite. Why not put graphite under pressure and force the atoms composing it into the tight packing characteristic of diamonds?

Many efforts were made and, like the alchemists, a number of experimenters reported successes. The most famous was the claim of the French chemist Ferdinand Frédéric Henri Moissan. In 1893, he dissolved graphite in molten cast iron and reported that he found small diamonds in the mass after it cooled. Most of the

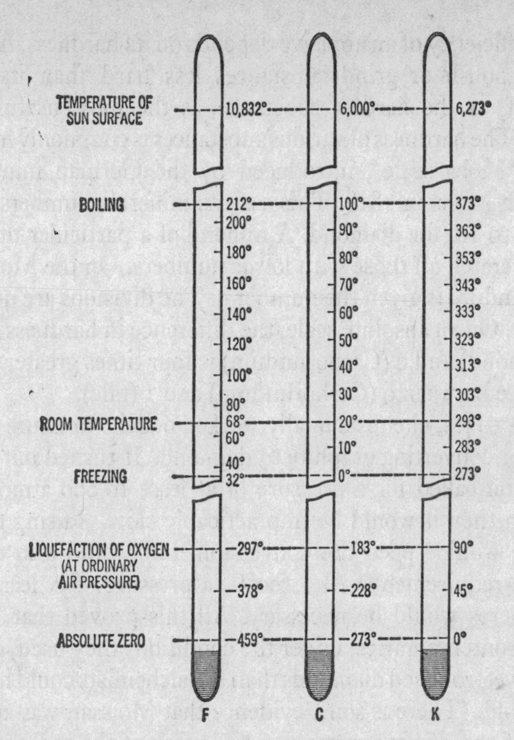

A comparison of the Fahrenheit, Centigrade, and Kelvin thermometric scales.

objects were black, impure, and tiny, but one was colourless and almost a millimetre long. These results were widely accepted, and, for a long time, Moissan was considered to have manufactured synthetic diamonds. However, his results were never successfully repeated.

The search for synthetic diamonds was not without its side victories, however. In 1891, the American inventor Edward Goodrich Acheson stumbled upon silicon carbide, to which he gave the trade name Carborundum. This proved harder than any substance then known but diamond, and it has been a much-used abrasive, that is, a substance used for grinding and polishing, ever since.

The efficiency of an abrasive depends on its hardness. An abrasive can polish or grind substances less hard than itself, and diamond, as the hardest substance, is the most useful in this respect. The hardness of various substances is commonly measured on the 'Mohs scale', introduced by the German mineralogist Friedrich Mohs in 1818. This assigns minerals numbers from 1, for talc, to 10, for diamond. A mineral of a particular number is able to scratch all those with lower numbers. On the Mohs scale, Carborundum is given the number 9. The divisions are not equal, however. On an absolute scale, the difference in hardness between 10 (diamond) and 9 (Carborundum) is four times greater than the difference between 9 (Carborundum) and 1 (talc).

In the 1930s, chemists finally worked out the pressure requirements for converting graphite to diamond. It turned out that the conversion called for a pressure of at least 10,000 atmospheres, and even then it would be impracticably slow. Raising the temperature would speed the conversion, but would also raise the pressure requirements. At 1,500°C, a pressure of at least 30,000 atmospheres would be necessary. All this proved that Moissan and his contemporaries, under the conditions they used, could no more have produced diamonds than the alchemists could have produced gold. (There is some evidence that Moissan was actually a victim of one of his assistants, who, tiring of the tedious experiments, decided to end them by planting a real diamond in the cast-iron mixture.)

Aided by Bridgman's pioneering work in attaining the necessary high temperatures and pressures, scientists at the General Electric Company finally accomplished the feat in 1955. Pressures of 100,000 atmospheres or more were produced, along with temperatures of up to 2,500°C. In addition, a small quantity of metal, such as chromium, was used to form a liquid film across the graphite. It was on this film that the graphite turned to diamond. In 1962, a pressure of 200,000 atmospheres and a temperature of 5,000°C could be attained. Graphite was then turned to diamond directly, without the use of a catalyst.

The synthetic diamonds are too small and impure to be used as gems, but they are now produced commercially as abrasives and cutting tools and, indeed, are a major source of such products. By the end of the decade, an occasional small diamond of gem quality could be produced.

A newer product made by the same sort of treatment can supplement the use of diamond. A compound of boron and nitrogen (boron nitride) is very similar in properties to graphite (except that boron nitride is white instead of black). Subjected to the high temperatures and pressures that convert graphite to diamond, the boron nitride undergoes a similar conversion. From a crystal arrangement like that of graphite, the atoms of boron nitride are converted to one like that of diamond. In its new form it is called 'borazon'. Borazon is about four times as hard as Carborundum. In addition it has the great advantage of being more resistant to heat. At a temperature of 900°C diamond burns up but borazon comes through unchanged. Over twenty new materials, in addition to diamond and borazon, had been formed by pressure work in the 1960s.

Metals

Most of the elements in the periodic table are metals. As a matter of fact, only about twenty of the 102 elements can be considered definitely non-metallic. Yet the use of metals came relatively late in the history of the human species. One reason is that, with rare exceptions, the metallic elements are combined in nature with other elements and are not easy to recognize or extract. Primitive man at first used only materials that could be manipulated by simple treatments such as carving, chipping, hacking, and grinding. This restricted his materials to bones, stones and wood.

His introduction to metals may have come in the form of discoveries of meteorites, or of small nuggets of gold, or metallic copper in the ashes of fires built on rocks containing a copper ore. In any case, people who were curious enough (and lucky enough) to find these strange new substances and look into ways of handling them would discover many advantages in them. Metal differed from rock in that it had an attractive lustre when polished. It could be beaten into sheets and drawn into wire. It could be melted and poured into a mould to solidify. It was much more beautiful and adaptable than rock and ideal for ornaments. Metals probably were fashioned into ornaments long before they were put to any other use.

Because they were rare, attractive, and did not alter with time, these metals were valued and bartered until they became a recognized medium of exchange. Originally, pieces of metal (gold, silver, or copper) had to be weighed separately in trading transactions, but, by 700 B.C., standardized weights of metal stamped in some official government fashion were issued in the Asia Minor kingdom of Lydia and the Aegean island of Aegina. Coins are still with us today.

What really brought metals into their own was the discovery that some of them would take a sharper cutting edge than stone could, and they would maintain that edge under conditions that would ruin a stone axe. Moreover, metal was tough. A blow that would splinter a wooden club or shatter a stone axe would only slightly deform a metal object of similar size. These advantages more than compensated for the fact that metal was heavier than stone and harder to obtain.

The first metal obtained in reasonable quantity was copper, which was in use by 4000 B.C. Copper itself is too soft to make useful weapons or armour (though it will make pretty ornaments), but it was often found alloyed with a little arsenic or antimony, which resulted in a substance that was harder than the pure metal. Then samples of copper ore must have been found that contained tin. The copper-tin alloy (bronze) was hard enough for purposes of weaponry. Men soon learned to add the tin deliberately. The Bronze Age replaced the Stone Age in Egypt and western Asia about 3000 B.C. and in south-eastern Europe by 2000 B.C. Homer's *Iliad* and *Odyssey* commemorate that period of culture.

Iron was known as early as bronze, but for a long time meteorites were the only source. It remained no more than a precious metal, limited to occasional use, until methods were discovered for smelting iron ore and thus obtaining iron in unlimited quantities. The difficulty lay in working with fires hot enough and suitable methods to add carbon to the iron and harden it into the form we now call 'steel'. Iron smelting began somewhere in Asia Minor about 1400 B.C. and developed and spread slowly.

An iron-weaponed army could rout a bronze-armed one, for iron swords would cut through bronze. The Hittites of Asia Minor were the first to use iron weapons to any extent, and they had a period of power in western Asia. Then the Assyrians succeeded the Hittites. By 800 B.C., they had a completely ironized army which was to dominate western Asia and Egypt for two and a half centuries. At about the same time, the Dorians brought the Iron Age to Europe by invading Greece and defeating the Achaeans, who committed the error of clinging to the Bronze Age.

Iron is obtained essentially by heating iron ore (usually a ferric oxide) with carbon. The carbon atoms carry off the oxygen of the ferric oxide, leaving a lump of pure iron behind. In ancient times, the temperatures used did not melt the iron, and the product was a tough metal that could be worked into the desired shape by hammering – that is, 'wrought iron'. Iron metallurgy on a larger scale came into being in the Middle Ages. Special furnaces were used, and higher temperatures that melted the iron. The molten iron could be poured into moulds to form castings, so it was called 'cast iron'. This was much less expensive than wrought iron and much harder, too, but it was brittle and could not be hammered. Increasing demand for iron of either form helped to de-forest England, for instance, which consumed its wood in the iron-smelting furnaces. But then, in 1780, the English iron-worker Abraham Darby showed that coke (carbonized coal) would work as well as, or better than, charcoal (carbonized wood). The pressures on the forests eased in this direction, and the more-than-century-long domination of coal as an energy source began.

It was not until late in the eighteenth century that chemists, thanks to the French physicist René Antoine Ferchault de Réaumur, finally realized that it was the carbon content that dictated the toughness and hardness of iron. To maximize those properties the carbon content ought to be between 0·2 and 1·5 per cent; the steel that then results is harder and tougher and generally stronger than either cast iron or wrought iron. But until the mid nineteenth century, high-quality steel could be made only by the complicated procedure of carefully adding the appropriate quantity of carbon to wrought iron (itself comparatively expensive). Steel remained therefore a luxury metal, used only where no substitute could be found – as in swords and springs.

The Age of Steel was ushered in by a British engineer named Henry Bessemer. Originally interested primarily in cannon and projectiles, Bessemer invented a system of rifling intended to enable cannon to shoot farther and more accurately. Napoleon III of France was interested and offered to finance further experiments. But a French artillerist killed the idea by pointing out that the propulsive explosion Bessemer had in mind would shatter the cast-iron cannons used in those days. Bessemer, chagrined, turned to the problem of creating stronger iron. He knew nothing of metallurgy, so he could approach the problem with a fresh mind. Cast iron was brittle because of its carbon content. Therefore the problem was to reduce the carbon.

Why not burn the carbon away by melting the iron and sending a blast of air through it? This seemed at first thought a ridiculous idea. Would not the air blast cool the molten metal and cause it to solidify? Bessemer tried it anyway, and he found quite the reverse was true. As the air burned the carbon, the combustion gave off heat and the temperature of the iron rose rather than fell. The carbon burned off nicely. By proper controls, steel could be produced in quantity and comparatively cheaply.

In 1856, Bessemer announced his 'blast furnace'. Ironmakers adopted the method with enthusiasm, then dropped it in anger when they found that inferior steel was being formed. Bessemer discovered that the iron ore used by the industry contained phosphorus (which had been absent from his own ore samples). Although Bessemer explained to the ironmakers that phosphorus had betrayed them, they refused to be twice bitten. Bessemer therefore had to borrow money to set up his own steel works in Sheffield. Importing phosphorus-free iron ore from Sweden, he speedily produced steel at a price that undersold the other ironmakers.

In 1875, the British metallurgist Sidney Gilchrist Thomas discovered that by lining the interior of the furnace with limestone and magnesia, he could easily remove the phosphorus from the molten iron. After this, almost any iron ore could be used in the manufacture of steel. Meanwhile, the German-British inventor Karl Wilhelm Siemens developed the 'open-hearth method' in 1868, in which pig iron was heated with iron ore; this process also could take care of the phosphorus content.

The Age of Steel then got under way. The name is no mere phrase. Without steel, skyscrapers, suspension bridges, great ships, railways, and many other modern constructions would be almost unthinkable, and, despite the rise of other metals, steel still remains the preferred metal in a host of everyday uses, from car bodies to knives.

(It is a mistake, of course, to think that any single advance can bring about a major change in the way of life of humanity. This is always the result of a whole complex of interrelated advances. For instance, all the steel in the world could not make skyscrapers practical without the existence of that too-often-taken-for-granted device, the lift. In 1861, the American inventor Elisha Graves Otis patented a hydraulic lift and in 1889, the company he founded installed the first electrically run lifts in a New York commercial building.)

With steel cheap and commonplace, it became possible to experiment with the addition of other metals ('alloy steel') to see if steel could be still further improved. The British metallurgist Robert Abbott Hadfield pioneered in this direction. In 1882, he found that adding manganese to steel to the extent of 13 per cent produced a harder alloy, which could be used in machinery for particularly brutal jobs, such as rock-crushing. In 1900, a steel alloy containing tungsten and chromium was found to retain its hardness well at high temperatures, even red heat, and this alloy proved a boon for high-speed tools. Today there are innumerable other alloy steels for particular jobs, employing such metals as molybdenum, nickel, cobalt, and vanadium.

The great difficulty with steel is its vulnerability to corrosion – a process that returns iron to the crude state of the ore whence it came. One way of combating this is to shield the metal by painting it or by plating it with a metal less likely to corrode, such as nickel, chromium, cadmium, or tin. A more effective method is to form an alloy that does not corrode. In 1913, the British metallurgist Harry Brearley discovered such an alloy by accident. He was looking for steel alloys that would be particularly suitable for gun barrels. Among the samples he discarded as unsuitable was a nickel-chromium alloy. Months later, he happened to notice that these particular pieces in his scrap heap were as bright as ever, although the rest were rusted. That was the birth of 'stainless steel'. It is too soft and too expensive for use in large-scale construction, but it serves admirably in cutlery and small appliances where non-rusting is more important than hardness.

Since something like a thousand million dollars a year is spent over the world in the not too successful effort to keep iron and steel from corroding, the search for a general rust inhibitor goes on unabated. One interesting recent discovery is that pertechnetates, compounds containing technetium, protect iron against rusting. Of course, this rare, man-made element may never be common enough to be used on any substantial scale, but it offers an invaluable research tool. Its radioactivity allows chemists to follow its fate and to observe what happens to it on the iron surface. If this use of technetium leads to a new understanding which will help solve the corrosion problem, that achievement alone will pay back in a matter of months all the money invested in research on the synthetic elements over the last quarter century.

One of iron's most useful properties is its strong ferromagnetism. Iron itself is an example of a 'soft magnet'. It is easily magnetized under the influence of an electric or magnetic field; that is, its magnetic domains (see p. 190) are easily lined up. It is also easily demagnetized when the field is removed, and the domains fall into random orientation again. This ready loss of magnetism can be useful, as in electromagnets, where the iron core is magnetized easily with the current on, but *should* be as easily demagnetized when the current goes off.

Since the Second World War, a new class of soft magnets has been developed. These are the 'ferrites', an example being nickel ferrite ($NiFe_2O_4$) and manganese ferrite ($MnFe_2O_4$), which are used in computers as elements which must gain or lose magnetism with the utmost ease and rapidity.

'Hard magnets', with domains which are difficult to orient or, once oriented, to disorient, will, once magnetized, retain the property over long periods. Various steel alloys are the commonest examples, though particularly strong, hard magnets have been found among alloys that contain little or no iron. The best-known example of this is 'alnico', discovered in 1931, one variety of which is made of aluminium, nickel, and cobalt (the name of the alloy being derived from the first two letters of each of the substances), plus a bit of copper.

In the 1950s, techniques were developed to use powdered iron as a magnet, the particles being so small as to consist of individual domains. These could be oriented in molten plastic, which would then be allowed to solidify, holding the domains fixed in their orientation. Such 'plastic magnets' are very easy to shape and mould, but can be made adequately strong as well.

We have seen in recent decades the emergence of enormously useful new metals – metals that were almost useless and even unknown up to a century or so ago and in some cases up to our own generation. The most striking example is aluminium. Aluminium is the most common of all metals – 60 per cent more common than iron. But it is also exceedingly difficult to extract from its ores. In 1825, Hans Christian Oersted (who had discovered the connection between electricity and magnetism) separated a little aluminium in impure form. Thereafter, many chemists tried unsuccessfully to purify the metal, until the French chemist Henri Étienne Sainte-Clair Deville in 1854 finally devised a method of obtaining aluminium in reasonable quantities. Aluminium is so active chemically that he had to use metallic sodium (even more active) to break aluminium's grip on its neighbouring atoms. For a while aluminium sold for a hundred dollars a pound, making it practically a precious metal. Napoleon III indulged himself in aluminium cutlery and had an aluminium rattle fashioned for his infant son; and in the United States, as a mark of the nation's great esteem for

George Washington, the Washington Monument was capped with a slab of solid aluminium.

In 1886, Charles Martin Hall, a young student of chemistry at Oberlin College, was so impressed by his professor's statement that anyone who could discover a cheap method of making aluminium would make a fortune that he decided to try his hand at it. In a home laboratory in his woodshed, Hall set out to apply Humphry Davy's early discovery that an electric current sent through a molten metal could separate the metal ions by depositing them on the cathode plate. Looking for a material that could dissolve aluminium, he stumbled across cryolite, a mineral found in reasonable quantity only in Greenland. (Nowadays synthetic cryolite is available.) Hall dissolved aluminium oxide in cryolite, melted the mixture, and passed an electric current through it. Sure enough, pure aluminium collected on the cathode. Hall rushed to his professor with his first few ingots of the metal. (To this day they are treasured by the Aluminum Company of America.)

As it happened, a young French chemist named Paul Louis Toussaint Héroult, who was just Hall's age (twenty-two), discovered the same process in the same year. (To complete the coincidence, Hall and Héroult both died in 1914.)

The Hall–Héroult process made aluminium an inexpensive metal, though it was never to be as cheap as steel, because useful aluminium ore is less common than useful iron ore, and electricity (the key to aluminium) is more expensive than coal (the key to steel). Nevertheless, aluminium has two great advantages over steel. First, it is light – only one third the weight of steel. Second, in aluminium's case corrosion merely takes the form of a thin, transparent film over its surface, which protects deeper layers from corrosion without affecting the metal's appearance.

Pure aluminium is rather soft, but alloying can take care of that. In 1906, the German metallurgist Alfred Wilm made a tough alloy by adding a bit of copper and a smaller bit of magnesium to the aluminium. He sold his patent rights to the Durener Metal Works in Germany, and they gave the alloy the name Duralumin.

Engineers quickly realized how valuable a light but strong metal could be in aircraft. After the Germans introduced Duralumin in zeppelins during the First World War and the British learned its composition by analysing the alloy in a crashed zeppelin, use of this new metal spread over the world. Because Duralumin was not quite as corrosion-resistant as aluminium itself, metallurgists covered it with thin sheets of pure aluminium, forming the product called Alclad.

Today there are aluminium alloys which, weight for weight, are stronger than some steels. Aluminium has tended to replace steel wherever lightness and corrosion resistance are more important than brute strength. It has become, as everyone knows, almost a universal metal, used in aeroplanes, rockets, railway trains, cars, doors, screens, paint, kitchen utensils, foil wrapping and what not.

And now we have magnesium, a metal even lighter than aluminium. Its main use is in aeroplanes, as you might expect; as early as 1910, Germany was making use of magnesium-zinc alloys for that purpose. After the First World War, magnesium-aluminium alloys came into increasing use.

Only about one fourth as abundant as aluminium and more active chemically, magnesium is harder to obtain from ores. But fortunately there is a rich source in the ocean. Magnesium, unlike aluminium or iron, is present in sea water in quantity. The ocean carries dissolved matter to the amount of 3·5 per cent of its mass. Of this dissolved material, 3·7 per cent is magnesium ion. The ocean as a whole, therefore, contains about 2,000 billion (2,000,000,000,000,000) tons of magnesium, or all we could use for the indefinite future.

The problem was to get it out. The method chosen was to pump sea water into large tanks and add calcium oxide (also obtained from the sea, i.e., from oyster shells). The calcium oxide reacts with the water and the magnesium ion to form magnesium hydroxide, which is insoluble and therefore precipitates out of solution. The magnesium hydroxide is converted to magnesium chloride by treatment with hydrochloric acid, and the magnesium metal is then separated from the chlorine by means of an electric current.

In January of 1941, the Dow Chemical Company produced the first ingots of magnesium from sea water, and the stage was laid for a tenfold increase in magnesium production during the war years.

As a matter of fact, any element that can be extracted profitably from sea water may be considered in virtually limitless supply since, after use, it eventually returns to the sea. It has been estimated that if 100 million tons of magnesium were extracted from sea water each year for a million years, the magnesium content of the ocean would drop from its present figure of 0·13 to 0·12 per cent.

If steel was the 'wonder metal' of the mid nineteenth century, aluminium of the early twentieth century, and magnesium of the mid twentieth century, what will the next new wonder metal be? The possibilities are limited. There are only seven really common metals in the earth's crust. Besides iron, aluminium, and magnesium, they are sodium, potassium, calcium, and titanium. Sodium, potassium, and calcium are far too active chemically to be used as construction metals. (For instance, they react violently with water.) That leaves titanium, which is about one eighth as abundant as iron.

Titanium has an extraordinary combination of good qualities. It is only a little more than half as heavy as steel, stronger, weight for weight, than aluminium or steel, resistant to corrosion, and able to withstand high temperatures. For all these reasons, titanium is now being used in aircraft, ships, and guided missiles wherever these properties can be put to good use.

Why was mankind so slow to discover the value of titanium? The reason is much the same as for aluminium and magnesium. It reacts too readily with other substances, and in its impure forms – combined with oxygen or nitrogen – it is an unprepossessing metal, brittle and seemingly useless. Its strength and other fine qualities emerge only when it is isolated in really pure form (in a vacuum or under an inert gas). The effort of metallurgists has succeeded to the point where a pound of titanium which would have cost $3,000 in 1947, cost $2 in 1969.

The search need not, however, be for new wonder metals. The older metals (and some non-metals, too) can be made far more 'wonderful' than they are now.

In Oliver Wendell Holmes' poem 'The Deacon's Masterpiece', the story is told of a 'one-hoss shay' (one-horse buggy) which was carefully made in such a way as to have no weakest point. In the end, the shay went all at once – decomposing into a powder. But it had lasted a hundred years.

The atomic structure of crystalline solids, both metal and non-metal, is rather like the one-hoss shay situation. A metal's crystals are riddled with sub-microscopic clefts and scratches. Under pressure, a fracture will start at one of these weak points and spread through the crystal. If, like the deacon's wonderful one-hoss shay, a crystal could be built with no weak points, it would have great strength.

Such no-weak-point crystals do form as tiny fibres called 'whiskers' on the surface of crystals. Tensile strengths of carbon whiskers have been found to run as high as 1,400 tons per square inch, which is from fifteen to seventy times the tensile strength of steel. If methods could be designed for manufacturing defect-free metal in quantity, we would find ourselves with materials of astonishing strength. In 1968, for instance, Soviet scientists produced a tiny defect-free crystal of tungsten that would sustain a load of 1,635 tons per square inch, as compared to 213 tons per square inch for the best steel. And even if defect-free substances were not available in bulk, the addition of defect-free fibres to ordinary metals would reinforce and strengthen them.

Then, too, as late as 1968, an interesting new method was found for combining metals. The two methods of historic interest were alloying, where two or more metals are melted together and form a more-or-less-homogeneous mixture, and plating, where one metal is bound firmly to another (a thin layer of expensive metal is usually bound to the surface of a bulky volume of cheaper metal, so that the surface is, for instance, as beautiful and corrosion-resistant as gold but the whole nearly as cheap as copper).

The American metallurgist Newell C. Cook and his associates were attempting to plate a silicon layer on a platinum surface, using molten alkali fluoride as the liquid in which the platinum was immersed. The expected plating did not occur. What happened, apparently, was that the molten fluoride removed the very thin film of bound oxygen ordinarily present on even the most resistant metals and presented the platinum surface 'naked' to the silicon atoms. Instead of binding themselves to the surface on the other side of the oxygen atoms, they worked their way *into* the surface. The result was that a thin outer layer of the platinum became an alloy.

Cook followed this new direction and found that many substances could be combined in this way to form a 'plating' of alloy on pure metal (or on another alloy). Cook called the process 'metalliding' and quickly showed its usefulness. Thus, copper to which 2 to 4 per cent of beryllium is added in the form of an ordinary alloy becomes extraordinarily strong. The same result can be achieved if copper is 'beryl-lined' at the cost of much less of the relatively rare beryllium. Again, steel metallided with boron ('boriding') is hardened. The addition of silicon, cobalt, and titanium, also produces useful properties.

Wonder metals, in other words, if not found in nature can be created by human ingenuity.

6 The Particles

The Nuclear Atom

As I pointed out in the preceding chapter, it was known by 1900 that the atom was not a simple, indivisible particle, but contained at least one sub-atomic particle – the electron, identified by J. J. Thomson. Thomson suggested that electrons were stuck like raisins in the positively charged main body of the atom.

But very shortly it developed that there were also other sub-particles within the atom. When Becquerel discovered radioactivity, he identified some of the radiation emitted by radioactive substances as consisting of electrons, but other emissions were discovered as well. The Curies in France and Ernest Rutherford in England found one that was less penetrating than the electron stream. Rutherford called this radiation 'alpha rays' and gave the electron emission the name 'beta rays'. The flying electrons making up the latter radiation are, individually, 'beta particles'. The alpha rays were also found to be made up of particles and these were called 'alpha particles'. 'Alpha' and 'beta' are, of course, the first two letters of the Greek alphabet.

Meanwhile the French chemist Paul Ulrich Villard discovered a third form of radioactive emission, which was named 'gamma rays' after the third letter of the Greek alphabet. The gamma rays were quickly identified as radiation resembling X-rays, but with shorter wavelengths.

Rutherford learned by experiment that a magnetic field deflected alpha particles much less than it did beta particles. Furthermore, they were deflected in the opposite direction, which meant that the alpha particle had a positive charge, as opposed to the electron's negative one. From the amount of deflection, it could be calculated that the alpha particle must have at least twice the mass of the hydrogen ion, which possessed the smallest known positive charge. The amount of deflection would be affected both by the particle's mass and by its charge. If the alpha particle's positive charge was equal to that of the hydrogen ion, its mass would be two times that of the hydrogen ion; if its charge was double that, it would be four times as massive as the hydrogen ion, and so on.

Rutherford settled the matter in 1909 by isolating alpha particles. He put some radioactive material in a thin-walled glass tube surrounded by a thick-walled glass tube, with a vacuum between. The alpha particles could penetrate the thin inner wall but not the thick outer one. They bounced back from the outer wall, so to speak, and in doing so lost energy and therefore were no longer able to penetrate the thin walls either. Thus they were trapped between. Now Rutherford excited the alpha particles by means of an electric discharge so that they glowed. They then showed the spectral lines of helium. (It has become evident that alpha particles produced by radioactive substances in the soil are the source of the helium in natural gas wells.) If the alpha particle is helium, its mass must be four times that of hydrogen. This, in turn, means that its positive charge amounts to two units, taking the hydrogen ion's charge as the unit.

Rutherford later identified another positive particle in the atom. This one had actually been detected, but not recognized, many years before. In 1886, the German physicist Eugen Goldstein, using a cathode-ray tube with a perforated cathode, had discovered a new radiation that streamed through the holes of the cathode in the direction opposite to the cathode rays themselves. He called it 'Kanalstrahlen' ('channel rays'). In 1902, this radiation served as the first occasion when the Doppler–Fizeau effect (see p. 49) was detected in any earthly source of light. The German physicist Johannes Stark placed a spectroscope in such a fashion that the rays raced towards it and demonstrated the violet shift.

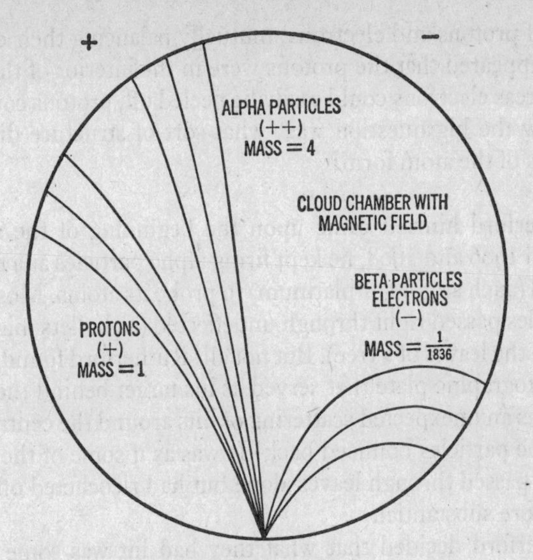

Deflection of particles by a magnetic field.

For this research, he was awarded the Nobel Prize for physics in 1919.

Since channel rays move in a direction opposite to the negatively charged cathode rays, Thomson suggested that this radiation be called 'positive rays'. It turned out that the particles of the 'positive rays' could easily pass through matter. They were therefore judged to be much smaller in volume than ordinary ions or atoms. The amount of their deflection by a magnetic field indicated that the smallest of these particles had the same charge and mass as a hydrogen ion, assuming that this ion carried the smallest possible unit of positive charge. The positive-ray particle was therefore deduced to be the fundamental positive particle – the opposite number of the electron. Rutherford named it the 'proton' (from the Greek word for 'first').

The proton and the electron do indeed carry equal, though opposite, electric charges, although the proton is 1,836 times as massive as the electron. It seemed likely, then, that an atom was composed of protons and electrons, mutually balancing their charges. It also appeared that the protons were in the interior of the atom, for whereas electrons could easily be peeled off, protons could not. But now the big question was: what sort of structure did these particles of the atom form?

Rutherford himself came upon the beginning of the answer. Between 1906 and 1908, he kept firing alpha particles at a thin foil of metal (such as gold or platinum) to probe its atoms. Most of the projectiles passed right through undeflected (as bullets might pass through the leaves of a tree). But not all: Rutherford found that on the photographic plate that served as his target behind the metal, there was an unexpected scattering of hits around the central spot, and some particles bounced back! It was as if some of the bullets had not passed through leaves alone but had ricocheted off something more substantial.

Rutherford decided that what they had hit was some sort of dense core, which occupied only a very small part of the volume of the atom. Most of an atom's volume, it seemed, must be occupied by electrons. As alpha particles charged through the foil of metal, they usually encountered only electrons, and they brushed aside this froth of light particles, so to speak, without being deflected. But once in a while an alpha particle might happen to hit an atom's denser core, and then it was deflected. That this happened only very occasionally showed that the atomic cores must be very small indeed, because a projectile passing through the metal foil must encounter many thousands of atoms.

It was logical to suppose that the hard core was made up of protons. Rutherford pictured the protons of an atom as crowded into a tiny 'atomic nucleus' at the centre. (It has since been demonstrated that this nucleus has a diameter of little more than 1/100,000 that of the whole atom.)

This, then, is the basic model of the atom: a positively charged nucleus taking up very little room, but containing almost all the mass of the atom, surrounded by a froth of electrons taking up nearly all the volume of the atom, but containing practically none of its mass. For his extraordinary pioneering work on the ultimate nature of matter, Rutherford received the Nobel Prize in chemistry in 1908.

It now became possible to describe specific atoms and their behaviour in more definite terms. For instance, the hydrogen atom possesses but a single electron. If this is removed, the proton that remains immediately attaches itself to some neighbouring molecule. But when the bare hydrogen nucleus does not find an electron to share in this fashion, it acts as a proton – that is to say, a subatomic particle – and in that form it can penetrate matter and react with other nuclei if it has enough energy.

Helium, with two electrons, does not give one up so easily. As I mentioned in the preceding chapter, its two electrons form a closed shell, and the atom is therefore inert. If helium is stripped of both electrons, however, it becomes an alpha particle – that is, a subatomic particle carrying two units of positive charge.

The third element, lithium, has three electrons in its atom. Stripped of one or two, it is an ion. If all three of its electrons are removed, it, too, becomes a bare nucleus, carrying a three-unit positive charge.

The number of units of positive charge in the nucleus of an atom has to be exactly equal to the number of electrons it normally contains, for the atom as a whole is ordinarily neutral. And, in fact, the atomic numbers of the elements are based on their units of positive charge rather than of negative charge, because the number of an atom's electrons may easily be made to vary in ion formation, whereas the number of its protons can be altered only with great difficulty.

This scheme of the construction of atoms had hardly been worked out when a new conundrum arose. The number of units of positive charge on a nucleus did not balance at all with the nucleus's weight, or mass, except in the case of the hydrogen atom. The helium nucleus, for instance, had a positive charge of two but was known to have four times the mass of the hydrogen nucleus. And the situation got worse and worse as one went down the table of elements, until, by the time uranium was reached, one had a nucleus with a mass equal to 238 protons but a charge equal to only 92.

How could a nucleus containing four protons (as the helium nucleus was supposed to) have only two units of positive charge? The first, and simplest, guess was that two units of its charge were neutralized by the presence in the nucleus of negatively charged particles of negligible weight. Naturally the electron sprang to mind. The puzzle might be straightened out if one assumed that the helium nucleus consisted of four protons and two neutralizing electrons, leaving a net positive charge of two – and so on all the way to uranium, whose nucleus would have 238 protons and 146 electrons, netting 92 units of positive charge. The whole idea was given encouragement by the fact that radioactive nuclei were actually known to emit electrons – i.e., beta particles.

This view of matter prevailed for more than a decade, until a better answer came in a roundabout way from other investigations. But, in the meantime, some serious objections to the hypothesis arose. For one thing, if the nucleus were built essentially of protons, with the light electrons contributing practically nothing to the mass, how was it that the relative masses of the various nuclei did not come to whole numbers? According to the measured atomic weights, the nucleus of the chlorine atom, for instance, had a mass of 35½ times that of the hydrogen nucleus. Did that mean it contained 35½ protons? No scientist (then or now) could accept the idea of half a proton.

Actually, this particular question had an answer that was discovered even before the main issue was solved. It makes an interesting story in itself.

Isotopes

As early as 1816, an English physician named William Prout had suggested that all atoms were built up from the hydrogen atom. As time went on and the atomic weights were worked out, Prout's theory fell by the wayside, because it developed that many elements had fractional weights (taking oxygen as the standard at 16). Chlorine, as I have mentioned, has an atomic weight of about

35·5 – 35·453, to be more exact. Other examples are antimony, 121·75; barium, 137·34; boron, 10·811; cadmium, 112·40.

Around the turn of the century there came a series of puzzling observations that was to lead to the explanation. The Englishman William Crookes (he of the Crookes tube) separated from uranium a small quantity of a substance that proved much more radioactive than uranium itself. He suggested that uranium was not radioactive at all – only this impurity, which he called 'uranium X'. Henri Becquerel, on the other hand, discovered that the purified, feebly radioactive uranium somehow increased in radioactivity with time. After it was left standing for a while, the active uranium X could be extracted from it, again and again. In other words, uranium was converted by its own radioactivity to the still more active uranium X.

Then Rutherford similarly separated a strongly radioactive 'thorium X' from thorium and found that thorium, too, went on producing more thorium X. It was already known that the most famous radioactive element of all, radium, broke down to the radioactive gas radon. So Rutherford and his assistant, the chemist Frederick Soddy, concluded that radioactive atoms, in the process of emitting their particles, generally transformed themselves into other varieties of radioactive atoms.

Chemists began searching for such transformations and came up with quite an assortment of new substances, giving them such names as radium A, radium B, mesothorium I, mesothorium II, and actinium C. All of them were grouped into three series, depending on their atomic ancestry. One series arose from the breakdown of uranium, another from that of thorium, and a third from that of actinium (later it turned out that actinium itself had a predecessor, named 'protactinium'). Altogether, some forty members of these series were identified, each distinguished by its own peculiar pattern of radiation. But the end product of all three series was the same: each chain of substances eventually broke down to the same stable element – lead.

Now obviously these forty substances could not all be separate elements; between uranium (92) and lead (82) there were only ten places in the periodic table, and all but two of these belonged to known elements. The chemists found, in fact, that though the substances differed in radioactivity, some of them were identical with one another in chemical properties. For instance, as early as 1907 the American chemists Herbert Newby McCoy and W. H. Ross showed that 'radiothorium', one of the disintegration products of thorium, showed precisely the same chemical behaviour as thorium. 'Radium D' behaved chemically exactly like lead; in fact, it was often called 'radiolead'. All this suggested that the substances in question were actually varieties of the same element: radiothorium a form of thorium, radiolead a member of a family of leads, and so on.

In 1913, Soddy gave clear expression of this idea and developed it further. He showed that when an atom emitted an alpha particle it changed into an element two places lower in the list of elements; when it emitted a beta particle it changed into an element one place higher. On this basis, 'radiothorium' would indeed fall in thorium's place in the table, and so would the substances called 'uranium X_1' and 'uranium Y': all three would be varieties of element 90. Likewise, 'radium D', 'radium B', 'thorium B', and 'actinium B' would all share lead's place as varieties of element 82.

To the members of a family of substances sharing the same place in the periodic table Soddy gave the name 'isotope' (from Greek words meaning 'same position'). Soddy received the Nobel Prize in chemistry in 1921.

The proton–electron model of the nucleus fitted in beautifully with Soddy's isotope theory. Removal of an alpha particle from a nucleus would reduce the positive charge of that nucleus by two – exactly what was needed to move it two places down in the periodic table. On the other hand, the ejection of an electron (beta particle) from a nucleus would leave an additional proton unneutralized and thus increase the nucleus's positive charge by one unit. That amounted to raising the atomic number by one, so the element would move to the next higher position in the periodic table.

How is it that when thorium breaks down to 'radiothorium', after going through not one but three disintegrations, the product is still thorium? Well, in the process the thorium atom loses an alpha particle, then a beta particle, then a second beta particle. If we accept the proton building-block idea, this means it has lost four electrons (two supposedly contained in the alpha particle) and four protons. (The actual situation differs from this picture, but in a way that does not affect the result.) The thorium nucleus started with 232 protons and 142 electrons (supposedly). Having lost four protons and four electrons, it is reduced to 228 protons and 138 electrons. This still leaves the atomic number 90, the same as before. So 'radiothorium', like thorium, has ninety planetary electrons circling around the nucleus. Since the chemical properties of an atom are controlled by the number of its planetary electrons, thorium and 'radiothorium' behave the same chemically, regardless of their difference in atomic weight (232 against 228).

The isotopes of an element are identified by their atomic weight, or 'mass number'. Thus ordinary thorium is called thorium 232, while 'radiothorium' is thorium 228. Similarly, the radioactive isotopes of lead are known as lead 210 ('radium D'), lead 214 ('radium B'), lead 212 ('thorium B'), and lead 211 ('actinium B').

The notion of isotopes was found to apply to stable elements as well as to radioactive ones. For instance, it turned out that the three radioactive series I have mentioned ended in three different forms of lead. The uranium series ended in lead 206, the thorium series in lead 208, and the actinium series in lead 207. Each of these was an 'ordinary', stable isotope of lead, but the three leads differed in atomic weight.

Proof of the existence of stable isotopes came from a device invented by an assistant of J. J. Thomson named Francis William Aston. It was an arrangement that separated isotopes very sensitively by virtue of the difference in deflection of their ions by a magnetic field; Aston called it a 'mass spectrograph'. In 1919, using an early version of this instrument, Thomson showed that neon was made up of two varieties of atom, one with a mass number of 20, the other with a mass number of 22. Neon 20 was the common isotope; neon 22 came with it in the ratio of one atom in ten. (Later a third isotope, neon 21, was discovered, amounting to only one atom in 400 in the neon of the atmosphere.)

Now the reason for the fractional atomic weights of the elements at last became clear. Neon's atomic weight of 20·183 represented the composite weight of the three differently weighted isotopes making up the element as it was found in nature. Each individual atom had an integral mass number, but the average mass number – the atomic weight – was fractional.

Aston proceeded to show that several common stable elements were indeed mixtures of isotopes. He found that chlorine, with a fractional atomic weight of 35·453, was made up of chlorine 35 and chlorine 37, in the 'abundance ratio' of four to one. Aston was awarded the Nobel Prize in chemistry in 1922.

In his address accepting the Prize, Aston clearly forecast the possibility of making use of the energy bound in the atomic nucleus, foreseeing both nuclear power plants and nuclear bombs (see Chapter 9). In 1935, the Canadian-American physicist Arthur Jeffrey Dempster used Aston's instrument to take a long step in that direction. He showed that, although 993 of every 1,000 uranium atoms were uranium 238, the remaining seven were uranium 235. This was a discovery fraught with a significance soon to be realized.

Thus, after a century of false trails, Prout's idea was finally vindicated. The elements *were* built of uniform building blocks – if not of hydrogen atoms, at least of units with hydrogen's mass. The reason the elements did not bear this out in their weights was that they were mixtures of isotopes containing different numbers of building blocks. In fact, even oxygen, whose atomic weight of sixteen was used as the standard for measuring the relative weights of the elements, was not a completely pure case. For every 10,000 atoms of common oxygen 16, there were twenty atoms of an isotope with a weight equal to 18 units and four with the mass number 17.

Actually there are a few elements consisting of a 'single isotope'. (This is a misnomer: to speak of an element as having only one isotope is like saying a woman has given birth to a 'single twin'.) The elements of this kind include beryllium, all of whose atoms have the mass number 9; fluorine, made up solely of fluorine 19; aluminium, solely aluminium 27; and a number of others. A nucleus with a particular structure is now called a 'nuclide', following the suggestion made in 1947 by the American chemist Truman Paul Kohman. One can properly say that an element such as aluminium is made up of a single nuclide.

Ever since Rutherford identified the first nuclear particle (the alpha particle), physicists have busied themselves poking around in the nucleus, trying either to change one atom into another or to break it up to see what it is made of. At first they had only the alpha particle to work with. Rutherford made excellent use of it.

One of the fruitful experiments Rutherford and his assistants carried out involved firing alpha particles at a screen coated with zinc sulphide. Each hit produced a tiny scintillation (an effect first discovered by Crookes in 1903), so that the arrival of single particles could be witnessed and counted with the naked eye. Pursuing this technique, the experimenters put up a metal disc that would block the alpha particles from reaching the zinc sulphide screen so that the scintillations stopped. When hydrogen was introduced into the apparatus, scintillations appeared on the screen despite the blocking metal disc. Moreover, these new scintillations differed in appearance from those produced by alpha particles. Since the metal disc stopped alpha particles, some other radiation must be penetrating it to reach the screen. The radiation, it was decided, must consist of fast protons. In other words, the alpha particles would now and then make a square hit on the nucleus of a hydrogen atom and send it careering forwards, as one billiard ball might send another forward on striking it. The struck protons, being relatively light would shoot forward at great velocity and so could penetrate the metal disc and strike the zinc sulphide screen.

This detection of single particles by scintillation is an example of a 'scintillation counter'. To make such counts, Rutherford and his assistants first had to sit in the dark for fifteen minutes in order to sensitize their eyes and then make their painstaking counts. Modern scintillation counters do not depend on the human eye and mind. Instead, the scintillations are converted to electric pulses that are then counted electronically. The result need merely be read off from appropriate dials. The counting may be made more practical where scintillations are numerous, by using electric circuits that allow only one in two or in four (or even more) scintillations to be recorded. Such 'scalers' (which scaled down the counting, so to speak) were first devised by the English physicist C. E. Wynn-Williams in 1931. Since the Second World War, organic substances have substituted for zinc sulphide and have proved preferable.

In Rutherford's original scintillation experiments, there came an unexpected development. When his experiment was performed with nitrogen instead of hydrogen as the target for the alpha-particle bombardment, the zinc sulphide screen still showed scintillations exactly like those produced by protons. Rutherford could only conclude that the bombardment had knocked protons out of the nitrogen nucleus.

To try to find out just what had happened, Rutherford turned to the 'Wilson cloud chamber'. This device had been invented in 1895 by the Scottish physicist Charles Thomson Rees Wilson. A glass container fitted with a piston is filled with moisture-saturated air. When the piston is pulled outwards, the air abruptly expands and therefore cools. At the reduced temperature, it is supersaturated with the moisture. Now any charged particle will cause the water vapour to condense on it. If a particle dashes through the chamber, ionizing atoms in it, a foggy line of droplets will mark its wake.

The nature of this track can tell a great deal about the particle. The light beta particle leaves a faint, wavering path; the particle is knocked about even in passing near electrons. The much more massive alpha particle makes a straight, thick track. If it strikes a nucleus and rebounds, the path has a sharp bend in it. If it picks up two electrons and becomes a neutral helium atom, its track ends. Aside from the size and character of its track, there are other ways of identifying a particle in the cloud chamber. Its response to an applied magnetic field tells whether it is positively or negatively charged, and the amount of curve indicates its mass and energy. By now physicists are so familiar with photographs of all sorts of tracks that they can read them off as if they were primer print. For the development of his cloud chamber, Wilson shared the Nobel Prize in physics in 1927.

The cloud chamber has been modified in several ways since its invention and cousin instruments have been devised. The original cloud chamber was not usable after expansion until the chamber had been reset. In 1939, A. Langsdorf, in the United States, devised a 'diffusion cloud chamber', in which warm alcohol vapour diffused into a cooler region in such a way that there was always a super-saturated region and tracks could be observed continuously.

Then came the 'bubble chamber', a device similar in principle. In it, super-heated liquids under pressure are used rather than super-saturated gas. The path of the charged particle is marked by a line of vapour bubbles in the liquid rather than liquid droplets in vapour. The inventor, the American physicist Donald Arthur Glaser, is supposed to have got the idea by studying a glass of beer in 1953. If so, it was a most fortunate glass of beer for the world of physics and for him, for Glaser received the Nobel Prize for physics in 1960 for the invention of the bubble chamber.

The first bubble chamber had only been a few inches in diameter. Within the decade, bubble chambers six feet long were being used. Bubble chambers, like diffusion cloud chambers, are constantly set for action. In addition, since many more atoms are present in a given volume of liquid than of gas, more ions are produced in a bubble chamber, which is thus particularly well adapted to the study of fast and short-lived particles. Within a decade of its invention, bubble chambers were producing hundreds of thousands of photographs per week. Ultra-shortlived particles were discovered in the 1960s that would have gone undetected without the bubble chamber.

Liquid hydrogen is an excellent liquid with which to fill bubble chambers, because the hydrogen nuclei are so simple (consisting of single protons) as to introduce a minimum of added complications. Bubble chambers 12 feet across and 7 feet high, using as much as 6,400 gallons of liquid hydrogen, now exist, and a 200-litre liquid-helium bubble chamber is in operation in Great Britain.

Although the bubble chamber is more sensitive to short-lived particles than the cloud chamber, it has its shortcomings. Unlike the cloud chamber, the bubble chamber cannot be triggered by desired events. It must record everything wholesale, and uncounted numbers of tracks must be searched through for those of significance. The search was on, then, for some method of detecting tracks that combined the selectivity of the cloud chamber with the sensitivity of the bubble chamber.

This need was met eventually by the 'spark chamber', in which incoming particles ionize gas and set off electric currents through neon gas that is crossed by many metal plates. The currents show up as a visible line of sparks, marking the passage of the particles, and the device can be adjusted to react only to those particles under study. The first practical spark chamber was constructed in 1959 by the Japanese physicists S. Fukui and S. Miyamoto. In 1963, Soviet physicists improved it further, heightening its sensitivity and flexibility. Short streamers of light are produced that, seen on end, make a virtually continuous line (rather than the separate sparks of the spark chamber). The modified device is therefore a 'streamer chamber'. It can detect events that take place within the chamber, and particles that streak off in any direction, where the original spark chamber fell short in both respects.

But, leaving modern sophistication in studying the flight of subatomic particles, we must turn back half a century to see what happened when Rutherford bombarded nitrogen nuclei with alpha particles within one of the original Wilson cloud chambers. The alpha particle would leave a track that would end suddenly in a fork. Plainly this represented a collision with a nitrogen nucleus. One branch of the fork would be comparatively thin, representing a proton shooting off. The other branch, a short, heavy track, represented what was left of the nitrogen nucleus, rebounding from the collision. But there was no sign of the alpha particle itself. It seemed that it must have been absorbed by the nitrogen nucleus, and this supposition was later verified by the British physicist Patrick Maynard Stuart Blackett, who is supposed to have taken more than 20,000 photographs in the process of collecting eight such collisions (surely an example of superhuman patience, faith, and persistence). For this and other work in the field of nuclear physics, Blackett received the Nobel Prize in physics in 1948.

The fate of the nitrogen nucleus could now be deduced. When it absorbed the alpha particle, its mass number of 14 and positive charge of 7 were raised to 18 and 9, respectively. But since the combination immediately lost a proton, the mass number dropped to 17 and the positive charge to 8. Now the element with a positive charge of 8 is oxygen, and the mass number 17 belongs to the isotope oxygen 17. In other words, Rutherford had, in 1919, transmuted nitrogen into oxygen. This was the first man-made transmutation in history. The dream of the alchemists had been fulfilled, though in a manner they could not possibly have foreseen.

Alpha particles from radioactive sources had limits as projectiles: they were not nearly energetic enough to break into nuclei of the heavier elements, whose high positive charges exercise a strong repulsion against positively charged particles. But the nuclear fortress had been breached, and more energetic attacks were to come.

New Particles

The matter of attacks on the nucleus brings us back to the question of the make-up of the nucleus. In 1930, two German physicists, Walther Bothe and H. Becker, reported that they had released from the nucleus a mysterious new radiation of unusual penetrating power. They had produced it by bombarding beryllium atoms with alpha particles. The year before Bothe had devised methods for using two or more counters in conjunction – 'coincidence counters'. These could be used to identify nuclear events taking place in a millionth of a second. For this and other work he shared in the Nobel Prize for physics in 1954.

Two years later the Bothe–Becker discovery was followed up by the French physicists Frédéric and Irène Joliot-Curie. (Irène was the daughter of Pierre and Marie Curie, and Joliot had added her name to his on marrying her.) They used the new-found radiation from beryllium to bombard paraffin, a waxy substance composed of hydrogen and carbon. The radiation knocked protons out of the paraffin.

The English physicist James Chadwick quickly suggested that the radiation consisted of particles. To determine their size, he bombarded boron atoms with them, and from the increase in mass of the new nucleus he calculated that the particle added to the boron had a mass about equal to the proton. Yet the particle itself

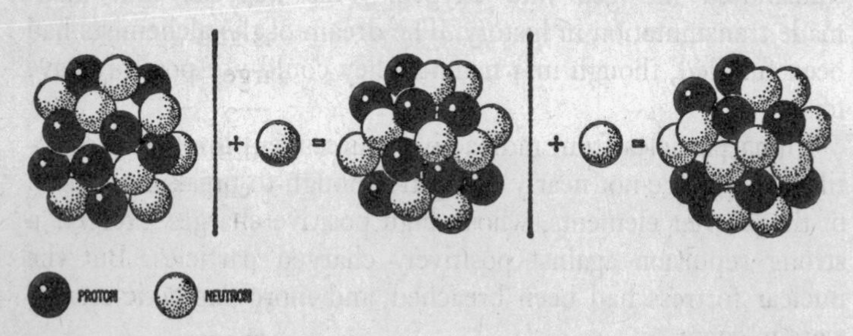

Nuclear make-up of oxygen 16, oxygen 17, and oxygen 18. They contain eight protons each and, in addition, eight, nine, and ten neutrons, respectively.

could not be detected in a Wilson cloud chamber. Chadwick decided that the explanation must be that the particle had no electric charge (an uncharged particle produces no ionization and therefore condenses no water droplets).

So Chadwick concluded that a completely new particle had turned up – a particle with just about the same mass as a proton but without any charge, or, in other words, electrically neutral. The possibility of such a particle had already been suggested, and a name had even been proposed – the 'neutron'. Chadwick accepted that name. For his discovery of the neutron, he was awarded the Nobel Prize in physics in 1935.

The new particle at once solved certain doubts that theoretical physicists had had about the proton–electron model of the nucleus. The German theoretical physicist Werner Heisenberg announced that the concept of a nucleus consisting of protons and neutrons, rather than protons and electrons, gave a much more satisfactory picture – more in accord with what the mathematics of the case said the nucleus should be like.

Furthermore, the new model fitted the facts of the periodic table of elements just as neatly as the old one had. The helium nucleus, for instance, would consist of two protons and two neutrons, which explained its mass of four and nuclear charge of two units. And the concept accounted for isotopes in very simple fashion. For example, the chlorine 35 nucleus would have 17 protons and 18 neutrons; the chlorine 37 nucleus, 17 protons and 20 neutrons. This would give both the same nuclear charge, and the extra weight of the heavier isotope would lie in its two extra neutrons. Likewise, the three isotopes of oxygen would differ only in their numbers of neutrons: oxygen 16 would have eight protons and eight neutrons; oxygen 17, eight protons and nine neutrons; oxygen 18, eight protons and ten neutrons.

In short, every element could be defined simply by the number of protons in its nucleus, which is equivalent to the atomic number. All the elements except hydrogen, however, also had neutrons in the nucleus, and the mass number of a nuclide was the sum of its protons and neutrons. Thus the neutron joined the proton as a basic building block of matter. For convenience, both are now lumped together under the general term 'nucleons', a term first used in 1941 by the Danish physicist Christian Moller. From this came 'nucleonics', suggested in 1944 by the American engineer Zay Jeffries to represent the study of nuclear science and technology.

This new understanding of nuclear structure has resulted in additional classifications of nuclides. Nuclides with equal numbers of protons are, as has just been explained, isotopes. Similarly, nuclides with equal numbers of neutrons (as, for instance, hydrogen 2 and helium 3, each containing one neutron in the nucleus) are 'isotopes'. Nuclides with equal total number of nucleons, and therefore of equal mass numbers, as calcium 40 and argon 40, are 'isobars'.

The proton–neutron model of the nucleus is not likely to be seriously upset in the future. At first it left unexplained the fact that radioactive nuclei emitted electrons, but that question was soon cleared up, as I shall explain shortly.

Nevertheless, in a very important respect the discovery of the neutron disappointed physicists. They had been able to think of the universe as being built of just two fundamental particles – the proton and the electron. Now a third had to be added. To scientists, every retreat from simplicity is regrettable.

The worst of it was that, as things turned out, this was only the beginning. Simplicity's backward step quickly became a headlong rout. There were more particles to come.

For many years physicists had been studying the mysterious 'cosmic rays' from space, first discovered in 1911 by the Austrian physicist Victor Francis Hess on balloon flights high in the atmosphere.

The presence of such radiation was detected by an instrument so simple as to hearten those who sometimes feel that modern science can progress only by use of unbelievably complex devices. The instrument was an 'electroscope', consisting of two pieces of thin gold foil attached to a metal rod within a metal housing fitted with windows. (The ancestor of this device was constructed as long ago as 1706 by the English physicist Francis Hauksbee.)

If the metal rod is charged with static electricity, the pieces of gold foil separate. Ideally, they would remain separated forever, but ions in the surrounding atmosphere slowly conduct away the charge so that the leaves gradually collapse towards each other. Energetic radiation, such as X-rays, gamma rays, or streams of charged particles, produces the ions necessary for such charge leakage. Even if the electroscope is well shielded there is still a slow leakage, indicating the presence of a very penetrating radiation not directly related to radioactivity. It was this penetrating radiation, which increased in intensity, that Hess noticed as he rose high in the atmosphere. Hess shared the Nobel Prize for physics in 1936 for this discovery.

The American physicist Robert Andrews Millikan, who collected a great deal of information on this radiation (and gave it the name 'cosmic rays'), decided that it must be a form of electromagnetic radiation. Its penetrating power was such that some of it could even pass through several feet of lead. To Millikan this suggested that the radiation was like the penetrating gamma rays, but with an even shorter wavelength.

Others, notably the American physicist Arthur Holly Compton, contended that the cosmic rays were particles. There was a way to investigate the question. If they were charged particles, they should be deflected by the earth's magnetic field as they approached the earth from outer space. Compton studied the measurements of cosmic radiation at various latitudes and found that it did indeed curve with the magnetic field: it was weakest near the magnetic equator and strongest near the poles, where the magnetic lines of force dipped down to the earth.

The 'primary' cosmic particles, as they enter our atmosphere, carry fantastically high energies. Most of them are protons, but some are nuclei of heavier elements. In general, the heavier the nucleus, the rarer it is among the cosmic particles. Nuclei as complex as those making up iron atoms were detected quickly enough, and in 1968, nuclei as complex as those of uranium were detected. The uranium nuclei make up only one particle in 10 million. A few very high-energy electrons are also included.

When the primary particles hit atoms and molecules of the air, they smash these nuclei and produce all sorts of 'secondary' particles. It is this secondary radiation (still very energetic) that we detect near the earth, but balloons sent to the upper atmosphere have recorded the primary radiation.

Now it was as a result of cosmic-ray research that the next new particle – after the neutron – was discovered. This discovery had actually been predicted by a theoretical physicist. Paul Adrien Maurice Dirac had reasoned, from a mathematical analysis of the properties of sub-atomic particles, that each particle should have an 'anti-particle'. (Scientists like nature to be not only simple but also symmetrical.) Thus there ought to be an 'anti-electron', exactly like the electron except that it had a positive instead of a negative charge, and an 'anti-proton' with a negative instead of a positive charge.

Dirac's theory did not make much of a splash in the scientific world when he proposed it in 1930. But sure enough, two years later the 'anti-electron' actually turned up. The American physicist Carl David Anderson was working with Millikan on the problem of whether cosmic rays were electromagnetic radiation or particles. By then most people were ready to accept Compton's evidence that they were charged particles, but Millikan was an extraordinarily hard loser, and he was not satisfied that the issue was settled. Anderson undertook to find out whether cosmic rays entering a Wilson cloud chamber would be bent by a strong magnetic field. To slow down the rays sufficiently so that the curvature, if any, could be detected, Anderson placed in the chamber a lead barrier about a quarter of an inch thick. He found that the cosmic radiation crossing the chamber after it came through the lead did make a curved track. But he also found something else. In their passage through the lead, the energetic cosmic rays knocked particles out of the lead atoms. One of these particles made a track just like that of an electron. But it curved in the wrong direction! Same mass but opposite charge. There it was – Dirac's 'anti-electron'. Anderson called his discovery the 'positron'. It is an example of the secondary radiation produced by cosmic rays, but in 1963 it was found that positrons were included among the primary radiations as well.

Left to itself, the positron is as stable as the electron (why not, since it is identical with the electron except for electric charge?) and could exist indefinitely. It is not, however, left to itself, for it comes into existence in a universe filled with electrons. As it streaks along, it almost immediately (say, within a millionth of a second) finds itself in the neighbourhood of one.

For a moment, there may be an electron–positron association – a situation in which the two particles circle each other about a mutual centre of force. In 1945, the American physicist Arthur Edward Ruark suggested that this two-particle system be called 'positronium', and in 1951, the Austrian-American physicist Martin Deutsch was able to detect positronium through the characteristic gamma radiation it gave up.

However, even if a positronium system forms, it remains in existence for only a 10-millionth of a second, at most. The dance ends in the combination of the electron and positron. When the two opposite bits of matter combine they cancel each other, leaving no matter at all ('mutual annihilation'); only energy, in the form of gamma rays, is left behind. This confirmed Albert Einstein's suggestion that matter could be converted into energy and vice versa. Indeed, Anderson soon succeeded in detecting the reverse phenomenon: gamma rays suddenly disappearing and giving rise to an electron–positron pair. This is called 'pair production'. (Anderson, along with Hess, received the Nobel Prize in physics in 1936.)

The Joliot-Curies shortly afterwards came across the positron in another connection, and in so doing made an important discovery. Bombarding aluminium atoms with alpha particles, they found that the procedure produced not only protons but also positrons. This in itself was interesting but not fabulous. When they stopped the bombardment, however, the aluminium kept right on emitting positrons! The emission faded off with time. Apparently they had created a new radioactive substance in the target.

The Joliot-Curies interpreted what had happened in this way. When an aluminium nucleus absorbed an alpha particle, the addition of two protons changed aluminium (atomic number 13) to phosphorus (atomic number 15). Since the alpha particle contained four nucleons altogether, the mass number would go up by four – from aluminium 27 to phosphorus 31. Now if the reaction knocked a proton out of this nucleus, the reduction of its atomic number and mass number by one would change it to another element – namely, silicon 30.

Since an alpha particle is the nucleus of helium and a proton the nucleus of hydrogen, we can write the following equation of this 'nuclear reaction':

$$\text{aluminium } 27 + \text{helium } 4 \rightarrow \text{silicon } 30 + \text{hydrogen } 1$$

Notice that the mass numbers balance: 27 plus 4 equals 30 plus 1. So do the atomic numbers, for aluminium's is 13 and helium's 2, making 15 together, while silicon's atomic number of 14 and hydrogen's 1 also add up to 15. This balancing of both mass numbers and atomic numbers is a general rule of nuclear reactions.

The Joliot-Curies assumed that neutrons as well as protons had been formed in the reaction. If phosphorus 31 emitted a neutron instead of a proton, the atomic number would not change, though the mass number would go down one. In that case the element would remain phosphorus but become phosphorus 30. This equation would read:

aluminium 27+helium 4→phosphorus 30+neutron 1

Since the atomic number of phosphorus is 15 and that of the neutron is 0, again the atomic numbers on both sides of the equation also balance.

Both processes – alpha absorption followed by proton emission, and alpha absorption followed by neutron emission – take place when aluminium is bombarded by alpha particles. But there is one important distinction between the two results. Silicon 30 is a perfectly well-known isotope of silicon, making up a little more than 3 per cent of the silicon in nature. But phosphorus 30 does not exist in nature. The only known natural form of phosphorus is phosphorus 31. Phosphorus 30, in short, is a radioactive isotope with a brief lifetime that exists today only when it is produced artificially; in fact, it was the first such isotope made by man. The Joliot-Curies received the Nobel Prize in chemistry in 1935 for their discovery of artificial radioactivity.

The unstable phosphorus 30 that the Joliot-Curies had produced by bombarding aluminium quickly broke down by emitting positrons. Since the positron, like the electron, has practically no mass, this emission did not change the mass number of the nucleus. However, the loss of one positive charge did reduce its atomic number by one, so that it was converted from phosphorus to silicon.

Where does the positron come from? Are positrons among the components of the nucleus? The answer is no. What happens is that a proton within the nucleus changes to a neutron by shedding its positive charge, which is released in the form of a speeding positron.

Now the emission of beta particles – the puzzle we encountered earlier in the chapter – can be explained. This comes about as the result of a process just the reverse of the decay of a proton into a neutron. That is, a neutron changes into a proton. The proton-to-neutron change releases a positron and, to maintain the symmetry, the neutron-to-proton change releases an electron (the beta particle). The release of a negative charge is equivalent to the gain of a positive charge and accounts for the formation of a positively charged proton from an uncharged neutron. But how does the uncharged neutron manage to dig up a negative charge and send it flying outwards?

Actually, if it were just a negative charge, the neutron could not do so. Two centuries of experience have taught physicists that neither a negative electric charge nor a positive electric charge can be created out of nothing. Neither can either type of charge be destroyed. This is the law of 'conservation of electric charge'.

However, a neutron does not create only an electron in the process of producing a beta particle; it creates a proton as well. The uncharged neutron disappears, leaving in its place a positively charged proton and a negatively charged electron. The two new particles, *taken together*, have an over-all electric charge of zero. No *net* charge has been created. Similarly, when a positron and electron meet and engage in mutual annihilation, the charge of the positron and electron, *taken together*, is zero to begin with.

When a proton emits a positron and changes into a neutron, the original particle (the proton) is positively charged, and the final particles (the neutron and positron), taken together, have a positive charge.

It is also possible for a nucleus to absorb an electron. When this happens, a proton within the nucleus changes to a neutron. An electron plus a proton (which, taken together, have a charge of zero) form a neutron, which has a zero charge. The electron captured is from the innermost electron shell of the atom, since the electrons of that shell are closest to the nucleus and most easily gathered in. The innermost shell is the K-shell (see p. 251) and the process is therefore called 'K-capture'. An electron from the L-shell then drops into the vacant spot, and an X-ray is emitted. It is by these X-rays that K-capture can be detected. This was first accomplished in 1938 by the American physicist Luis W. Alvarez. Ordinary nuclear reactions involving the nucleus alone are usually not affected by chemical change, which affects electrons only. Since K-capture affects electrons as well as nuclei, the chance of its occurring can be somewhat altered as a result of chemical change.

All of these particle interactions satisfy the law of conservation of electric charge and must also satisfy quite a number of other conservation laws. Any particle interaction that violates none of the conservation laws will eventually occur, physicists suspect, and an observer with the proper tools and proper patience will detect it. Those events that violate a conservation law are 'forbidden' and will not take place. Nevertheless, physicists are occasionally surprised to find that what had seemed a conservation law is not as rigorous or as universal as had been thought. We shall come across examples of that.

Once the Joliot-Curies had created the first artificial radioactive isotope, physicists proceeded merrily to produce whole tribes of them. In fact, radioactive varieties of every single element in the periodic table have now been formed in the laboratory. In the modern periodic table, each element is really a family, with stable and unstable members, some found in nature, some only in the laboratory.

For instance, hydrogen comes in three varieties. First there is ordinary hydrogen, containing a single proton. In 1932, the chemist Harold Urey succeeded in isolating a second. He did it by slowly evaporating a large quantity of water, working on the theory that he would be left in the end with a concentration of the heavier form of hydrogen that was suspected to exist. Sure enough, when he examined the last few drops of unevaporated water spectroscopically, he found a faint line in the spectrum in exactly the position predicted for 'heavy hydrogen'.

Heavy hydrogen's nucleus is made up of one proton and one neutron. Having a mass number of two, the isotope is hydrogen 2. Urey named the atom 'deuterium', from a Greek word meaning 'second', and the nucleus a 'deuteron'. A water molecule containing deuterium is called 'heavy water'. Because deuterium has twice the mass of ordinary hydrogen, heavy water has higher boiling and freezing points than ordinary water. Whereas ordinary water boils at 100°C, and freezes at 0°C, heavy water boils at 101·42°C and freezes at 3·79°C. Deuterium itself has a boiling point of 23·7°K as compared with 20·4°K for ordinary hydrogen. Deuterium occurs in nature in the ratio of one part to 6,000 parts of ordinary hydrogen. For his discovery of deuterium, Urey received the Nobel Prize in chemistry in 1934.

The deuteron turned out to be a valuable particle for bombarding nuclei. In 1934, the Australian physicist Marcus Lawrence Elwin Oliphant and the Austrian chemist Paul Harteck, attacking deuterium itself with deuterons, produced a third form of hydrogen, made up of one proton and two neutrons. The reaction went:

$$\text{hydrogen 2} + \text{hydrogen 2} \rightarrow \text{hydrogen 3} + \text{hydrogen 1}$$

The new 'super-heavy' hydrogen was named 'tritium', from the Greek word for 'third', and its nucleus is a 'triton'. Its boiling point is 25·0°K, and its melting point 20·5°K. Pure tritium oxide ('super-heavy water') has been prepared, and its melting point is 4·5°C. Tritium is radioactive and breaks down comparatively rapidly. It exists in nature, being formed as one of the products of the bombardment of the atmosphere by cosmic rays. In breaking down, it emits an electron and changes to helium 3, a stable but rare isotope of helium.

Helium 3 differs from ordinary helium 4 in some interesting ways, particularly in the fact that it does not display the same properties of super-conductivity and super-fluidity in the liquid state, discussed in the preceding chapter. Atmospheric helium contains only 0·00013 per cent of helium 3, all originating, no doubt, from the breakdown of tritium. (Tritium, because it is unstable, is even rarer. It is estimated that only three and a half pounds exist all told in the atmosphere and oceans.) The helium 3 content of helium obtained in natural gas wells, where cosmic rays have had less opportunity to form tritium, is even smaller in percentage.

These two isotopes, helium 3 and helium 4, are not the only heliums. Physicists have created two radioactive forms: helium 5, one of the most unstable nuclei known, and helium 6, also very unstable.

And so it goes. By now the list of known isotopes has grown to about 1,400 altogether. Over 1,100 of these are radioactive and many of them have been created by new forms of atomic

Nuclei of ordinary hydrogen, deuterium, and tritium.

artillery far more potent than the alpha particles from radioactive sources which were the only projectiles at the disposal of Rutherford and the Joliot-Curies.

The sort of experiment performed by the Joliot-Curies in the early 1930s seemed a matter of the scientific ivory-tower at the time, but it has come to have a highly practical application. Suppose a set of atoms of one kind, or of many, are bombarded with neutrons. A certain percentage of each kind of atom will absorb a neutron, and a radioactive atom will generally result. This radioactive element will decay, giving off sub-atomic radiation in the form of particles or gamma rays.

Every different type of atom will absorb neutrons to form a different type of radioactive atom, giving off different and characteristic radiation. The radiation can be detected with great delicacy. From its type and from the rate at which its production declines, the radioactive atom giving it off can be identified and, therefore, so can the original atom before it absorbed a neutron. Substances can be analysed in this fashion ('neutron-activation analysis') with unprecedented precision: amounts as small as a billionth of a gram of a particular nuclide are detectable.

Neutron-activation analysis can be used to determine delicate differences in impurities-content in samples of particular pigments from different centuries and in this way can determine the authenticity of a supposedly old painting, using only the barest fragment of its pigment. Other delicate decisions of this sort can be made: even hair from Napoleon's century-and-a-half-old corpse was studied and found to contain suspicious quantities of arsenic (which perhaps he took medicinally).

Particle Accelerators

Dirac had predicted not only an anti-electron (the positron) but also an anti-proton. But to produce an anti-proton would take vastly more energy. The energy needed was proportional to the mass of the particle. Since the proton was 1,836 times as massive as the electron, the formation of an anti-proton called for at least 1,836 times as much energy as the formation of a positron. The feat had to wait for the development of a device for accelerating sub-atomic particles to sufficiently high energies.

At the time of Dirac's prediction, the first step in this direction had just been taken. In 1928, the English physicists John D. Cockcroft and Ernest Walton, working in Rutherford's laboratory, developed a 'voltage multiplier', a device for building up electric potential, which could drive the charged proton up to an energy of nearly 400,000 electron volts. (One electron volt is equal to the energy developed by an electron accelerated across an electric field with a potential of one volt.) With protons accelerated in this machine they were able to break up the lithium nucleus, and for this work they were awarded the Nobel Prize for physics in 1951.

Meanwhile the American physicist Robert Jemison Van de Graaff was creating another type of accelerating machine. Essentially, it operated by separating electrons from protons and depositing them at opposite ends of the apparatus by means of a moving belt. In this way the 'Van de Graaff electrostatic generator' developed a very high electric potential between the opposite ends; Van de Graaff got it up to 8 million volts. Electrostatic generators can easily accelerate protons to a speed amounting to 24 million electron volts (physicists now invariably abbreviate million electron volts to 'Mev').

The dramatic pictures of the Van de Graaff electrostatic generator producing huge sparks caught the popular imagination and introduced the public to 'atom smashers'. It was popularly viewed as a device to produce 'man-made lightning', although, of course, it was much more than that. (A generator designed to produce artificial lightning and nothing more had actually been built in 1922 by the German-American electrical engineer Charles Proteus Steinmetz.)

The energy that can be reached in such a machine is restricted by practical limits on the attainable potential. However, another scheme for accelerating particles shortly made its appearance. Suppose that, instead of firing particles with one big shot, you accelerated them with a series of small pushes. If each successive push was timed just right, it would increase the speed each time, just as pushes on a child's swing will send it higher and higher if they are applied 'in phase' with the swing's oscillations.

This idea gave birth, in 1931, to the 'linear accelerator'. The particles are driven down a tube divided into sections. The driving force is an alternating electric field, so managed that as the particles enter each successive section, they get another push. Since the particles speed up as they go along, each section must be longer than the one before, so that the particles will take the same time to get through it and will be in phase with the timing of the pushes.

It is not easy to keep the timing just right, and anyway there is a limit to how long a tube you can make practicably, so the linear accelerator did not catch on in the 1930s. One of the things that pushed it into the background was that Ernest Orlando Lawrence of the University of California conceived a better idea.

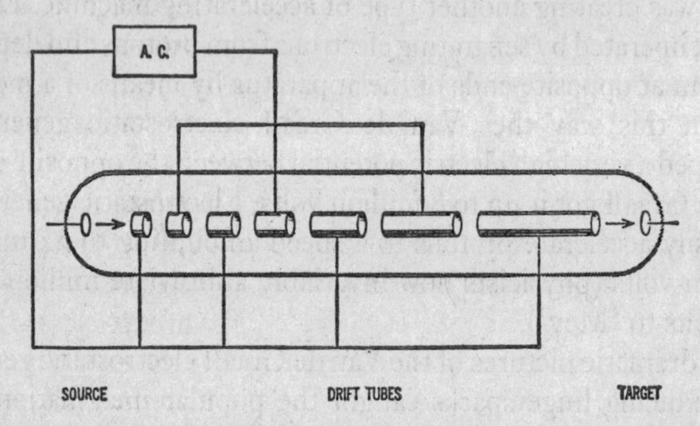

Principle of the linear accelerator. A high-frequency alternating charge alternately pushes and pulls the charged particles in the successive drive tubes, accelerating them in one direction.

Instead of driving the particles down a straight tube, why not whirl them around in a circular path? A magnet could bend them in such a path. Each time they completed a half circle, they would be given a kick by the alternating field, and in the set-up the timing would not be so difficult to control. As the particles speeded up, their path would be bent less sharply by the magnet, so they would move in ever wider circles and perhaps take the same time for each round trip. At the end of their spiralling flight, the particles would emerge from the circular chamber (actually divided into semi-circular halves, called 'dees') and attack their target.

Lawrence's compact new device was named the 'cyclotron'. His first model, less than a foot in diameter, could accelerate protons to energies of nearly 1·25 Mev. By 1939 the University of California had a cyclotron, with magnets five feet across, capable of raising particles to some 20 Mev, twice the speed of the most energetic alpha particles emitted by radioactive sources. In that year Lawrence received the Nobel Prize in physics for his invention.

The cyclotron itself had to stop at about 20 Mev, because at that energy the particles were travelling so fast that the mass increase with velocity – an effect predicted by Einstein's Theory of Relativity – became appreciable. This increase in mass caused the particles to start lagging and falling out of phase with the electrical kicks. But there was a cure for this, and it was worked out in 1945 independently by the Soviet physicist Vladimir Iosifovich Veksler and the California physicist Edwin Mattison McMillan. The cure was simply to synchronize the alternations of the electric field with the increase in mass of the particles. This modification of the cyclotron was called the 'synchrocyclotron'. By 1946 the University of California had built one which accelerated particles to energies of 200 to 400 Mev. Later larger synchrocyclotrons in the United States and in the Soviet Union raised the energies to 700 to 800 Mev.

Meanwhile the acceleration of electrons had been getting separate attention. To be useful in smashing atoms, the light electrons had to be raised to much higher speeds than protons (just as a ping-

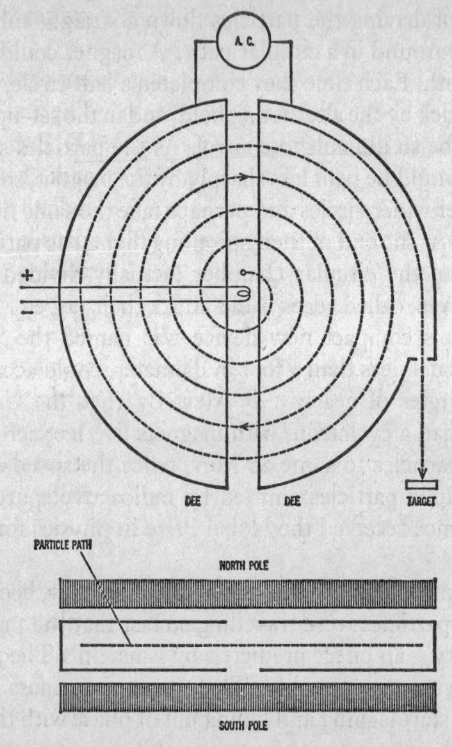

Principle of the cyclotron, shown in top view (*above*) and side view (*below*). Particles injected from the source are given a kick in each dee by the alternating charge and are bent in their spiral path by a magnet.

pong ball has to be moved much faster than a golf ball to do as much damage). The cyclotron would not work for electrons, because at the high velocities needed to make the electrons effective, their increase in mass was too great. In 1940 the American physicist Donald William Kerst designed an electron-accelerating device which balanced the increasing mass with an electric field of increasing strength. The electrons were kept in the same circular path instead of spiralling outwards. This instrument was named the 'betatron', after beta particles. Betatrons now generate electron velocities up to 340 Mev.

They have been joined by another instrument of slightly different design called the 'electron synchrotron'. The first of these was built in England in 1946 by F. K. Goward and D. E. Barnes. These raise electron energies to the 1,000 Mev mark, but cannot go higher because electrons moving in a circle radiate energy at increasing rates as velocity is increased. This radiation produced by an accelerating particle is called 'Bremsstrahlung', a German word meaning 'braking radiation'.

Taking a leaf from the betatron and electron synchrotron, physicists working with protons began about 1947 to build 'proton synchrotrons', which likewise kept their particles in a single circular path. This helped save on weight. Where particles move in outwardly spiralling paths, a magnet must extend the entire width of the spiral to keep the magnetic force uniform throughout. With the path held in a circle, the magnet need be only large enough to cover a narrow area.

Because the more massive proton does not lose energy with motion in a circular path as rapidly as does the electron, physicists set out to surpass the 1,000-Mev mark with a proton synchrotron. This value of 1,000 Mev is equal to a thousand million electron volts – abbreviated to Gev, the G coming from 'giga', Greek for 'giant'.

In 1952, the Brookhaven National Laboratory on Long Island completed a proton synchrotron that reached 2 to 3 Gev. They called it the 'cosmotron', because it had arrived at the main energy range of particles in the cosmic rays. Two years later the University of California brought in its 'bevatron', capable of producing particles of between 5 and 6 Gev. Then, in 1957, the Soviet Union announced its 'phasotron' had got to 10 Gev.

But by now these machines seem puny in comparison with accelerators of a newer type, called the 'strong-focusing synchrotron'. The limitation on the bevatron type is that particles in the stream fly off into the walls of the channel in which they travel. The new type counteracts this tendency by means of alternating magnetic fields of different shape which keep focusing the particles in a narrow stream. The idea was first suggested by Christofilos, whose 'amateur' abilities outshone the professionals here as well as in the case of the Christofilos effect. This, incidentally, further decreased the size of the magnet required for the energy levels attained. Where particle energy was increased fiftyfold, the weight of the magnet involved was less than doubled.

In November 1959, the European Committee for Nuclear Research (CERN), a cooperative agency of twelve nations, completed in Geneva a strong-focusing synchrotron which reached 24 Gev and produced large pulses of particles (containing 10,000 million protons) every three seconds. This synchrotron is nearly three city blocks in diameter, and one round trip through it is two fifths of a mile. In the three-second period during which the pulse builds up, the protons travel half a million times around that track. The instrument has a magnet weighing 3,500 tons and costs 30 million dollars.

It is not, however, the last word. Brookhaven has completed an even larger machine that has moved well above the 30-Gev mark, and the Soviet Union now has one that is over a mile in diameter and that attained over 70 Gev when put into operation in 1967. American physicists are supervising the construction of one that will be three miles in diameter and will attain 300 Gev, while others of 1,000 Gev are dreamed of.

The linear accelerator, or 'linac', has also undergone a revival. Improvements in technique have removed the difficulties that plagued the early models. For extremely high energies, a linear accelerator has some advantages over the cyclic type. Since electrons do not lose energy when travelling in a straight line, a linac can accelerate electrons more powerfully and focus beams on targets more sharply. Stanford University has built a linear accelerator two miles long which can reach energies of perhaps 45 Gev.

Nor is sheer size the only answer to greater power. The notion has been repeatedly broached of having two accelerators in tandem so that a stream of energetic particles collides head-on with another stream moving in the opposite direction. This will quadruple the energies involved over the collision of one such stream with a stationary object. Such a tandem-accelerator may be the next step.

With merely the bevatron, man at last came within reach of creating the anti-proton. The California physicists set out deliberately to produce and detect it. In 1955, Owen Chamberlain and Emilo G. Segrè, after bombarding copper with protons of 6·2 Gev hour after hour, definitely caught the anti-proton – in fact, sixty of them. It was far from easy to identify them. For every anti-proton produced, 40,000 particles of other types came into existence. But by an elaborate system of detectors, so designed and arranged that only an anti-proton could touch all the bases, they recognized the particle beyond question. For their achievement, Chamberlain and Segrè received the Nobel Prize in physics in 1959.

The anti-proton is as evanescent as the positron – at least in our universe. Within a tiny fraction of a second after it is created, the particle is snatched up by some normal, positively charged nucleus. There the anti-proton and one of the protons of the nucleus annihilate each other, turning into energy and minor particles. In 1965, enough energy was concentrated to reverse the process and produce a proton–anti-proton pair.

Once in a while a proton and an anti-proton have only a near collision instead of a direct one. When that happens, they mutually neutralize their respective charges. The proton is converted to a neutron, which is fair enough. But the anti-proton becomes an 'anti-neutron'! What can an 'anti-neutron' be? The positron is the opposite of the electron by virtue of its opposite charge, and the anti-proton is likewise 'anti' by virtue of its charge. But what gives the uncharged anti-neutron the quality of oppositeness?

Here we have to digress a little into the subject of the spin of particles. A particle may usually be viewed as spinning on its axis, like a top or the earth or the sun of our Galaxy or, for all we know, the universe itself. This particle spin was first suggested in 1925 by the Dutch physicists George Eugene Uhlenbeck and Samuel Abraham Goudsmit. In spinning, the particle generates a tiny magnetic field; such fields have been measured and thoroughly explored, notably by the German physicist Otto Stern and the

American physicist Isidor Isaac Rabi who received the Nobel Prizes in physics in 1943 and 1944, respectively, for their work on this phenomenon.

Spin is measured in such a way that the spin of an electron or a proton is said to be equal to one half. When this is doubled, it becomes an odd integer (1). The energies of particles whose spin, on being doubled, becomes an odd integer in this fashion can be dealt with according to a system of rules worked out independently, in 1926, by Fermi and Dirac. These rules make up the 'Fermi–Dirac statistics', and particles (such as the electron and proton) that obey these are called 'fermions'. The neutron is also a fermion.

There also exist particles whose spin, when doubled, is an even number. Their energies can be dealt with by another set of rules devised by Einstein and by the Indian physicist S. N. Bose. Particles that follow the 'Bose–Einstein statistics' are 'bosons'. The alpha particle, for instance, is a boson.

These classes of particles have different properties. For instance, the Pauli exclusion principle (see p. 251) applies not only to electrons, but to all fermions. It does not, however, apply to bosons.

It is easy to understand how a charged particle sets up a magnetic field, but not so easy to see why the uncharged neutron should. Yet it unquestionably does. The most direct evidence of this is that when a neutron beam strikes magnetized iron, it behaves differently from the way it does when the iron is not magnetized. The neutron's magnetism remains a mystery; physicists suspect that the neutron contains positive and negative charges which add up to zero, but which somehow manage to set up a magnetic field when the particle spins.

In any case, the spin of the neutron gives us the answer to the question as to what the anti-neutron is. It is simply a neutron with its spin direction reversed; its south magnetic pole, say, is up instead of down. Actually the proton and anti-proton and the electron and positron show exactly the same pole-reversed phenomenon.

Anti-particles can undoubtedly combine to form 'anti-matter', as ordinary particles form ordinary matter. The first actual example of anti-matter was produced at Brookhaven in 1965. There the bombardment of a beryllium target with 7-Gev protons produced combinations of anti-protons and anti-neutrons, something that was an 'anti-deuteron'. 'Anti-helium 3' has since been produced and undoubtedly, if enough pains are taken, still more complicated anti-nuclei can be formed. The principle is clear, however, and no physicist doubts it. Anti-matter can exist.

But does it exist in actuality? Are there masses of anti-matter in the universe? If there were, they would not betray themselves from a distance. Their gravitational effects and the light they produce would be exactly like that of ordinary matter. If, however, they encountered ordinary matter, the massive annihilation reactions that result ought to be most noticeable. Astronomers have therefore taken to looking speculatively at distant galaxies to see if any unusual activity might betray matter–anti-matter interactions. What about the exploding galaxies? What about Messier 87 with a bright jet of luminosity sticking out of its globular main body? What about the enormous energies pouring out of quasars? For that matter, what about the meteoric strike in Siberia in 1908 and its destructive consequences? Was it just a meteor or was its destructiveness the result of the fact that it was a piece (and perhaps a not-very-large piece) of anti-matter?

There are some questions that are even more fundamental. Why should there be some chunks of anti-matter in a universe composed mainly of matter? Since matter and anti-matter are equivalent in all respects but that of electromagnetic oppositeness, any force which would create one would have to create the other, and the universe should be made up of equal quantities of each. If they were intimately mixed, particles of matter and anti-matter would annihilate each other, but what if some effect served to separate them after their creation?

In this connection, there is a suggestion made by the Swedish physicist Oskar Klein which has been popularized by his compatriot Hannes Alfven. Suppose the universe came into being in the form of a very rarefied collection of particles spread out over a sphere a billion light-years or more in diameter. The particles might indeed be half ordinary ones and half anti-particles, but under such conditions of rarefaction, they would move freely and hardly ever collide and annihilate one another.

If there were magnetic fields in this thin-universe, particles in a particular region would tend to curve in one direction and anti-particles in the other, so that they would tend to separate. Eventually, after separation, they would collect into galaxies and anti-galaxies. If sizeable chunks of matter and anti-matter met and began to interact – either while the galaxies were forming or after they had formed – the radiation liberated at their volumes of junction would, by its pressure, drive them apart. Finally, then, there would be evenly spread out galaxies and anti-galaxies.

Mutual gravitational attraction would now draw them together in a 'contracting universe'. The further the universe contracted, the greater the chance of galaxy–anti-galaxy collisions, and the greater the energetic radiation produced. Finally, when the universe shrank to a diameter of a thousand million light-years or so, the radiation produced would have sufficient pressure to explode it and drive the galaxies and anti-galaxies forcibly apart. It is this expansion stage in which we now find ourselves, according to Klein and Alfven, and what we interpret as evidence in favour of the 'big bang' is evidence in favour of the crucial moment when the universe–anti-universe produces enough energy to blow itself up, so to speak.

If this is so, then half the galaxies we can see are anti-galaxies. But which half? And how can we tell? There are no answers to that – so far.

A consideration of anti-matter on a somewhat less majestic scale brings up the question of cosmic rays again. Most of the cosmic-ray particles have energies between 1 and 10 Gev. This

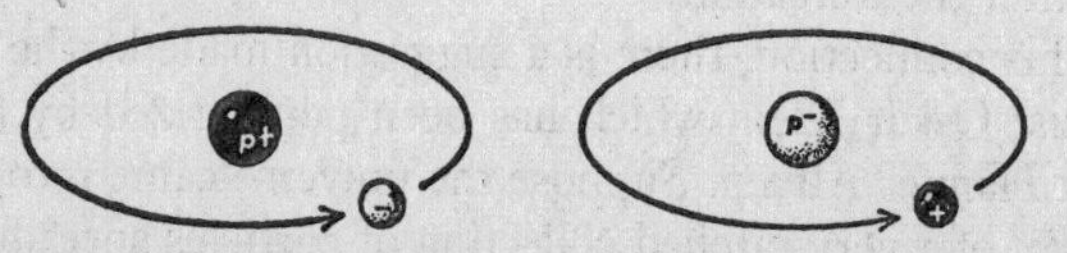

An atom of hydrogen and an atom of its anti-matter counterpart, consisting of an anti-proton and a positron.

might be accounted for by matter–anti-matter interaction, but a few cosmic particles run much higher: 20 Gev, 30 Gev, 40 Gev. Physicists at the Massachusetts Institute of Technology have even detected some with the colossal energy of 20,000 million Gev! Numbers such as this are more than the mind can grasp, but we may get some idea of what that energy means when we calculate that the amount of energy represented by 10 billion Gev would be enough to enable a single sub-microscopic particle to raise a one-ton weight two inches.

Ever since cosmic rays were discovered, people have wondered where they came from and how they arise. The simplest concept is that somewhere in the Galaxy, perhaps in our sun, perhaps farther away, there are nuclear reactions going on which shoot forth particles with the huge energies we find them possessing. Indeed, bursts of mild cosmic rays occur every other year or so (as was first discovered in 1942) in connection with flares from the sun. What then of such sources as supernovae, pulsars, and quasars? But there is no known nuclear reaction that could produce anything like thousand millions and billions of Gev. The most energetic one we can conceive of would be the mutual annihilation of heavy nuclei of matter and anti-matter, and this would liberate at most 250 Gev.

The alternative is to suppose, as Fermi did, that some force in space accelerates the cosmic particles. They may come originally with moderate energies from explosions such as supernovae and gradually be accelerated as they travel through space. The most popular theory at present is that they are accelerated by cosmic magnetic fields, acting like gigantic synchrotrons. Magnetic fields do exist in space, and our Galaxy as a whole is thought to possess one, although this can at best be but 1/20,000 as intense as the magnetic field associated with the earth.

Travelling through this field, the cosmic particles would be slowly accelerated in a curved path. As they gained energy, their paths would swing out wider and wider until the most energetic ones would whip right out of the Galaxy. Although most of the particles would never reach this escape trajectory, because they would lose energy by collisions with other particles or with large bodies, some would. Indeed, the most energetic cosmic particles that reach us may be passing through our Galaxy after having been hurled out of other galaxies in this fashion.

More New Particles

The discovery of the anti-particles did not really disturb physicists; on the contrary, it was a pleasing confirmation of the symmetry of the universe. What did disturb them was a quick succession of discoveries showing that the proton, the electron, and the neutron were not the only 'elementary particles' they had to worry about.

The first of these complications had arisen even before the neutron was discovered. It had to do with the emission of beta particles by radioactive nuclei. The particle emitted by a radioactive nucleus generally carries a considerable amount of energy. Where does the energy come from? It is created by conversion of a little of the nucleus's mass into energy; in other words, the nucleus always loses a little mass in the act of expelling the particle. Now physicists had long been troubled by the fact that often the beta particle emitted in a nucleus's decay did not carry enough energy to account for the amount of mass lost by the nucleus. In fact, the electrons weren't all equally deficient. They emerged with a wide spectrum of energies, the maximum (attained by very few electrons) being almost right, but all the others falling short to a smaller or greater degree. Nor was this a necessary concomitant of sub-atomic particle emission. Alpha particles emitted by a particular nuclide possessed equal energies in expected quantities. What, then, was wrong with beta-particle emission? What had happened to the missing energy?

Lise Meitner, in 1922, was the first to ask this question with suitable urgency and by 1930, Niels Bohr, for one, was ready to abandon the great principle of conservation of energy, at least as far as it applied to sub-atomic particles. In 1931, however, Wolfgang Pauli, in order to save conservation of energy, suggested a solution to the riddle of the missing energy. His solution was very simple: another particle carrying the missing energy came out of the nucleus along with the beta particle. This mysterious second particle had rather strange properties. It has no charge and no mass; all it had, as it sped along at the velocity of light, was a certain amount of energy. It looked, in fact, like a fictional item created just to balance the energy books.

And yet, no sooner had it been proposed than physicists were sure that the particle existed. When the neutron was discovered and found to break down into a proton, releasing an electron which, as in beta decay, also carried a deficiency of energy, they were still surer. Enrico Fermi in Italy gave the putative particle a name – 'neutrino', Italian for 'little neutral one'.

The neutron furnished physicists with another piece of evidence for the existence of the neutrino. As I have mentioned, almost every particle has a spin. The amount of spin is expressed in multiples of one half, plus or minus, depending on the direction of the spin. Now the proton, the neutron, and the electron have each a spin of one half. If, then, the neutron, with spin one half, gives rise to a proton and an electron, each with spin one half, what happens to the law of conservation of angular momentum? There is something wrong here. The proton and the electron may total their spins to one (if both spin in the same direction) or to zero (if their spins are opposite), but any way you slice it their spins cannot add up to one half. Again, however, the neutrino comes to the rescue. Let the spin of the neutron be $+\frac{1}{2}$. Let the proton's spin be $+\frac{1}{2}$ and the electron's $-\frac{1}{2}$, for a net of 0. Now give the neutrino the spin $+\frac{1}{2}$, and the books are neatly balanced.

$$+\tfrac{1}{2}\,(\text{n}) = +\tfrac{1}{2}\,(\text{p}) - \tfrac{1}{2}\,(\text{e}) + \tfrac{1}{2}\,(\text{neutrino}).$$

There is still more balancing to do. A single particle (the neutron) has formed two particles (the proton and the electron), and if we include the neutrino, actually three particles. It seems more reasonable to suppose that the neutron is converted into two particles and an anti-particle, or a net of one particle. In other words, what we really need to balance is not a neutrino but an anti-neutrino.

The neutrino itself would arise from the conversion of a proton into a neutron. There the products would be a neutron (particle), a positron (anti-particle), and a neutrino (particle). This, too, balances the books.

The most important proton-to-neutron conversions are those involved in the nuclear reactions that go on in the sun and other stars. Stars therefore emit fast floods of neutrinos, and it is estimated that perhaps 6 to 8 per cent of their energy is carried off in this way. This, however, is only true for such stars as our sun. In 1961, the American physicist Hong Yee Chiu suggested that, as the central temperatures of a star rise, additional neutrino-producing reactions become important. As a star progresses in its evolutionary course towards a hotter and hotter core (see Chapter 2), a larger and larger proportion of its energy is carried off by neutrinos.

There is crucial importance in this. The ordinary method of transmitting energy, by photons, is slow. Photons interact with matter, and they make their way out from the sun's core to the surface only after uncounted myriads of absorptions and re-emissions. Consequently, although the sun's central temperature is 15,000,000°C, its surface is only 6,000°C. The substance of the sun is a good heat insulator.

Neutrinos, however, virtually do not interact with matter. It has been calculated that the average neutrino could pass through 100 light-years of solid lead with only a 50 per cent chance of being absorbed. This means that any neutrinos formed in the sun's core leave at once and at the speed of light, reaching the sun's surface, without interference, in less than three seconds and speeding off. (Any that move in our direction pass through us without affecting us in any way. This is true, day or night, for at night, when the bulk of the earth is between ourselves and the sun, the neutrinos can pass through the earth and ourselves as easily as through ourselves alone.)

By the time a central temperature of 6,000,000,000°K is reached, Chiu calculates, most of the star's energy is being pumped into neutrinos. The neutrinos leave at once, carrying the energy with them, and the sun's centre cools drastically. It is this, perhaps, which leads to the catastrophic contraction that then makes itself evident as a supernova.

Anti-neutrinos are produced in any neutron-to-proton conversion, but these do not go on (as far as is known) on the vast scale that leads to such floods of neutrinos from every star. The most important sources of anti-neutrinos are from natural radioactivity and uranium fission (which I shall discuss in more detail in Chapter 9).

Naturally physicists could not rest content until they had actually tracked down the neutrino; scientists are never happy to accept phenomena or laws of nature entirely on faith (as concepts such as the 'soul' must be). But how detect an entity as nebulous as the neutrino – an object with no mass, no charge, and practically no propensity to interact with ordinary matter?

Still, there was some slight hope. Although the probability of a neutrino reacting with any particle is exceedingly small, it is not quite zero. To be unaffected in passing through one hundred light-years of lead is just a measure of the average, but there will be some neutrinos that react with a particle before they go that far, and a few – an almost unimaginably small proportion of the total number – that will be stopped within the equivalent of 1/10 inch of lead.

In 1953, a group of physicists led by Clyde L. Cowan and Frederick Reines of the Los Alamos Scientific Laboratory set out to try the next-to-impossible. They erected their apparatus for detecting neutrinos next to a large fission reactor of the Atomic Energy Commission on the Savannah River in Georgia. The reactor would furnish streams of neutrons, which, hopefully, would release floods of anti-neutrinos. To catch them, the experimenters used large tanks of water. The plan was to let the anti-neutrinos bombard the protons (hydrogen nuclei) in the water and detect the results of the capture of an anti-neutrino by a proton.

What would happen? When a neutron breaks down, it yields a proton, an electron, and an anti-neutrino. Now a proton's absorption of an anti-neutrino should produce essentially the reverse. That is to say, the proton should be converted to a neutron, emitting a positron in the process. So there were two things to be looked for: (1) the creation of neutrons, and (2) the creation of positrons. The neutrons could be detected by dissolving a cadmium compound in the water, for when cadmium absorbs neutrons, it emits gamma rays of a certain characteristic energy. And the positrons could be identified by their annihilating interaction with electrons, which would yield certain other gamma rays. If the experimenters' instruments detected gamma rays of exactly these two telltale energies and separated by the proper time interval, they could be certain that they had caught anti-neutrinos.

The experimenters arranged their ingenious detection devices, waited patiently, and, in 1956, exactly a quarter century after Pauli's invention of the particle, they finally trapped the anti-neutrino. The newspapers and even some learned journals called it simply the 'neutrino'.

To get the real neutrino, we need some source that is rich in neutrinos. The obvious one is the sun. What system can be used to detect the neutrino as opposed to the anti-neutrino? One possibility (following a suggestion of the Italian physicist Bruno Pontecorvo) begins with chlorine 37, which makes up about a quarter of all chlorine atoms. Its nucleus contains 17 protons and 20 neutrons. If one of those neutrons absorbs a neutrino, it becomes a proton (and emits an electron). The nucleus will then have 18 protons and 19 neutrons and will be argon 37.

To form a sizeable target of chlorine neutrons, one might use liquid chlorine, but that is a very corrosive and toxic substance, and to keep it liquid would present a problem in refrigeration. Instead, chlorine-containing organic compounds can be used; one called tetrachloroethylene is a good one for the purpose.

The American physicist Raymond R. Davis made use of such a neutrino trap in 1956 to show that there really was a difference between the neutrino and the anti-neutrino. Assuming the two particles were different, the trap would detect only neutrinos and not anti-neutrinos. When it was set up near a fission reactor in 1956 under conditions where it would certainly detect anti-neutrinos (if anti-neutrinos were identical to neutrinos), it did *not* detect them.

The next step was to try to detect neutrinos from the sun. A huge tank containing 100,000 gallons of tetrachloroethylene was used for the purpose. It was set up in a deep mine in South Dakota. There was enough earth above it to absorb any particles emerging from the sun except neutrinos. (Consequently, we have the odd situation that in order to study the sun we must burrow deep, deep into the bowels of the earth.) The tank was then exposed to the solar neutrinos for several months to allow enough argon 37 to accumulate to be detectable. The tank was then flushed with helium for 22 hours and the tiny quantity of argon 37 in the helium gas determined. By 1968, solar neutrinos were indeed detected, but in less than half the amounts expected from current theories as to what is going on inside the sun. However, the experimental techniques involved here are fantastically difficult and it is, as yet, early days.

Our list of particles has grown, then, to eight: proton, neutron, electron, neutrino, and their respective anti-particles. This, however, did not exhaust the list. Further particles began to seem necessary to physicists if they were to explain how the particles in the nucleus hung together.

Ordinary attractions between protons and electrons, between one atom and another, between one molecule and another, could be explained by electromagnetic forces – the mutual attraction of opposite electric charges. This would not suffice for the nucleus, where the only charged particles present were protons. Indeed, by electromagnetic reasoning, we would suppose that the protons, all positively charged, should repel one another violently and that any atomic nucleus should explode with shattering force the instant it was formed (if it ever could be formed in the first place).

Clearly, some other force must be involved, something much stronger than the electromagnetic force and capable of overpowering it. The superior strength of this 'nuclear force' can be easily demonstrated by the following consideration. The atoms of a strongly bound molecule, such as that of carbon monoxide, can be pried apart by the application of only eleven electron volts of energy. That quantity of energy suffices to handle a strong manifestation of electromagnetic force.

On the other hand, the proton and neutron making up a

deuteron, one of the most weakly bound of all nuclei, require 2 million electron volts for disruption. Making allowance for the fact that particles within the nucleus are much closer to one another than atoms within a molecule, it is still fair to conclude that the nuclear force is 130 times as strong as the electromagnetic force.

But what is the nature of this nuclear force? The first fruitful lead came in 1932, when Werner Heisenberg suggested that the protons were held together by 'exchange forces'. He pictured the protons and neutrons in the nucleus as continually interchanging identity, so that any given particle was first a proton, then a neutron, then a proton, and so on. This might keep the nucleus stable in the same way that you might be able to hold a hot potato by tossing it quickly from hand to hand. Before a proton could 'realize' (so to speak) that it was a proton and try to flee its neighbour protons, it had become a neutron and could stay where it was. Naturally it could get away with this only if the changes took place exceedingly quickly, say within a millionth of a trillionth of a second.

Another way of looking at it is to imagine two particles, exchanging a third. Each time particle A emits the exchange particle it moves backwards to conserve momentum. Each time particle B accepts the exchange particle it is pushed backwards for the same reason. As the exchange particle bounces back and forth, particles A and B move farther and farther apart so that they seem to experience a repulsion. If, on the other hand, the exchange particle moves around boomerang-fashion, from the rear of particle A to the rear of particle B, then the two particles would be pushed closer together and seem to experience an attraction.

It would seem by Heisenberg's theory that all forces of attraction and repulsion would be the result of exchange particles. In the case of electromagnetic attraction and repulsion the exchange particle is the photon, which, as we shall see in the next chapter, is a mass-less particle associated with light and electromagnetic radiation generally. It can be argued that it is because the photon is mass-less that electromagnetic attraction and repulsion is long-range, lessening in intensity only as the square of the distance, and therefore important over interstellar and even intergalactic distances.

By this reasoning, the gravitational force, which also is long-range and also lessens in intensity as the square of the distance, should involve the continual exchange of mass-less particles. Physicists have named such a particle the 'graviton'.

The gravitational force is much, much weaker than the electromagnetic force. A proton and an electron attract each other gravitationally with only about $1/10^{39}$ as much force as they attract each other electromagnetically. The graviton must be correspondingly less energetic than the photon and must therefore be unimaginably difficult to detect.

Nevertheless, the American physicist Joseph Weber has been trying to detect the graviton since 1957. His most recent attempts have made use of a pair of aluminium cylinders 153 centimetres long and 66 centimetres wide, suspended by a wire in a vacuum chamber. The gravitons (which would be detected in wave-form) would displace those cylinders slightly, and a measuring system for detecting a displacement of a hundred billionth of a centimetre is used. The feeble waves of the gravitons, coming from deep in space, ought to wash over the entire planet, and cylinders separated by great distances ought to be affected simultaneously. In 1969, Weber announced he had detected the effects of gravitational waves. If so, the question is what – even out in space – could represent fluctuations in gravitational force sufficient to produce detectable waves? Are they events involving neutron stars, black holes, or what? We don't know.

But back to the nuclear force. Unlike the electromagnetic and the gravitational fields, it was short-range. Although extremely strong within the nucleus, it vanished almost completely outside the nucleus. For that reason, no mass-less exchange particle would do.

In 1935, the Japanese physicist Hideki Yukawa mathematically analysed the problem. An exchange particle possessing mass would produce a short-range force-field. The mass would be in inverse ratio to the range: the greater the mass, the shorter the range. It turned out that the mass of the appropriate particle lay somewhere between that of the proton and the electron; Yukawa estimated it to be between 200 and 300 times the mass of an electron.

Barely a year later, this very kind of particle was discovered. At the California Institute of Technology, Carl Anderson (the discoverer of the positron), investigating the tracks left by secondary cosmic rays, came across a short track that was more curved than a proton's and less curved than an electron's. In other words, the particle had an intermediate mass. Soon more such tracks were detected, and the particles were named 'mesotrons', or 'mesons' for short.

Eventually other particles in the intermediate mass range were discovered, and this first one was distinguished as the 'mu meson' or the 'muon'. ('Mu' is one of the letters of the Greek alphabet; almost all of them have now been used in naming subatomic particles.) As in the case of the particles mentioned earlier, the muon comes in two varieties, a negative and a positive.

The negative muon, 206·77 times as massive as the electron (and therefore about one ninth as massive as a proton) is the particle; the positive muon is the anti-particle. The negative muon and positive muon correspond to the electron and positron, respectively. Indeed, by 1960 it had become evident that the negative muon was identical with the electron in almost every way except mass. It was a 'heavy electron'. Similarly, the positive muon was a 'heavy positron'.

There is no explanation, so far, for the identity, but it carries through to the point where, as was discovered in 1953, negative muons can replace electrons in atoms to form 'muonic atoms'. Similarly, positive muons could replace positrons in anti-matter.

Positive and negative muons will undergo mutual annihilation and may briefly circle about a mutual centre of force before doing so – just as is true of positive and negative electrons. A more interesting situation, however, was discovered in 1960 by the American physicist Vernon Willard Hughes. He detected a system in which the electron circled a positive muon, a system he called 'muonium'. (A positron circling a negative muon would be 'anti-muonium'.)

The muonium atom (if it may be called that) is quite analogous to hydrogen 1, in which an electron circles a positive proton, and the two are similar in many of their properties. Although muons and electrons seem to be identical except for mass, that mass-difference is enough to keep the electron and the positive muon from being true opposites, so that one will not annihilate the other. Muonium, therefore, doesn't have the kind of instability that positronium has. Muonium endures longer and would endure for ever (if undisturbed from without) were it not for the fact that the muon itself does not. After two millionths of a second or so, the muon decays and the muonium atom ceases to exist.

Another similarity between muons and electrons is this: just as heavy particles may produce electrons plus anti-neutrinos (as when a neutron is converted to a proton) or positrons plus neutrinos (as when a proton is converted to a neutron) so heavy particles can interact to form negative muons plus anti-neutrinos or positive muons plus neutrinos. For years, physicists took it for granted that the neutrinos that accompanied electrons and positrons and those that accompanied negative and positive muons were identical. In 1962, however, it was learned that the neutrinos never crossed over, so to speak; the electron's neutrino was never involved in any interaction that would form a muon, and the muon's neutrino was never involved in any interaction that would form an electron or positron.

In short, physicists found themselves with two pairs of chargeless, mass-less particles, the electron's anti-neutrino and the positron's neutrino plus the negative muon's anti-neutrino and the positive muon's neutrino. What the difference between the two neutrinos and between the two anti-neutrinos might be is more than anyone can tell at the moment, but they are different.

The muons differ from the electron and positron in another respect, that of stability. The electron or positron, left to itself, will remain unchanged indefinitely. The muon is unstable, however, and breaks down after an average lifetime of a couple of millionths of a second. The negative muon breaks down to an electron (plus an anti-neutrino of the electron variety and a neutrino of the muon variety), while the positive muon does the same in reverse, producing a positron, an electron-neutrino, and a muon-anti-neutrino.

Since the muon is a kind of heavy electron, it cannot very well be the nuclear cement Yukawa was looking for. Electrons are not found within the nucleus and therefore neither should the muon be. This was discovered to be true on a purely experimental basis, long before the near identity of muon and electron was suspected; muons simply showed no tendency to interact with nuclei. For a while, Yukawa's theory seemed to be tottering.

In 1947, however, the British physicist Cecil Frank Powell discovered another type of meson in cosmic-ray photographs. It was a little more massive than the muon and proved to possess about 273 times the mass of an electron. The new meson was named a 'pi meson' or a 'pion'.

The pion was found to react strongly with nuclei and to be just the particle predicted by Yukawa. (Yukawa was awarded the Nobel Prize in physics in 1949, and Powell received it in 1950.) Indeed, there was a positive pion that acted as the exchange force between protons and neutrons, and there was a corresponding anti-particle, the negative pion, which performed a similar service for anti-protons and anti-neutrons. Both are even shorter lived than muons; after an average lifetime of about one fortieth of a microsecond, they break up into muons plus neutrinos of the muon variety. (And, of course, the muon breaks down further to electrons and additional neutrinos.) There is also a neutral pion, which is its own anti-particle. (There is, in other words, only one variety of that particle.) It is extremely unstable, breaking down in less than a trillionth of a second to form a pair of gamma rays.

Despite the fact that a pion 'belongs' within the nucleus, it will fleetingly circle a nucleus before interacting with it, sometimes, to form a 'pionic atom'. This was detected in 1952. Indeed, any pair of negative and positive particles or particle-systems can be made to circle each other, and in the 1960s, physicists have studied a number of evanescent 'exotic atoms' in order to gain some notion as to the details of particle structure.

Since its discovery, the pion has grown quite important to the physicists' view of the sub-atomic world. Even free protons and neutrons may be surrounded by tiny clouds of pions and may even be composed of them. The American physicist Robert Hofstadter investigated nuclei with extremely energetic electrons produced by a linear accelerator. He has suggested that both proton and neutron consist of cores made up of mesons. As a result of his work in this field, he shared in the Nobel Prize in physics in 1961.

As the number of known particles grew, it became necessary for physicists to divide them into groups. The lightest were called 'leptons', from a Greek word meaning 'small'. These include the electrons, the muons, their anti-particles, and their neutrinos. The photon is usually included also, as a particle with zero mass and zero charge but with a spin equal to 1. (In this way, the photon differs from the various neutrinos, which also have zero mass and charge, but which have a spin of one half.) The photon is considered to be its own anti-particle. The graviton is also included, differing from the other mass-less, charge-less particles by having a spin of 2.

The fact that photons and gravitons are their own anti-particles helps explain why it is so difficult to detect whether a distant galaxy is matter or anti-matter. Much of what we receive from a distant galaxy is photons and gravitons; a galaxy of anti-matter emits exactly the same photons and gravitons that a galaxy of matter would. There are no anti-photons and anti -gravitons that might act as distinctive fingerprints of anti-matter. We ought to receive neutrinos, however – or anti-neutrinos. A preponderance of neutrinos would mark matter; one of anti-neutrinos, anti-matter. With the development and improvement of techniques for detecting neutrinos or anti-neutrinos from outer space, it may become possible some day to pin down this matter of the existence and location of anti-galaxies.

Among the leptons, those that do not carry an electric charge also do not possess mass. These charge-less, mass-less particles are all stable. Left to themselves, each will endure unchanged (as far as we know) for ever. For some reason, charge can only exist when mass exists, but particles with mass tend to break down to particles with lesser mass. Thus a muon tends to break down to an electron. An electron (or positron) is, as far as we know, the least massive particle that can exist. For it to break down further would mean the loss of mass altogether, and this would also require the loss of electric charge. Since the law of conservation of electric charge makes it impossible to lose charge, the electron cannot break down. An electron and a positron can undergo mutual annihilation, for the opposite charges will cancel each other, but either one left to itself would, as far as we know, exist eternally.

More massive than the leptons are the 'mesons', which are a family that no longer includes the muon even though that particle was the original meson. Among the mesons now are the pions and a newer variety, the 'K-mesons' or 'kaons'. These were first detected in 1952 by two Polish physicists, Marian Danysz and Jerzy Pniewski. These are about 970 times as massive as an electron and, therefore, about half the mass of a proton or neutron. The kayon comes in two varieties, a positive kayon and a neutral kayon, and each has an anti-particle associated with it. They are unstable, of course, breaking down in about a microsecond to pions.

Above the mesons are the 'baryons' (from a Greek word meaning 'heavy'). Until the 1950s, the proton and the neutron were the only specimens known. Beginning in 1954, however, a series of still more massive particles (sometimes called 'hyperons') were discovered. It is the baryon particles that have particularly proliferated in recent years, in fact, and the proton and neutron are but the lightest of a large variety.

There is a 'law of conservation of baryon number', physicists have discovered, for in all particle breakdowns the net number of baryons (that is, baryons minus anti-baryons) remains the same. The breakdown is always from a more massive to a less massive particle and that explains why the proton is stable and is the *only* baryon to be stable. It happens to be the lightest baryon. If it broke down it would have to cease being a baryon and that would break the law of conservation of baryon number. For the same reason, an anti-proton is stable, because it is the lightest anti-baryon. Of course, a proton and an anti-proton can engage in mutual annihilation since, taken together, they make up one baryon plus one anti-baryon for a net baryon number of zero.

The first baryons beyond the proton and neutron to be discovered were given Greek names. There was the 'lambda particle', the 'sigma particle', and the 'xi particle'. The first came in one variety, a neutral particle; the second in three varieties, positive, negative, and neutral; the third in two varieties, negative and neutral. Every one of these had an associated anti-particle, making a dozen particles altogether. All were exceedingly unstable; none could live for more than a hundredth of a microsecond or so; and some, such as the neutral sigma particle, broke down after a hundred billionth of a microsecond.

The lambda particle, which is neutral, can replace a neutron in a nucleus to form a 'hypernucleus' – an entity which endures less than a thousand millionth of a second. The first to be discovered was a hyper-tritium nucleus made up of a proton, a neutron, and a lambda particle. This was located among the products of cosmic radiation by Danysz and Pniewski in 1952. In 1963, Danysz reported hypernuclei containing two lambda particles. What's more, negative hyperons can be made to replace electrons in atomic structure, as was first reported in 1968. Such massive electron-replacements circle the nucleus at such close quarters as to spend their time actually within the nuclear outer regions.

But all these are the comparatively stable particles; they live long enough to be directly detected and to be easily awarded a lifetime and personality of their own. In the 1960s, the first of a whole series of particles was detected by Alvarez (who received the Nobel Prize in physics in 1968 as a result). These were so short-lived that their existence could only be deduced from the necessity of accounting for their breakdown products. Their half-lives are something of the order of a few billionths of a billionth of a second, and one might wonder whether they are really individual particles or merely a combination of two or more particles, pausing to nod at each other before flashing by.

These ultra-short-lived entities are called 'resonance particles' and, counting them, over 150 different sub-atomic particles are now known; physicists are uncertain as to how many remain to be found. The situation among the particles is about what it was a century ago among the elements, before Mendeleev had proposed the periodic table.

Some physicists believe that the various mesons and baryons are not truly independent particles – that baryons can absorb and emit mesons to attain various levels of excitement, and that each excited baryon might easily be mistaken for a different particle. (We have a similar situation among atoms, where a particular atom may absorb or emit photons to reach various levels of electronic excitement – except that an excited hydrogen atom is still recognized as a hydrogen atom.)

What is needed, then, is some sort of periodic table for sub-atomic particles – something that would group them into families consisting of a basic member or members with other particles that are excited states of that basic member or members.

Something of the sort was proposed in 1961 by the American physicist Murray Gell-Mann and the Israeli physicist Yuval Ne'emen, who were working independently. Groups of particles were put together in a pattern that depended on their various properties into a beautifully symmetric fashion, which Gell-Mann called the 'eightfold way' but which is more formally referred to as 'SU 3'. In particular, one such grouping needed one more particle for completion. That particle, if it was to fit into the group, had to have a particular mass and a particular set of other properties. The combination was not a likely one for a particle to have; yet, in 1964, a particle (the 'omega-minus') was detected with just the predicted set of properties, and in succeeding years it was detected dozens of times. In 1971 its anti-particle, the 'anti-omega-minus', was detected.

Even if baryons are divided into groups and a sub-atomic periodic table is set up, there would still be enough different particles to give physicists the urge to find something still simpler and more fundamental. In 1964, Gell-Mann – having endeavoured to work out the simplest way of accounting for all the baryons with a minimum number of more fundamental 'sub-baryonic particles' – came up with the notion of 'quarks'. He got this name because he found that only three different quarks were necessary and that different combinations of the three quarks were needed to make up all the known baryons. This reminded him of a line from *Finnegans Wake* by James Joyce, which goes 'Three quarks for Musther Mark'.

In order to account for the known properties of baryons, the three different quarks had to have specific properties of their own. The most astonishing property was a fractional electric charge. All known particles had either no electric charge, an electric charge exactly equal to that of the electron (or positron), or an electric charge equal to some exact multiple of the electron (or positron). The known charges, in other words, were 0, +1, −1, +2, −2, and so on. One quark, however (the 'p-quark'), had a charge of +2/3, while the other two (the 'n-quark' and the 'lambda-quark') had a charge of −1/3 apiece. The n-quark and the lambda-quark were distinguished from each other by something called the 'strangeness number'. Whereas the n-quark (and the p-quark, too) had a strangeness number of 0, the lambda-quark had a strangeness number of −1.

Each quark had its 'anti-quark'. There was the anti-p-quark, with a charge of −2/3 and a strangeness number of 0; the anti-n-quark, with a charge of +1/3 and a strangeness number of 0; and the anti-lambda-quark, with a charge of +1/3 and a strangeness number of +1.

Now then, one can imagine a proton built up of two p-quarks and an n-quark, while a neutron is built up of two n-quarks and a p-quark (which is why we have the 'p' and 'n' prefixes to the quarks). A lambda particle is made up of a p-quark, an n-quark, and a lambda-quark (which accounts for the last-named prefix), an omega-minus particle is made up of three lambda-quarks, and so on. One can even combine quarks by twos to form the different mesons.

The question is, though, however convenient the quarks may

THE LONGER-LIVED SUB-ATOMIC PARTICLES

FAMILY NAME	PARTICLE NAME	SYMBOL	MASS	SPIN	ELECTRIC CHARGE
	PHOTON	γ (GAMMA RAY)	0	1	NEUTRAL
	GRAVITON	———	0	2	NEUTRAL
ELECTRON FAMILY	ELECTRON'S NEUTRINO	ν_e	0	½	NEUTRAL
	ELECTRON	e^-	1	½	NEGATIVE
MUON FAMILY	MUON'S NEUTRINO	ν_μ	0 (?)	½	NEUTRAL
	MUON	μ^-	206.77	½	NEGATIVE
MESONS	PION	π^+	273.2	0	POSITIVE
		π^-	273.2	0	NEGATIVE
		π^0	264.2		
	KAON	K^+	966.6	0	POSITIVE
		K^0	974	0	NEUTRAL
BARYONS	NUCLEON	p (PROTON)	1836.12	½	POSITIVE
		n (NEUTRON)	1838.65	½	NEUTRAL
	LAMBDA	Λ^0	2128.8	½	NEUTRAL
	SIGMA	Σ^+	2327.7	½	POSITIVE
		Σ^-	2340.5	½	NEGATIVE
		Σ^0	2332	½	NEUTRAL
	XI	Ξ^-	2580	½	NEGATIVE
		Ξ^0	2570	½	NEUTRAL

* The K^0 meson has two different lifetimes. All other particles have only one.

After a table in K. W. Ford, *The World of Elementary Particles*. Reprinted through the courtesy of Blaisdell Publishing Company, a division of Ginn and Company.

ANTI-PARTICLE	NO. OF DISTINCT PARTICLES	AVERAGE LIFETIME (SECONDS)	TYPICAL MODE OF DECAY
SAME PARTICLE	1	INFINITE	——
SAME PARTICLE	1	INFINITE	——
$\overline{\nu_e}$	2	INFINITE	——
e^+ (POSITRON)	2	INFINITE	——
$\overline{\nu_\mu}$	2	INFINITE	——
μ^+	2	2.212×10^{-6}	$\mu^- \rightarrow e^- + \nu_e + \nu_\mu$
π^- SAME AS		2.55×10^{-8}	$\pi^+ \rightarrow \mu^+ + \nu_\mu$
π^+ THE	3	2.55×10^{-8}	$\pi^- \rightarrow \mu^- + \nu_\mu$
π^0 PARTICLES		1.9×10^{-16}	$\pi^0 \rightarrow \gamma + \gamma$
$\overline{K^+}$ (NEGATIVE)		1.22×10^{-8}	$K^+ \rightarrow \pi^+ + \pi^0$
$\overline{K^0}$	4	1.00×10^{-10} and* 6×10^{-8}	$K^0 \rightarrow \pi^+ + \pi^-$
$\overline{p}$ (NEGATIVE)	4	INFINITE	
$\overline{n}$		1013	$n \rightarrow p + e^- + \nu_e$
$\overline{\Lambda^0}$	2	2.51×10^{-10}	$\Lambda^0 \rightarrow p + \pi^-$
$\overline{\Sigma^+}$ (NEGATIVE)		8.1×10^{-11}	$\Sigma^+ \rightarrow n + \pi^+$
$\overline{\Sigma^-}$ (POSITIVE)	6	1.6×10^{-10}	$\Sigma^- \rightarrow n + \pi^-$
$\overline{\Sigma^0}$		ABOUT 10^{-20}	$\Sigma^0 \rightarrow \Lambda^0 + \gamma$
$\overline{\Xi^-}$ (POSITIVE)	4	1.3×10^{-10}	$\Xi^- \rightarrow \Lambda^0 + \pi^-$
$\overline{\Xi^0}$		ABOUT 10^{-10}	$\Xi^0 \rightarrow \Lambda^0 + \pi^0$
	33		

be mathematically, do they really exist? That is, we may agree that there are four quarters to a dollar, but does that mean that somewhere in a paper dollar bill there are four metallic quarters? One way of answering the question would be to stroke a proton, neutron, or other particle, with such energy as to cause it to fly apart into its constituent quarks. Unfortunately, the binding forces holding quarks together are far higher than those holding baryons together (just as those are far higher than the forces holding atoms together), and man does not yet dispose of enough energy to break up the baryon. There are cosmic-ray particles with enough energy to do so (if it can be done), but though some reports of quark-like particles among cosmic-ray products have been heard, these are not widely accepted.

At the moment of writing, the quark hypothesis must be regarded as interesting – but speculative.

The K-mesons and the hyperons introduced physicists to a fourth field of force different from the three already known: gravitational, electromagnetic, and nuclear.

Of these three, the nuclear force is by far the strongest, but it acts only over an extremely short distance. Whereas the electromagnetic and gravitational forces decrease only as the square of the distance, nuclear forces drop off so rapidly with distance that the force between two nucleons falls almost to zero if they are separated by a distance greater than their own diameter. (This is as if the earth's gravity were to become practically zero 4,000 miles above the surface.) Consequently interactions between particles under the influence of nuclear forces must take place very quickly.

For instance, imagine a pi meson and a proton approaching each other. If the nuclear force is to cause them to interact, it must do so while they are within a proton's width of each other. A proton's width is about 0·0000000000001 centimetre. Flying mesons are travelling at almost the speed of light, which is 30,000 million centimetres a second. Thus the pi meson will be within the influence of the nuclear force for only about 0·00000000000000000000001 second (a hundred thousandth of a trillionth of a second). And yet, even in this short time, the nuclear force brings about an interaction. The pi meson and the proton can react to produce a lambda hyperon and a K-meson.

This is an example of what physicists call a 'strong interaction'. Baryons and those mesons subject to strong interactions are lumped together as 'hadrons', from a Greek word meaning 'bulky'. A 'weak interaction' is one that requires a considerably longer time. The theory of such interactions was first worked out in 1934 by Fermi. An example of such an interaction is the breakdown of a K-meson or a hyperon. This takes one thousand millionth of a second or so. That may seem a breathlessly short time, but compared to the time it takes for a pi meson and proton to interact, it is very long. It is, in fact, about a billion times as long as the physicists had expected, considering the speed of most nuclear interactions.

They concluded that the 'weak interactions' were governed by forces much weaker than the nuclear forces, and they took to calling some of the particles that broke down as a result of weak interactions 'strange particles'. This name applied primarily to K-mesons and hyperons.

Gell-Mann gave different particles something called 'strangeness number' (which we met earlier in connection with quarks), awarding the numbers in such a way that strangeness is conserved in all particle interactions. That is, the net strangeness number is the same afterwards as before.

This fourth type of interaction should be mediated by way of an exchange particle, too. The gravitational, electromagnetic, and strong interactions, have the graviton, the photon, and the pion as theirs; the weak interaction should have something called the 'W-particle'. It is also called an 'intermediate boson', because it ought to obey Bose–Einstein statistics and because it ought to have an intermediate rate of decay. So far the W-particle has not actually been detected.

Nuclear physicists now deal with about a dozen conservation laws. Some are the familiar conservation laws of nineteenth-century physics: the conservation of energy, the conservation of momentum, the conservation of angular momentum, and the conservation of electric charge. Then there are conservation laws that are less familiar: the conservation of strangeness, the conservation of baryon number, the conservation of isotopic spin, and so on.

The strong interactions seem to obey all these conservation laws; in the early 1950s, physicists took it for granted that the laws were universal and irrevocable. But they were not. In the case of weak interactions, some of the conservation laws are not obeyed.

The particular conservation law that was shattered was the 'conservation of parity'. Parity is a strictly mathematical property that cannot be described in concrete terms; suffice it to say that the property refers to a mathematical function that has to do with the wave characteristics of a particle and its position in space. Parity has two possible values – 'odd' and 'even'. The key point we must bear in mind is that parity has been considered a basic property that, like energy or momentum, is subject to the law of conservation: in any reaction or change, parity must be conserved. That is to say, when particles interact to form new particles, the parity on both sides of the equation (so it was thought) must balance, just as mass numbers must, or atomic numbers, or angular momentum.

Let me illustrate. If an odd-parity particle and an even-parity particle interact to form two other particles, one of the new particles must be odd parity and the other even parity. If two odd-parity particles form two new particles, both of the new ones must be odd or both even. Conversely, if an even-parity particle breaks down to form two particles, both must be even parity or both must be odd parity. If it forms three particles, either all three have even parity or one has even parity and the other two have odd parity. (You may be able to see this more clearly if you consider the odd and even numbers, which follow similar rules. For instance, an even number can only be the sum of two even numbers or of two odd numbers, but never the sum of an even number and an odd one.) This is what is meant by the 'conservation of parity'.

The beginning of the trouble came when it was found that K-mesons sometimes broke down to two pi mesons (which, since the pi meson has odd parity, added up to even parity) and sometimes gave rise to three pi mesons (adding up to odd parity). Physicists concluded that there were two types of K-meson, one of even parity and one of odd parity; they named the two the 'theta meson' and the 'tau meson', respectively.

Now in every respect except the parity result, the two mesons were identical: the same mass, the same charge, the same stability, the same everything. It was hard to believe that there could be two particles with exactly the same properties. Was it possible that the two were actually the same and that there was something wrong with the ideas of the conservation of parity? In 1956, two young Chinese physicists working in the United States, Tsung Dao Lee and Chen Ning Yang, made precisely that suggestion. They proposed that, although the conservation of parity held in strong interactions, it might break down in weak interactions, such as the decay of K-mesons.

As they worked out this possibility mathematically, it seemed to them that if the conservation of parity broke down, the particles involved in weak interactions should show a 'handedness', something first pointed out in 1927 by the Hungarian physicist Eugene Wigner. Let me explain.

Your right hand and left hand are opposites. One can be considered the mirror image of the other: in a mirror the right hand looks like a left hand. If all hands were symmetrical in every respect, the mirror image would be no different from the direct image, and there would be no such distinction as 'right' and 'left' hand. Very well then, let us apply this to a group of particles emitting electrons. If electrons come out in equal numbers in all directions, the particle in question has no 'handedness'. But if most of them tend to go in a preferred direction – say up rather than down – then the particle is not symmetrical. It shows a 'handedness': if we look at the emissions in a mirror, the preferred direction will be reversed.

The thing to do, therefore, was to observe a collection of particles that emit electrons in a weak interaction (say some particle that decays by beta emission) and see if the electrons came out in a preferred direction. Lee and Yang asked an experimental physicist at Columbia University, Chien-Shiung Wu, to perform the experiment.

She set up the necessary conditions. All the electron-emitting

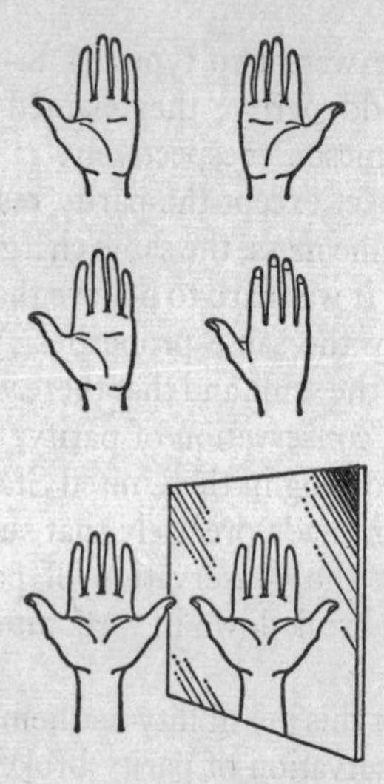

Mirror-image asymmetry and symmetry illustrated by hands.

atoms had to be lined up in the same direction if a uniform direction of emission was to be detected; this was done by means of a magnetic field and the material was kept at a temperature near absolute zero.

Within forty-eight hours the experiment yielded the answer. The electrons were indeed emitted asymmetrically. The conservation of parity did break down in weak interactions. The 'theta meson' and the 'tau meson' were one and the same particle, breaking down with odd parity in some cases, with even parity in others. Other experimenters soon confirmed the overthrow of parity, and for their bold conjecture the theoretical physicists Lee and Yang received the Nobel Prize in physics in 1957.

If symmetry breaks down with respect to weak interactions, perhaps it will break down elsewhere. The universe as a whole may be left-handed (or right-handed) after all. Alternatively, there may be two universes, one left-handed, the other right-handed; one composed of matter, the other of anti-matter.

Physicists are now viewing the conservation laws in general with a new cynicism. Any one of them might, like conservation of parity, apply under some conditions and not under others.

Parity, after its fall, was combined with 'charge conjugation', another mathematical property assigned to sub-atomic particles, which governed its status as a particle or anti-particle, and the two together were thought to be conserved. Added also was still another symmetry, which implied that the law governing sub-atomic events was the same whether time runs forwards or backwards. The whole may be called 'CPT-conservation'. In 1964, however, nuclear reactions were discovered that violated CPT-conservation. This affects 'time reversal'. It means you can distinguish between time running forwards (on the sub-atomic scale) and time running backwards – something physicists had thought to be impossible. To avoid this dilemma, a possible fifth force, weaker even than the gravitational force, was postulated. Such a fifth force, however, ought to produce certain detectable effects. In 1965, those effects were sought and not found. Physicists were left with their time dilemma. But there is no reason to despair. Such problems seem always to lead, eventually, to new and deeper understanding of the universe.

Inside the Nucleus

Now that so much has been learned about the general make-up and nature of the nucleus, there is great curiosity as to its structure, particularly the fine structure inside. First of all, what is its shape? Because it is so small and so tightly packed with neutrons and protons, physicists naturally assume that it is spherical. The fine details of the spectra of atoms suggest that many nuclei have a spherical distribution of charge. Some do not: they behave as if they have two pairs of magnetic poles, and these nuclei are said to have 'quadrupole moments'. But their deviation from the spherical is not very large. The most extreme case is that of the nuclei of the lanthanides, in which the charge distribution seems to make up a prolate spheroid (rugby-ball-shaped, in other words). Even here the long axis is not more than 20 per cent greater than the short axis.

As for the internal structure of the nucleus, the simplest model pictures it as a tightly packed collection of particles much like a drop of liquid, where the particles (molecules) are packed closely with little space between, where the density is virtually even throughout, and where there is a sharp surface boundary.

This 'liquid-drop model' was first worked out in detail in 1936 by Niels Bohr. It suggests a possible explanation of the absorption and emission of particles by some nuclei. When a particle enters the nucleus, one can suppose, it distributes its energy of motion among all the closely packed particles, so that no one particle receives enough energy immediately to break away. After perhaps a thousand billionth of a second, when there has been time for thousands of millions of random collisions, some particle accumulates sufficient energy to fly out of the nucleus.

The model could also account for the emission of alpha particles by the heavy nuclei – that is, the unstable elements with atomic numbers above 83. In these large nuclei, the short-range nuclear forces may not reach all the way across the nucleus; hence the force of repulsion between positive particles can take effect. As a result, portions of the nucleus in the form of the two-proton, two-neutron alpha particle (a very stable combination) may break off spontaneously from the surface of the nucleus. After the nucleus has decayed to a size such that the nuclear force overwhelms the force of repulsion, the nucleus becomes stable.

The liquid-drop model suggests another form of nuclear instability. When a large drop of liquid suspended in another liquid is set wobbling by currents in the surrounding fluid, it tends to break up into smaller spheres, often into roughly equal halves. We can think of the fission of uranium as closely analogous to this process. The fissionable nucleus, when struck by a neutron, begins to wobble, in a manner of speaking. It may stretch out into the shape of a dumb-bell (as a liquid drop would), and in that case the nuclear attractive forces would not reach from one end of the dumb-bell to the other, with the result that the repulsive force would drive the two portions apart. Bohr offered this explanation when nuclear fission was discovered.

Other nuclei besides uranium 235 ought to be (and proved to be) subject to fission if they receive enough input of energy. In fact, if a nucleus is large enough for the repulsive forces to become important, it ought occasionally to fission even without the input of energy. (This is like saying that the drop-like nucleus is always vibrating and wobbling, and every once in a while the vibration is strong enough to produce the dumb-bell and bring about the break.)

In 1940, two Soviet physicists, G. N. Flerov and K. A. Petrjak, discovered indeed that the heavier isotope of uranium, U 238, sometimes fissions spontaneously, without the addition of any particle. Uranium exhibits instability mainly by emitting alpha particles, but in a pound of uranium there are four spontaneous fissions per second while about 8 million nuclei are emitting alpha particles.

Spontaneous fission also takes place in uranium 235, in protactinium, in thorium, and, more frequently, in the transuranium elements. As nuclei get larger and larger, the probability of spontaneous fission increases. In the heaviest elements of all – einsteinium, fermium, and mendelevium – it becomes the most important method of breakdown, far outweighing alpha-particle emission.

Another popular model of the nucleus likens it to the atom as a whole, picturing the nucleons within the nucleus, like the electrons around the nucleus, as occupying shells and sub-shells, each affecting the others only slightly. This is called the 'shell model'.

How can there be room for independent shells of nucleons in the tiny, tightly packed nucleus? Well, however it is managed, the evidence suggests that there is some 'empty space' there. For instance, in a mesonic atom the meson may actually circle in an orbit within the nucleus for a short time. And Robert Hofstadter has found that the nucleus consists of a high-density core surrounded by a 'skin' of gradually decreasing density. The thickness of the skin is about half the radius of the nucleus, so that it actually makes up seven eighths of the volume.

By analogy with the situation in the atom's electronic shells, one may suppose that the nuclei with filled outer nucleonic shells should be more stable than those whose outer shells are not filled. The simplest theory would indicate that nuclei with 2, 8, 20, 40, 70, or 112 protons or neutrons, would be particularly stable. This, however, does not fit observation. The German-American physicist Maria Goeppert Mayer took account of the spin of the protons and neutrons and showed how this would affect the situation. It turned out that nuclei containing 2, 8, 20, 50, 82, or 126 protons or neutrons would then be particularly stable, and that fit the observations. Nuclei with 28 or 40 protons or neutrons would be fairly stable. All others would be less stable, if stable at all. These 'shell numbers' are sometimes called 'magic numbers' (with 28 or 40 occasionally referred to as 'semi-magic numbers').

Among the magic-number nuclei are helium 4 (2 protons and 2 neutrons), oxygen 16 (8 protons and 8 neutrons), and calcium 40 (20 protons and 20 neutrons), all especially stable and more abundant in the universe than other nuclei of similar size.

As for the higher magic numbers, tin has ten stable isotopes, each with 50 protons, and lead has four, each with 82 protons. There are five stable isotopes (each of a different element) with 50 neutrons each, and seven stable isotopes with 82 neutrons each. In general, the detailed predictions of the nuclear-shell theory work best near the magic numbers. Midway between (as in the case of the lanthanides and actinides), the fit is poor. But just in the midway regions, nuclei are farthest removed from the spherical (and shell theory assumes spherical shape) and are most markedly ellipsoidal. The 1963 Nobel Prize for physics was awarded to Goeppert Mayer and to two others: Wigner, and the German physicist J. Hans Daniel Jensen, who also contributed to the theory.

In general, as nuclei grow more complex, they become rarer in the universe, or less stable, or both. The most complex stable isotopes are lead 208 and bismuth 209, each with the magic number of 126 neutrons, and lead, with the magic number of 82 protons in addition. Beyond that all nuclides are unstable and with (in general) shortening half-lives as the number of protons, neutrons, or both are increased. But the shortening is not invariable. Thorium 232 has a half-life of 14,000 million years and is all but stable. Even californium 251 has a half-life in the centuries.

Could very complex nuclides – those more complex than any

that have been observed or synthesized – be sufficiently stable to permit formation in relatively large quantities? There are, after all, magic numbers beyond 126, and these might result in certain relatively stable super-complex atoms. Calculations have shown that, in particular, an element with 114 protons and 184 neutrons might have a surprisingly long half-life; physicists who have now reached element number 105 are seeking out ways of attaining this isotope of element 114 which, chemically, ought to resemble lead.

7 The Waves

Light

Of all the helpful attributes of nature, the one that man probably appreciates most is light. According to the Bible, the first words of God were 'Let there be light', and the sun and the moon were created primarily to serve as sources of light: 'And let them be for lights in the firmament of the heaven to give light upon the earth.'

The scholars of ancient and medieval times were completely in the dark as to the nature of light. They speculated that it consisted of particles emitted by the glowing object or perhaps by the eye itself. The only facts about it that they were able to establish were that light travelled in a straight path, that it was reflected from a mirror at an angle equal to that at which the beam struck the mirror, and that a light beam was bent ('refracted') when it passed from air into glass, water, or some other transparent substance.

The first important experiments on the nature of light were conducted by Isaac Newton in 1666. He let a beam of sunlight, entering a dark room through a chink in a blind, fall obliquely on one face of a triangular glass prism. The beam was bent when it entered the glass and then bent still farther in the same direction when it emerged from a second face of the prism. Newton caught the emerging beam on a white screen to see the effect of the reinforced refraction. He found that instead of forming a spot of white light, the beam was now spread out in a band of colours – red, orange, yellow, green, blue, and violet, in that order.

Newton deduced that ordinary white light was a mixture of different kinds of light which, separately, affect our eyes so as to produce the sensation of different colours. The spread-out band of its components was called a 'spectrum', from a Latin word meaning 'ghost'.

Newton decided that light consisted of tiny particles ('corpuscles') travelling at enormous speed. This would explain why light travelled in straight lines and cast sharp shadows. It was reflected by a mirror because the particles bounced off the surface, and it was bent on entering a refracting medium (such as water or glass) because the particles travelled faster in such a medium than in air.

Still, there were some awkward questions. Why should the particles of green light, say, be refracted more than those of yellow light? Why was it that two beams of light could cross without affecting each other – that is, without the particles colliding?

In 1678, the Dutch physicist Christian Huyghens (a versatile scientist who had built the first pendulum clock and done important work in astronomy) suggested an opposing theory, namely, that light consisted of tiny waves. If it was made up of waves, there was no difficulty about explaining the different amount of refraction of different kinds of light through a refracting medium, provided it was assumed that light travelled more slowly through the refracting medium than through air. The amount of refraction would vary with the length of the waves: the shorter the wavelength, the greater the refraction. This meant that violet light (the most refracted) had a shorter wavelength than blue light, blue shorter than green, and so on. It was this difference in wavelength that distinguished the colours to the eye. And, of course, if light consisted of waves, two beams could cross without trouble. (After all, sound waves and water waves crossed without losing their identity.)

But Huyghens' wave theory was not very satisfactory either. It did not explain why light rays travelled in straight lines and cast sharp shadows, nor why light waves could not go round obstacles, as water waves and sound waves could. Furthermore, if light consisted of waves, how could it travel through a vacuum, as it certainly did in coming to us through space from the sun and stars? What medium was it waving?

For about a century, the two theories contended with each other. Newton's 'corpuscular theory' was by far the more popular, partly because it seemed on the whole more logical, and partly because it had the support of Newton's great name. But in 1801 an English physician and physicist, Thomas Young, performed an experiment that swung opinion the other way. He projected a narrow beam of light through two closely spaced holes towards a screen behind. If light consisted of particles, presumably the two beams emerging through the holes would simply produce a brighter region on the screen where they overlapped and less bright regions where they did not. But this was not what Young found. The screen showed a series of bands of light, each separated from the next by a dark band. It seemed that in these dark intervals, the light of the two beams together added up to darkness!

The wave theory would easily explain this. The bright band represented the reinforcement of waves of one beam by waves of the other; in other words, there the two sets of waves were 'in phase', both peaks together and strengthening each other. The dark bands, on the other hand, represented places where the waves were 'out of phase', the trough of one cancelling the peak of the other. Instead of reinforcing each other, the waves at these places interfered with each other, leaving the net light energy at those points zero.

From the width of the bands and the distance between the two holes through which the beams issued, it was possible to calculate the length of light waves, say of red light or violet or colours between. The wavelengths turned out to be very small indeed. The wavelength of red light, for example, came to about 0·000075 centimetre. (Nowadays the wavelengths of light are expressed in a convenient unit suggested by Ångström. The unit, called the ångström – abbreviated Å – is one hundred-millionth of a centimetre. Thus the wavelength of red light is about 7,500 ångström units, the wavelength of violet light is about 3,900 ångström units, and the colour wavelengths of the visible spectrum lie between these numbers.)

The shortness of the wavelengths is very important. The reason light waves travel in straight lines and cast sharp shadows is that they are incomparably smaller than ordinary objects; waves can curve round an obstruction only when that obstruction is not much larger than the wavelength. Even bacteria, for instance, are vastly wider than a wavelength of light, so light can define them sharply under a microscope. Only objects somewhere near a wavelength of light in size (for example, viruses and other sub-microscopic particles) are small enough for light waves to pass round them.

It was the French physicist Augustin Jean Fresnel who showed (in 1818) that if an interfering object is small enough, a light wave will indeed travel round it. In that case, the light produces what is called a 'diffraction' pattern. For instance, the very fine parallel lines of a 'diffraction grating' act as a series of tiny obstacles that reinforce one another. Since the amount of diffraction depends on the wavelength, a spectrum is produced. From the amount by which any colour or portion of the spectrum is diffracted and from the known separation of the scratches on the glass, the wavelength can again be calculated.

Fraunhofer pioneered in the use of such diffraction gratings, an advance generally forgotten in the light of his more famous discovery of spectral lines. The American physicist Henry Augustus Rowland invented concave gratings and developed techniques for ruling them with as many as 20,000 lines to the inch. It was this that made it possible for the prism to be supplanted in spectroscopy.

Between such experimental findings and the fact that Fresnel worked out the mathematics of wave motion systematically, the wave theory of light seemed established and the corpuscular theory smashed – seemingly for good.

Not only were light waves accepted as existing, their length was measured with increasingly good precision. By 1827, the French physicist Jacques Babinet was suggesting that the wavelength of light – an unalterable physical quantity – be used as the standard for measurement of length, instead of the various man-made standards that were in fact used. This suggestion did not become practicable, however, until the 1880s, when the German-American physicist Albert Abraham Michelson invented an instrument called the 'interferometer', which could measure the wavelengths of light with unprecedented accuracy. In 1893, Michelson measured the wavelength of the red line in the cadmium spectrum and found it to be 1/1,553,164 metre long.

A measure of uncertainty still existed when it was discovered that elements consisted of different isotopes, each contributing a line of slightly different wavelength. As the twentieth century progressed, however, the spectral lines of individual isotopes were measured. In the 1930s, the lines of krypton 86 were measured. This isotope, being that of a gas, could be dealt with at low temperatures where atomic motion was slowed, with less consequent thickening to the line.

In 1960, the krypton-86 line was adopted by the General Conference of Weights and Measures as the fundamental standard of length. The metre has been redefined as equal to 1,650,763·73 wavelengths of this spectral line. This has increased the precision of measurement of length a thousandfold. The old standard metre bar could be measured, at best, to within one part in a million, whereas the light wave can be measured to within one part in a thousand million.

Light obviously travels at tremendous speeds. If you put out a light, it gets dark everywhere at once, as nearly as can be made out. This is not quite true for sound, for instance. If you watch a man in the distance chopping wood, you do not hear the stroke until some moments after the axe strikes. Sound has clearly taken a certain amount of time to travel to the ear. In fact, its speed of travel is easy to measure: it amounts to 1,090 feet per second, or about 750 miles per hour, in the air at sea level.

Galileo was the first to try to measure the speed of light. Standing on one hill while an assistant stood on another, he would uncover a lantern; as soon as the assistant saw the flash, he would signal by uncovering a light of his own. Galileo did this at greater and greater distances, assuming that the time it took the assistant to make his response would remain uniform and therefore any increase in the interval between his uncovering his own lantern and seeing the responding flash would represent the time taken by the light to cover the extra distance. The idea was sound, but of course light travels much too fast for Galileo to have detected any difference by this crude method.

In 1676, the Danish astronomer Olaus Roemer did succeed in timing the speed of light – on an astronomical distance scale. Studying Jupiter's eclipses of its four large satellites, Roemer noticed that the interval between successive eclipses became longer when the earth was moving away from Jupiter and became shorter when it was moving towards Jupiter in its orbit. Presumably the difference in eclipse times reflected the difference in distance between the earth and Jupiter; that is, it would be a measure of the distance in the time that light took to travel between Jupiter and the earth. From a rough estimate of the size of the earth's orbit, and from the maximum discrepancy in the eclipse timing, which Roemer took to represent the time it took light to cross the full width of the earth's orbit, he calculated the speed of light. His estimate came to 140,000 miles per second, remarkably good for what might be considered a first try and high enough to evoke the disbelief of his contemporaries.

Roemer's results were, however, confirmed a half century later from a completely different direction. In 1728, the British Astro-

Fizeau's arrangement for measuring the speed of light. Light reflected by the semi-mirror near the source passes through a gap in the rapidly spinning toothed wheel to a distant mirror (*right*) and is reflected back to the next tooth or the next gap.

nomer Royal, James Bradley, found that stars seemed to shift position because of the earth's motion; not through parallax, but because the velocity of the earth's motion about the sun was a measurable (though small) fraction of the speed of light. The analogy usually used is that of a man under an umbrella striding through a rainstorm. Even though the drops are falling vertically, the man must tip the umbrella forwards, for he is stepping into the drops. The faster he walks, the farther he must tip the umbrella. Similarly, the earth moves into the light rays falling from the stars, and the astronomer must tip his telescope a little bit, and in different directions, as the earth changes its direction of motion. From the amount of tip (the 'aberration of light'), Bradley could estimate the value of the speed of light and got a higher, and better, value than Roemer had.

Eventually, scientists obtained still more accurate measurements by applying refinements of Galileo's original idea. In 1849, the French physicist Armand Hippolyte Louis Fizeau set up an arrangement whereby a light flashed in a mirror five miles away and reflected back to the observer. The elapsed time for the ten-mile round trip of the flash was not much more than 1/20,000 of a second, but Fizeau was able to measure it by placing a rapidly rotating toothed wheel in the path of the light beam. When the wheel turned at a certain speed, the flash going out between the two teeth would hit the next tooth when it came back from the mirror, and so Fizeau, behind the wheel, would not see it. When the wheel was speeded up, the returning flash would not be blocked but would come through the next gap between teeth. Thus, by controlling and measuring the speed of the turning wheel, Fizeau was able to calculate the elapsed time, and therefore the speed of travel, of the flash of light.

A year later, Jean Foucault (who was soon to perform his pendulum experiment; see p. 111) refined the measurement by using a rotating mirror instead of a toothed wheel. Now the elapsed time was measured by a slight shift in the angle of reflection by the rapidly turning mirror. Foucault got a value of 187,000 miles per second for the speed of light in air. In addition, Foucault used his method to determine the speed of light through various liquids.

He found the speed to be markedly less than the speed of light in air. This fitted Huyghens' wave theory, too.

Still greater precision in the measurement of light's velocity came with the work of Michelson, who over a period of more than forty years, starting in 1879, applied the Fizeau–Foucault approach with greater and greater refinement. He eventually sent

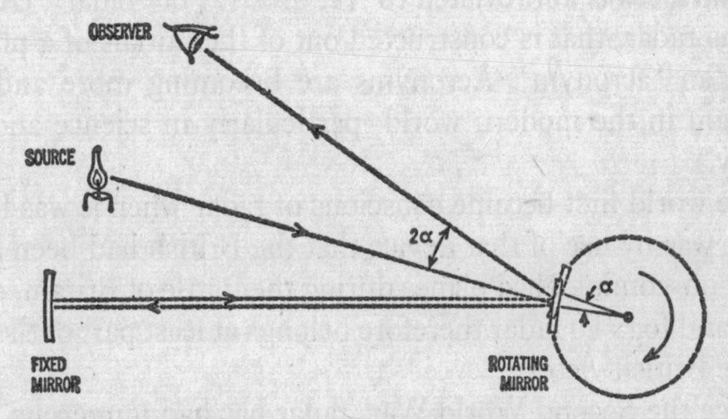

Foucault's method. The amount of rotation of the mirror, instead of Fizeau's toothed wheel, gave the speed of the light's travel.

light through a vacuum rather than through air (even air slows it up slightly), using evacuated steel pipes up to a mile long for the purpose. He measured the speed of light in a vacuum to be 186,284 miles per second. He was also to show that all wavelengths of light travel at the same speed in a vacuum.

In 1963, still more precise measurements have placed the speed of light at 186,281·7 miles per second. With the speed of light known with such amazing precision, it became possible to use light, or at least forms of light, to measure distance.

Imagine a short pulse of light moving outwards, striking some obstacle, being reflected backwards, and being received at the point where it had issued forth an instant before. What is needed is a wave form of low enough frequency to penetrate fog, mist, and cloud, but of high enough frequency to be reflected efficiently. The ideal range was found to be in the microwave (very short radio wave) region, with wavelengths of from 0·5 to 100 centimetres. From the time lapse between emission of the pulse and return of the echo, the distance of the reflected object can be estimated.

A number of physicists worked on devices making use of this principle, but the Scottish physicist Robert Alexander Watson-Watt was the first to make it thoroughly practicable. By 1935, he had made it possible to follow an aeroplane by the microwave reflections it sent back. The system was called '*ra*dio *d*etection *a*nd *r*anging', the word 'range' meaning 'to determine the distance of'. The phrase was abbreviated to 'ra. d. a. r.', or 'radar'. (A word, such as radar, that is constructed out of the initials of a phrase is called an 'acronym'. Acronyms are becoming more and more common in the modern world, particularly in science and technology.)

The world first became conscious of radar when it was learned that it was by use of that device that the British had been able to detect on-coming Nazi planes during the Battle of Britain, despite night and fog. To radar therefore belongs at least part of the credit for the British victory.

Since the Second World War, radar has had numerous peace-time uses. It has been used to detect rainstorms and has helped the weatherman in this respect. It has turned up mysterious reflections called 'angels', which turned out to be, not heavenly messengers, but flocks of birds, so that now radar is used in the study of bird migrations.

And, as described in Chapter 2, it was radar reflections from Venus and Mercury that gave astronomers new knowledge concerning the rotations of those planets and, with regard to Venus, information about the nature of the surface.

Through all the mounting evidence of the wave nature of light, a nagging question kept bothering the physicists. How was light transmitted through a vacuum? Other kinds of wave – sound, for instance – required a material medium. (From our observation platform here on earth, we could never hear an explosion on the moon or anywhere else in space, however loud, because sound waves cannot travel across space.) Yet here was light travelling through a vacuum more easily than through matter, and reaching us from galaxies thousands of millions of light-years away.

The classical scientists were always uncomfortable about the notion of 'action at a distance'. Newton, for instance, worried about how the force of gravity could operate through space. As a possible explanation, he revived the Greeks' idea of an 'ether' filling the heavens and speculated that perhaps the force of gravity might somehow be conducted by the ether.

Trying to account for the travel of light waves through space, physicists decided that light, too, must be conducted by the supposed ether. They began to speak of the 'luminiferous ether'. But this idea at once ran into a serious difficulty. Light waves are transverse waves: that is, they undulate at right angles to the direction of travel, like the ripples on the surface of water, in contrast to the 'longitudinal' motion of sound waves. Now physical theory said that only a *solid* medium could convey transverse waves. (Transverse water waves travel on the water surface, a special case, but cannot penetrate the body of the liquid.) Therefore the ether had to be solid, not gaseous or liquid. Not only must it be extremely rigid; to transmit waves at the tremendous speed of light, it had to be far more rigid than steel. What is more, this rigid ether had to permeate ordinary matter – not merely the vacuum of space but gases, water, glass, and the other transparent substances through which light could travel.

To cap it all, this solid, super-rigid material has to be so frictionless, so yielding, that it did not interfere in the slightest with the motion of the smallest planetoid or the flicker of an eyelid!

Yet despite the difficulties introduced by the notion of the ether, it seemed so useful. Faraday, who had no mathematical background at all but who had marvellous insight, worked out the concept of lines of force – lines along which a magnetic field had equal strength – and, visualizing these as elastic distortions of the ether, used them to explain magnetic phenomena.

In the 1860s, Clerk Maxwell, a great admirer of Faraday, set about supplying the mathematical analysis that would account for the lines of force. In doing so, he evolved a set of four simple equations that among them described almost all phenomena involving electricity and magnetism. These equations, advanced in 1864, not only described the interrelationship of the phenomena of electricity and magnetism, but showed the two could not be separated. Where an electric field existed, there had to be a magnetic field, too, at right angles, and vice versa. There was, in fact, only a single 'electromagnetic field'. Furthermore, in considering the implications of his equations, Maxwell found that a changing electric field had to induce a changing magnetic field, which in turn had to induce a changing electric field, and so on; the two leap-frogged, so to speak, and the field progressed outwards in all directions. The result was a radiation possessing the properties of a wave-form. In short, Maxwell predicted the existence of 'electromagnetic radiation' with frequencies equal to that in which the electromagnetic field waxed and waned.

It was even possible for Maxwell to calculate the velocity at which such an electromagnetic wave would have to move. He did this by taking into consideration the ratio of certain corresponding values in the equations describing the force between electric charges and the force between magnetic poles. This ratio turned out to be precisely equal to the velocity of light, and Maxwell could not accept that as a mere coincidence. Light was an electromagnetic radiation, and along with it were other radiations with wavelengths far longer, or far shorter, than that of ordinary light – and all these radiations involved the ether.

Maxwell's equations, by the way, introduced a problem that is still with us. They seemed to emphasize a complete symmetry between the phenomena of electricity and magnetism: what was true of one was true of the other. Yet in one fundamental way, the two seemed different. Particles existed which carried one or the other of the two opposed charges – positive or negative, but not both. Thus the electron carried a negative electric charge only, while the positron carried a positive electric charge only. Analogously, ought not there be particles with a north magnetic pole only, and others with a south magnetic pole only? These 'magnetic monopoles' have, however, never been found. Every particle involving a magnetic field has always possessed both a north and a south magnetic pole. Theory seems to indicate that the separation of the monopoles would require enormous energies of which only cosmic rays can dispose, but cosmic-ray research has as yet revealed no sign of them.

But back to the ether which, at the height of its power, met its Waterloo as a result of an experiment undertaken to test another classical question as knotty as 'action at a distance', namely, the question of 'absolute motion'.

By the nineteenth century, it had become perfectly plain that the earth, the sun, the stars, and in fact all objects in the universe were in motion. Where, then, could you find a fixed reference point, one that was at 'absolute rest', to determine 'absolute motion' – the foundation on which Newton's laws of motion were based? There was one possibility. Newton had suggested that the fabric of space itself (the aether, presumably) was at rest, so that one could speak of 'absolute space'. If the ether were motionless, perhaps one could find the 'absolute motion' of an object by determining its motion in relation to the ether.

In the 1880s, Albert Michelson conceived an ingenious scheme to do just that. If the earth was moving through a motionless ether, he reasoned, then a beam of light sent in the direction of its motion and reflected back should travel a shorter distance than one sent out at right angles and reflected back. To make the test, Michelson invented the 'interferometer', a device with a semi-mirror that lets half of a light beam through in the forward direction and reflects the other half at right angles. Both beams are then reflected back by mirrors to an eyepiece at the source. If one beam has travelled a slightly longer distance than the other, they arrive out of phase and form interference bands. This instrument is an extremely sensitive measurer of differences in length – so sensitive, in fact, that it can measure the growth of a plant from second to second and the diameter of some stars that seem to be dimensionless points of light in even the largest telescope.

Michelson's plan was to point the interferometer in various directions with respect to the earth's motion and detect the effect of the ether by the amount by which the split beams were out of phase on their return.

In 1887, with the help of the American chemist Edward Williams Morley, Michelson set up a particularly delicate version of the experiment. Stationing the instrument on a stone floating on mercury, so that it could be turned in any direction easily and smoothly, they projected their beam in various directions with respect to the earth's motion. They discovered practically no difference! The interference bands were virtually the same no matter in what direction they pointed the instrument or however

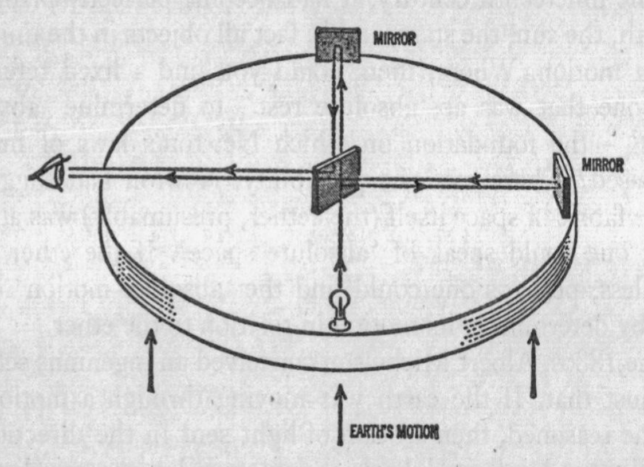

Michelson's interferometer. The semi-mirror (*centre*) splits the light beam, reflecting one half and letting the other half go straight ahead. If the two reflecting mirrors (*at right and straight ahead*) are at different distances, the returning beams of light will arrive at the observer out of phase.

many times they performed the experiment. (It should be said here that more recent experiments along the same line with still more delicate instruments have shown the same negative results.)

The foundations of physics tottered. Either the ether was moving with the earth, which made no sense at all, or there was no such thing as the ether. In either case there was no 'absolute motion' or 'absolute space'. 'Classical' physics – the physics of Newton – had had the rug pulled out from under it. Newtonian physics still held in the ordinary world: planets still moved in accordance with his law of gravitation, and objects on earth still obeyed his law of inertia and of action-and-reaction. It was just that the classical explanations were incomplete, and physicists must be prepared to find phenomena that did not obey the classical 'laws'. The observed phenomena, both old and new, would remain, but the theories accounting for them would have to be broadened and refined.

The 'Michelson–Morley experiment' is probably the most important experiment-that-did-not-work in the whole history of science. Michelson was awarded the Nobel Prize in physics in 1907 – the first American scientist to receive a Nobel Prize.

Relativity

In 1893, the Irish physicist George Francis FitzGerald came up with a novel explanation to account for the negative results of the Michelson–Morley experiment. He suggested that all matter contracted in the direction of its motion and that the amount of contraction increased with the rate of motion. According to this interpretation, the interferometer was always shortened in the direction of the earth's 'true' motion by an amount which exactly compensated for the difference in distance that the light beam would have to travel. Moreover, all possible measuring devices, including human sense organs, would be 'foreshortened' in just the same way. FitzGerald's explanation almost made it look as if nature conspired to keep man from measuring absolute motion by introducing an effect that just cancelled out any differences he might try to use to detect that motion.

This frustrating phenomenon became known as the 'FitzGerald contraction'. FitzGerald worked out an equation for it. An object moving at 7 miles per second (about the speed of our present fastest rockets) would contract by only about two parts per thousand million in the direction of flight. But at really high speeds the contraction would be substantial. At 93,000 miles per second (half the speed of light) it would be 15 per cent; at 163,000 miles per second ($\frac{7}{8}$ the speed of light), 50 per cent. That is, a one-foot ruler moving past us at 163,000 miles per second would seem only six inches long to us – provided we knew a method of measuring its length as it flew by. And at the speed of light, 186,282 miles per second, its length in the direction of motion would be nothing. Since presumably there can be no length shorter than nothing, it would follow that the speed of light in a vacuum is the greatest possible velocity in the universe.

The Dutch physicist Hendrik Antoon Lorentz soon carried FitzGerald's idea one step further. Thinking about cathode rays, on which he was working at the time, he reasoned that if the charge of a charged particle were compressed into a smaller volume, the mass of the particle should increase. Therefore a flying particle foreshortened in the direction of its travel by the FitzGerald contraction would have to increase in mass.

Lorentz presented an equation for the mass increase that turned out to be very similar to FitzGerald's equation for shortening. At 93,000 miles per second, an electron's mass would be increased by 15 per cent; at 163,000 miles per second, by 100 per cent (that is, its mass would be doubled); and at the speed of light, its mass would be infinite. Again it seemed that no speed greater than that of light could be possible, for how could mass be more than infinite?

The FitzGerald length effect and the Lorentz mass effect are so closely connected that the equations are often lumped together as the 'Lorentz–FitzGerald equations'.

If the FitzGerald contraction could not be measured, the Lorentz mass effect could be – indirectly. The ratio of an electron's mass to its charge can be determined from its deflection by a magnetic field. As an electron's velocity increased, the mass would increase, but there was no reason to think that the charge would; therefore, its mass-charge ratio should increase. By 1900, the German physicist W. Kauffman discovered that this ratio increased with velocity in such a way as to indicate that the electron's mass increased just as predicted by the Lorentz–FitzGerald equations. Later and better measurements showed the agreement to be just about perfect.

In discussing the speed of light as a maximum velocity, we must remember that it is the speed of light in a vacuum (186,282 miles per second) that is important here. In transparent material media, light moves more slowly. Its velocity in such a medium is equal to its velocity in a vacuum divided by the index of refraction of the medium. (The 'index of refraction' is a measure of the extent by which a light beam, entering the material obliquely from a vacuum, is bent.)

In water, with an index of refraction of about 1·3, the speed of light is 186,282 divided by 1·3, or about 143,000 miles per second. In glass (index of refraction about 1·5), the speed of light is 124,000 miles per second; while in diamond (index of refraction, 2·4) the speed of light is a mere 78,000 miles per second.

It is possible for sub-atomic particles to travel through a particular transparent medium at a velocity greater than that of light in that medium (though *not* greater than that of light in a vacuum). When particles travel through a medium in this fashion, they throw back a wake of bluish light much as an aeroplane travelling at supersonic velocities throws back a wake of sound.

The existence of such radiation was observed by the Russian physicist Paul Alekseyevich Cherenkov (his name is also spelled Cerenkov) in 1934; in 1937, the theoretical explanation was offered by the Russian physicists Ilya Mikhailovich Frank and Igor Yevgenevich Tamm. All three shared the Nobel Prize for physics in 1958 as a result.

Particle detectors have been devised that detect the 'Cerenkov radiation', and these 'Cerenkov counters' are particularly well adapted to study particularly fast particles such as those making up the cosmic rays.

While the foundations of physics were still rocking, a second explosion took place.

This time the innocent question that started all the trouble had to do with the radiation emitted by matter when it is heated. (Although the radiation in question is usually in the form of light, physicists speak of the problem as 'black-body radiation'. All this means is that they are thinking of an ideal body that absorbs light perfectly – without reflecting any of it away, as a perfectly black body would do – and also radiates perfectly.) The Austrian physicist Josef Stefan showed in 1879 that the total radiation emitted by a body depended only on its temperature (not at all on the nature of its substance) and that in ideal circumstances the radiation was proportional to the fourth power of the absolute temperature: i.e., doubling the absolute temperature would increase its total radiation sixteen-fold ('Stefan's law'). It was also known that as the temperature rose, the predominant radiation moved towards shorter wavelengths. As a lump of steel is heated, for instance, it starts by radiating chiefly in the invisible infra-red, then glows dim red, then bright red, then orange, then yellow-white, and finally, if it could somehow be kept from vaporizing at that point, it would be blue-white.

In 1893, the German physicist Wilhelm Wien worked out a theory of the energy distribution of black-body radiation, that is, of the amount of energy radiated at each particular wavelength range. It provided a formula that accurately described the distribution of energy at the violet end of the spectrum, but not at the red end. (For his work on heat he received the Nobel Prize in physics in 1911.) On the other hand, the English physicists Lord Rayleigh and James Jeans worked up an equation that described the distribution at the red end of the spectrum, but failed completely at the violet end. In short, the best theories available could explain one half of the radiation or the other, but not both at once.

The German physicist Max Karl Ernst Ludwig Planck tackled the problem. He found that, in order to make the equations fit the facts, he had to introduce a completely new notion. He suggested that radiation consisted of small units or packets, just as matter was made up of atoms. He called the unit of radiation the 'quantum' (after the Latin word for 'how much?'). Planck argued that radiation could be absorbed only in whole numbers of quanta. Furthermore, he suggested that the amount of energy in a quantum depended on the wavelength of the radiation. The shorter the wavelength, the more energetic the quantum; or, to put it another way, the energy content of the quantum is inversely proportional to the wavelength.

Now the quantum could be related directly to the frequency of a given radiation. Like the quantum's energy content, the frequency is inversely proportional to the radiation's wavelength. If both the frequency and the quantum's energy content were inversely proportional to the wavelength, then the two were directly proportional to each other. Planck expressed this by means of his now-famous equation:

$$e = h\nu$$

The symbol '*e*' stands for the quantum energy, 'ν' (the Greek letter 'nu') for the frequency, and '*h*' for 'Planck's constant', which gives the proportional relation between quantum energy and frequency.

The value of *h* is extremely small, and so is the quantum. The units of radiation are so small, in fact, that light looks continuous to us, just as ordinary matter seems continuous. But at the beginning of the twentieth century the same fate befell radiation as had befallen matter at the beginning of the nineteenth: both now had to be accepted as discontinuous.

Planck's quanta cleared up the connection between temperature and the wavelengths of emitted radiation. A quantum of violet light was twice as energetic as a quantum of red light, and naturally it would take more heat energy to produce violet quanta than red quanta. Equations worked out on the basis of the quantum explained the radiation of a black body very neatly at both ends of the spectrum.

Eventually Planck's quantum theory was to do a great deal more: it was to explain the behaviour of atoms, of the electrons in atoms, and of nucleons in the atoms' nuclei. Planck was awarded the Nobel Prize in physics in 1918.

Planck's theory made little impression on physicists when it was first announced in 1900. It was too revolutionary to be accepted at once. Planck himself seemed appalled at what he had done. But five years later a young German-born Swiss physicist named Albert Einstein verified the existence of his quanta.

The German physicist Philipp Lenard had found that when light struck certain metals, it caused the metal surface to emit electrons, as if the force of the light kicked electrons out of the atoms. The phenomenon acquired the name 'photoelectric effect', and for its discovery Lenard received the Nobel Prize for physics in 1905. When physicists began to experiment with it, they found, to their surprise, that increasing the intensity of the light did not give the kicked-out electrons any more energy. But changing the wavelength of light did affect them: blue light, for instance, caused the electrons to fly out at greater speed than yellow light did. A very dim blue light would kick out fewer electrons than a bright yellow light would, but those few 'blue-light' electrons would travel with greater speed than any of the 'yellow-light' electrons. On the other hand, red light, no matter how bright, failed to knock out any electrons at all from some metals.

None of this could be explained by the old theories of light. Why should blue light do something red light could not do?

Einstein found the answer in Planck's quantum theory. To absorb enough energy to leave the metal surface, an electron had to be hit by a quantum of a certain minimum size. In the case of an electron held only weakly by its atom (e.g., in cesium) even a quantum of red light would do. Where atoms held electrons more strongly, yellow light was required, or blue light, or even ultra-violet. And in any case, the more energetic the quantum, the more speed it would give to the electron it kicked out.

Here was a case where the quantum theory explained a physical phenomenon with perfect simplicity, whereas the pre-quantum view of light had remained helpless. Other applications of quantum mechanics followed thick and fast. For his explanation of the photoelectric effect (not for his theory of relativity) Einstein was awarded the Nobel Prize in physics in 1921.

In his 'Special Theory of Relativity', presented in 1905 and evolved in his spare time while he worked as examiner at the Swiss patent office, Einstein proposed a new fundamental view of the universe based on an extension of the quantum theory. He suggested that light travelled through space in quantum form (the 'photon'), and thus he resurrected the concept of light consisting of light particles. But this was a new kind of particle. It had properties of a wave as well as of a particle, and sometimes it showed one set of properties and sometimes the other.

This has been made to seem a paradox, or even a kind of mysticism, as if the true nature of light passes all possible understanding. That is not so. To illustrate with an analogy, a man may have many aspects: husband, father, friend, businessman. Depending on circumstances and on his surroundings, he behaves like a husband, father, friend, or businessman. You would not expect him to exhibit his husbandly behaviour towards a customer or his businesslike behaviour towards his wife, and yet that makes him neither a paradox nor more than one man.

In the same way, radiation has both corpuscular and wave properties. In some capacities, the corpuscular properties are particularly pronounced; in others, the wave properties. This dual character gives a more satisfactory account of radiation than either set of properties alone can.

The discovery of the wave nature of light had led to all the triumphs of nineteenth-century optics, including spectroscopy. But it had also required physicists to imagine the existence of the aether. Now Einstein's particle-wave view kept all the nineteenth-century victories (including Maxwell's equations), but made it unnecessary to assume that the aether existed. Radiation could travel through a vacuum by virtue of its particle attributes, and the aether idea, killed by the Michelson–Morley experiment, could now be buried.

Einstein introduced a second important idea in his Special Theory of Relativity: that the speed of light in a vacuum never varied, regardless of the motion of its source. In Newton's view of the universe, a light beam from a source moving towards an observer should seem to travel more quickly than one from a source moving in any other direction. In Einstein's view, this did not happen, and from that assumption he was able to derive the Lorentz–FitzGerald equations. He showed that the increase of mass with velocity, which Lorentz had applied only to charged particles, could be applied to all objects of any sort. He reasoned further that increases in velocity would not only foreshorten length and increase mass but also slow the pace of time: in other words, clocks would slow down along with the shortening of yardsticks.

The most fundamental aspect of Einstein's theory was its

denial of the existence of 'absolute space' and 'absolute time'. This may sound like nonsense: How can the human mind learn anything at all about the universe if it has no point of departure? Einstein answered that all we needed to do was to pick a 'frame of reference' to which the events of the universe could be related. Any frame of reference (the earth motionless, or the sun motionless, or we ourselves motionless, for that matter) would be equally valid, and we could simply choose the frame that was most convenient. It is more convenient to calculate planetary motions in a frame of reference in which the sun is motionless than in one in which the earth is motionless – but no more 'true'.

Thus measurements of space and time are 'relative' to some arbitrarily chosen frame of reference – and that is the reason for naming Einstein's idea the 'theory of relativity'.

To illustrate. Suppose we on the earth were to observe a strange planet ('Planet X'), exactly like our own in size and mass, go whizzing past us at 163,000 miles per second relative to ourselves. If we could measure its dimensions as it shot past, we would find that it was foreshortened by 50 per cent in the direction of its motion. It would be an ellipsoid rather than a sphere and would, on further measurement, seem to have twice the mass of the earth.

Yet to an inhabitant of Planet X, it would seem that he himself and his own planet were motionless. The earth would seem to be moving past *him* at 163,000 miles per second, and it would appear to have an ellipsoidal shape and twice the mass of *his* planet.

One is tempted to ask which planet would *really* be foreshortened and doubled in mass, but the only possible answer is: that depends on the frame of reference. If you find that frustrating, consider that a man is small compared to a whale and large compared to a beetle. Is there any point in asking what a man is *really*, large or small?

For all its unusual consequences, relativity explains all the known phenomena of the universe at least as well as pre-relativity theories do. But it goes further: it explains easily some phenomena that the Newtonian outlook explained poorly or not at all. Consequently, Einstein has been accepted over Newton, not as a replacement so much as a refinement. The Newtonian view of the universe can still be used as a simplified approximation that works well enough in ordinary life and even in ordinary astronomy, as in placing satellites in orbit. But when it comes to accelerating particles in a synchrotron, for example, we find that we must take account of the Einsteinian increase of mass with velocity to make the machine work.

Einstein's view of the universe so mingled space and time that either concept by itself became meaningless. The universe was four-dimensional, with time one of the dimensions (but behaving not quite like the ordinary spatial dimensions of length, breadth, and height). The four-dimensional fusion is often referred to as 'space-time'. This notion was first developed by one of Einstein's teachers, the Russian-German mathematician Hermann Minkowski, in 1907.

With time, as well as space, up to odd tricks in relativity, one aspect of relativity that still provokes arguments among physicists is Einstein's notion of the slowing of clocks. A clock in motion, he said, keeps time more slowly than a stationary one. In fact, all phenomena that change with time change more slowly when moving than when at rest, which is the same as saying that time itself is slowed. At ordinary speeds the effect is negligible, but at 163,000 miles per second a clock would seem (to an observer watching it fly past) to take two seconds to tick off one second. And at the speed of light, time would stand still.

The time-effect is more disturbing than those involving length and weight. If an object shrinks to half its length and then returns to normal, or if it doubles its weight and then returns to normal, no trace is left behind to indicate the temporary change, and opposing viewpoints need not quarrel.

Time, however, is a cumulative thing. If a clock on Planet X seems to be running at half-time for an hour because of its great speed, and if it is then brought to rest, it will resume its ordinary time-rate, but it will bear the mark of being half an hour slow! Well then, if two ships pass each other, and each considers the other to be moving at 163,000 miles per second and to be moving at half-time, when the two ships come together again, observers on each ship will expect the clock on the other ship to be half an hour slower than their own. But it isn't possible for each clock to be slower than the other. What, then, would happen? This problem is called the 'clock paradox'.

Actually, it isn't a paradox at all. If one ship just flashed by the other and both crews swore the other ship's clock was slow, it wouldn't matter which clock was 'really' slow, because the two ships would separate for ever. The two clocks would never be brought to the same place at the same time in order to be matched, and the clock paradox would never arise. Indeed, Einstein's Special Theory of Relativity only applies to uniform motion, so it is only the steady separation we are talking about.

Suppose, though, the two ships *did* come together after the flash-past, so that the clocks *could* be compared. In order for that to happen, something new must be added. At least one ship must accelerate. Suppose ship B did so: slowing down, travelling in a huge curve to point itself in the direction of A, then speeding up until it catches up with A. Of course, B might choose to consider itself at rest; by its chosen frame of reference, it is A that does all the changing, speeding up backwards to come to A. If the two ships were all there were to the universe, then indeed the symmetry would keep the clock paradox in being.

However, A and B are *not* all there is to the universe, and that upsets the symmetry. When B accelerates, it is doing so with reference not only to A, but to all the rest of the universe besides. If B chooses to consider itself at rest, it must consider not only A, but all the galaxies without exception, to be accelerating with respect to itself. It is B against the universe, in short. Under those circumstances, it is B's clock that ends up half an hour slow, not A's.

This affects notions of space travel. If astronauts leaving earth speed up to near the speed of light, their time would be much slower than ours. They might reach a distant destination and return in what seemed to them weeks, though on the earth many centuries would have passed. If time really slows in motion, a person might journey even to a distant star in his own lifetime. But of course he would have to say good-bye to his own generation and the world he knew. He would return to a world of the future.

In the Special Theory of Relativity Einstein did not deal with accelerated motion or gravitation. These were treated in his 'General Theory of Relativity', published in 1915. The General Theory presented a completely altered view of gravitation. It was viewed as a property of space, rather than as a force between bodies. As the result of the presence of matter, space became curved, and bodies followed the line of least resistance among the curves, so to speak. Strange as Einstein's idea seemed, it was able to explain something that the Newtonian law of gravity had not been able to explain.

The greatest triumph of Newton's law of gravity had come in 1846. The planet Uranus, discovered in 1781, had a slightly erratic orbit around the sun. A half century of observation made that unmistakable. Astronomers decided that some still undiscovered planet beyond it must be exerting a gravitational pull on it. The British astronomer John Couch Adams and the French astronomer Urbain Jean Joseph Leverrier calculated the position of this hypothetical planet, using Newton's theories as a basis. In 1846, the German astronomer Johann Gottfried Galle pointed his telescope at the spot indicated by Leverrier and, sure enough, there was a new planet – since named Neptune.

After that, nothing seemed capable of shaking Newton's law of gravity. And yet one planetary motion remained unexplained. The planet Mercury's point of nearest approach to the sun ('perihelion') changed from one trip to the next; it was never in the same place twice in the planet's 'yearly' revolutions around the sun. Astronomers were able to account for most of this irregularity as due to 'perturbations' of its orbit by the pull of the neighbouring planets.

Indeed, there had been some feeling in the early days of work with the theory of gravitation that perturbations arising from the shifting pull of one planet on another might eventually act to break up the delicate mechanism of the solar system. In the earliest decades of the nineteenth century, however, the French astronomer Pierre Simon Laplace showed that the solar system was not as delicate as all that. The perturbations were all cyclic, and orbital irregularities never increased to more than a certain amount in any direction. In the long run, the solar system is stable, and astronomers were more certain than ever that all particular irregularities could be worked out by taking perturbations into account.

This, however, did not work for Mercury. After all perturbations were allowed for, there was still an unexplained one-way shift of Mercury's perihelion by an amount equal to 43 seconds of arc per century. This motion, discovered by Leverrier in 1845, is not much: in 4,000 years it adds up only to the width of the moon. It was enough, however, to upset astronomers.

Leverrier suggested that this deviation might be caused by a small, undiscovered planet closer to the sun than Mercury. For decades astronomers searched for the supposed planet (called 'Vulcan'), and many were the reports of its discovery. All the reports turned out to be mistaken. Finally it was agreed that Vulcan did not exist.

Then Einstein's General Theory of Relativity supplied the answer. It showed that the perihelion of any revolving body should have a motion beyond that predicted by Newton's law. When this new calculation was applied to Mercury, the planet's shift of perihelion fitted it exactly. Planets farther from the sun than Mercury should show a progressively smaller shift of perihelion. In 1960, the perihelion of Venus's orbit had been found to be advancing about 8 seconds of arc per century; this shift fits Einstein's theory almost exactly.

More impressive were two unexpected new phenomena that only Einstein's theory predicted. First, Einstein maintained that an intense gravitational field should slow down the vibrations of atoms. The slow-down would be evidenced by a shift of spectral lines towards the red (the 'Einstein shift'). Casting about for a gravitational field strong enough to produce this effect, astronomers thought of the dense white-dwarf stars. They looked at the spectra of white dwarfs and did indeed find the predicted shift of lines.

The verification of Einstein's second prediction was even more dramatic. His theory said a gravitational field would bend light rays. Einstein calculated that a ray of light just skimming the sun's surface would be bent out of a straight line by 1·75 seconds of arc. How could that be checked? Well, if stars beyond the sun and just off its edge could be observed during an eclipse of the sun and their positions compared with what they were against the background when the sun did not interfere, any shift resulting from bending of their light should show up. Since Einstein had published his paper on general relativity in 1915, the test had to wait until after the end of the First World War. In 1919, the British Royal Astronomical Society organized an expedition to make the test by witnessing a total eclipse visible from the island of Principe, a small Portuguese-owned island off West Africa. The stars did shift position. Einstein had been verified again.

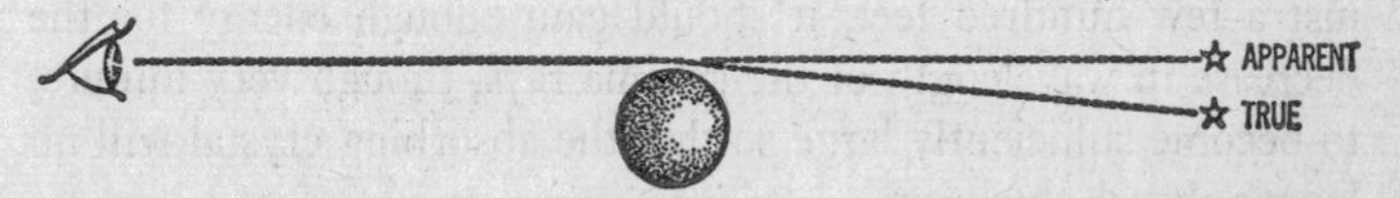

The gravitational bending of light waves, postulated by Einstein in the General Theory of Relativity.

By this same principle, if one star were directly behind another, the light of the farther star would bend about the nearer in such a way that the farther star would appear larger than it really is. The nearer star would act as a 'gravitational lens'. Unfortunately, the apparent size of stars is so minute that an eclipse of a distant star by a much closer one (as seen from earth) is extremely rare, although some astronomers have speculated that the puzzling properties of quasars may be due to gravitational lens-effects. An eclipse of this sort should take place in 1988. No doubt astronomers will be watching.

The three great victories of Einstein's General Theory were all astronomic in nature. Scientists longed to discover a way to check it in the laboratory under conditions they could vary at will. The key to such a laboratory demonstration arose in 1958, when the German physicist, Rudolf Ludwig Mössbauer, showed that, under certain conditions, a crystal could be made to produce a beam of gamma rays of sharply defined wavelength. Ordinarily, the atom emitting the gamma ray recoils, and this recoil broadens the band of wavelengths produced. In crystals under certain conditions, the crystal acts as a single atom: the recoil is distributed among all the atoms and sinks to virtually nothing, so that the gamma ray emitted is exceedingly sharp. Such a sharp-wavelength beam can be absorbed with extraordinary efficiency by a crystal similar to that which produced it. If the gamma rays are of even slightly different wavelength from that which the crystal would naturally produce, it would not be absorbed. This is called the 'Mössbauer effect'.

If such a beam of gamma rays is emitted downwards so as to fall with gravity, the General Theory of Relativity requires it to gain energy so that its wavelength becomes shorter. In falling just a few hundred feet, it should gain enough energy for the decrease in wavelength of the gamma rays, though very minute, to become sufficiently large so that the absorbing crystal will no longer absorb the beam.

Furthermore, if the crystal emitting the gamma ray is moved upwards while the emission is proceeding, the wavelength of the gamma ray is increased through the Doppler–Fizeau effect. The velocity at which the crystal is moved upwards can be adjusted so as to just neutralize the effect of gravitation on the falling gamma ray, which will then be absorbed by the crystal on which it impinges.

Experiments conducted in 1960 and later made use of the Mössbauer effect to confirm the General Theory with great exactness. They were the most impressive demonstration of its validity that has yet been seen; as a result, Mössbauer was awarded the 1961 Nobel Prize for physics.

Despite all this, the claim to validity of Einstein's General Theory remains tenuous. The confirmations remain borderline. In 1961, the American physicist Robert Henry Dicke evolved a more complex concept he calls the 'scalar-tensor theory', which treats gravitation not as a geometric effect, as Einstein's theory does, but as a combination of two fields of different properties. The two theories predict phenomena so nearly alike as to be virtually indistinguishable. In the summer of 1966, Dicke measured the sphericity of the sun and, by very delicate measurement, claimed to have detected a slight equatorial bulge. This bulge would account for 8 per cent of the observed advance of Mercury's perihelion and would destroy the excellent fit of the General Theory. This would weaken Einstein's theory but leave Dicke's unaffected.

On the other hand, although both theories predict that light (or radio) waves would be slowed when passing a massive object, they differ somewhat in the degree of slowing they predict. In 1970, radio signals were reflected from planetary probes at a time when they were about to pass behind the sun (as viewed from earth) at a known distance. The time it took for the radio waves to arrive back would measure the degree to which they were slowed as they skimmed by the sun in each direction. The results, as reported, were considerably closer to Einstein's prediction than to Dicke's, but the matter was still not conclusive.

Heat

So far in this chapter I have been neglecting a phenomenon that usually accompanies light in our everyday experience. Almost all luminous objects from a star to a candle give off heat as well as light.

Heat was not studied, other than qualitatively, before modern times. It was enough for a person to say, 'It is hot', or 'It is cold', or 'This is warmer than that'. To subject temperature to quantitative measure, it was first necessary to find some measurable change that seemed to take place uniformly with change in temperature. One such change was found in the fact that substances expand when warmed and contract when cooled.

Galileo was the first to try to make use of this fact to detect changes in temperature. In 1603, he inverted a glass tube of heated air into a bowl of water. As the air in the tube cooled to room temperature, it contracted and drew water up the tube, and there Galileo had his 'thermometer' (from Greek words meaning 'heat measure'). When the temperature of the room changed, the water level in the tube changed. If the room warmed, the air in the tube expanded and pushed the water level down; if it grew cooler, the air contracted and the water level moved up. The only trouble was that the basin of water into which the tube had been inserted was open to the air and the air pressure kept changing. That also shoved the water level up and down, independently of temperature, confusing the results. The thermometer was the first important scientific instrument to be made of glass.

By 1654, the Grand Duke of Tuscany, Ferdinand II, had evolved a thermometer which was independent of air pressure. It contained a liquid sealed into a bulb to which a straight tube was attached. The contraction and expansion of the liquid itself was used as the indication of temperature change. Liquids change their volume with temperature much less than gases do, but by using a sizeable reservoir of liquid and a filled bulb, so that the liquid could expand only up a very narrow tube, the rise and fall within that tube, for even tiny volume changes, could be made considerable.

The English physicist Robert Boyle did much the same thing about the same time, and he was the first to show that the human body had a constant temperature, markedly higher than the usual room temperature. Others demonstrated that certain physical phenomena always took place at some fixed temperature. Before the end of the seventeenth century this was found to be so in the case of the melting of ice and the boiling of water.

The first liquids used in thermometry were water and alcohol. Since water froze too soon and alcohol boiled away too easily, the French physicist Guillaume Amontons resorted to mercury. In his device, as in Galileo's, the expansion and contraction of air caused the mercury level to rise or fall.

Then in 1714 the German physicist Gabriel Daniel Fahrenheit combined the advances of the Grand Duke and of Amontons by enclosing mercury in a tube and using its own expansion and contraction with temperature as the indicator. Furthermore, Fahrenheit put a graded scale on the tube to allow the temperature to be read quantitatively.

There is some argument as to exactly how Fahrenheit arrived at the particular scale he used. He set zero, according to one account, at the lowest temperature he could get in his laboratory, attained by mixing salt and melting ice. He then set the freezing point of pure water at 32 and its boiling point at 212. This had two advantages. First, the range of temperature over which water was liquid came to 180°, which seemed a natural number to use in connection with 'degrees'. (It is the number of degrees in a semicircle.) Second, body temperature came near a round 100°; normally it is 98·6° Fahrenheit, to be exact.

So constant is body temperature normally, that if it is more than a degree or so above the average value, the body is said to run a fever and there is a clear feeling of illness. In 1858, the German physician Karl August Wunderlich introduced the procedure of frequent checks on body temperature as an indication of the course of disease. In the next decade, the British physician Thomas Clifford Allbutt invented the 'clinical thermometer' in which there is a constriction in the narrow tube containing the mercury. The mercury thread rises to a maximum when placed in the mouth, but does not fall when the thermometer is removed. The mercury thread simply divides at the constriction, leaving the portion above with its reading held constant. In Great Britain and the United States, the Fahrenheit scale is still used. We are familiar with it in everyday affairs such as weather-reporting and in the use of clinical thermometers.

In 1742, however, the Swedish astronomer Anders Celsius adopted a different scale. In its final form, this set the freezing point of water at 0 and its boiling point at 100. Because of the hundredfold division of the temperature range in which water was liquid, this is called the 'centigrade scale', from Latin words meaning 'hundred steps'. Most people still speak of measurements on this scale as 'degrees centigrade', but scientists, at an international conference in 1948, renamed the scale after the inventor, following the Fahrenheit precedent. Officially, then, one should speak of the 'Celsius scale' and of 'degrees Celsius'. The symbol C still holds. It was Celsius's scale that won out in most of the civilized world. Scientists, in particular, found the Celsius scale convenient.

Temperature measures the intensity of heat but not its quantity. Heat will always flow from a place of higher temperature to a place of lower temperature until the temperatures are equal, just as water will flow from a higher level to a lower one until the levels are equal. This is true regardless of the relative amounts of heat contained in the bodies involved. Although a bath-tub of lukewarm water contains far more heat than a burning match, when the match is placed near the water, heat goes from the match to the water, not vice versa.

Joseph Black, who had done important work on gases (see p. 181), was the first to make clear the distinction between temperature and heat. In 1760, he announced that various substances were raised in temperature by different amounts when a given amount of heat was poured into them. To raise the temperature of a gram of iron by one degree Celsius takes three times as much heat as to warm a gram of lead by one degree. And beryllium requires three times as much heat as iron.

Furthermore, Black showed it was possible to pour heat into a substance without raising its temperature at all. When ice is heated, it begins to melt, but it does not rise in temperature. Heat will eventually melt all the ice, but the temperature of the ice itself never goes above 0°C. The same thing happens in the case of boiling water at 100°C. As heat is poured into the water, more and more of it boils away as vapour, but the temperature of the liquid does not change.

The development of the steam engine (see pp. 410 ff.), which came at about the same time as Black's experiments, intensified the interest of scientists in heat and temperature. They began to speculate about the nature of heat, as earlier they had speculated about the nature of light.

In the case of heat, as of light, there were two theories. One held that heat was a material substance which could be poured or shifted from one substance to another. It was named 'caloric', from the Latin for 'heat'. According to this view, when wood was burned the caloric in the wood passed into the flame and from that into a kettle above the flame and from that into the water in the kettle. As water filled with caloric, it was converted to steam.

In the late eighteenth century, two famous observations gave rise to the theory that heat was a form of vibration. One was published by the American physicist and adventurer Benjamin Thompson, a Tory who fled the country during the American Revolution, was given the title Count Rumford and then proceeded to knock around Europe. While supervising the boring of cannon in Bavaria in 1798, he noticed that quantities of heat were being produced. He found that enough heat was being generated to bring eighteen pounds of water to the boiling point in less than three hours. Where was all the caloric coming from? Thompson decided that heat must be a vibration set up and intensified by the mechanical friction of the borer against the cannon.

The next year the chemist Humphry Davy performed an even more significant experiment. Keeping two pieces of ice below the freezing point, he rubbed them together, not by hand but by a mechanical contrivance, so that no caloric could flow into the ice. By friction alone, he melted some of the ice. He, too, concluded that heat must be a vibration and not a material. Actually, this experiment should have been conclusive, but the caloric theory, though obviously wrong, persisted to the middle of the nineteenth century.

Nevertheless, although the nature of heat was misunderstood, scientists learned some important things about it, just as the investigators of light turned up interesting facts about the reflection and refraction of light beams before they knew its nature. Jean Baptiste Joseph Fourier and Nicholas Léonard Sadi Carnot in France studied the flow of heat and made important advances. In fact, Carnot is usually considered the founder of the science of 'thermodynamics' (from Greek words meaning 'movement of heat'). He placed the working of steam engines on a firm theoretical foundation.

Carnot did his work in the 1820s. By the 1840s, physicists were concerned with the manner in which the heat that was put into steam could be converted into the mechanical work of moving a piston. Was there a limit to the amount of work that could be obtained from a given amount of heat? And what about the reverse process: How was work converted to heat?

Joule spent thirty-five years converting various kinds of work into heat, doing very carefully what Rumford had earlier done clumsily. He measured the amount of heat produced by an electric current. He heated water and mercury by stirring them with paddle wheels, or by forcing water through narrow tubes. He heated air by compressing it, and so on. In every case he calculated how much mechanical work had been done on the system and how much heat was obtained as a result. He found that a given amount of work, of any kind, always produced a given amount of heat, which was called the 'mechanical equivalent of heat'.

Since heat could be converted into work, it must be considered a form of 'energy' (from Greek words meaning 'containing work'). Electricity, magnetism, light, and motion could all be used to do work, so they, too, were forms of energy. And work itself, being convertible into heat, was a form of energy.

This emphasized something that had been more or less suspected since Newton's time: that energy was 'conserved' and could neither be created nor destroyed. Thus, a moving body has 'kinetic energy' ('energy of motion'), a term introduced by Lord Kelvin in 1856. Since a body moving upwards is slowed by gravity, its kinetic energy slowly disappears. However, as the body loses kinetic energy, it gains energy of position, for, by virtue of its location high above the surface of the earth, it can eventually fall and regain kinetic energy. In 1853, the Scottish physicist William John Macquorn Rankine named this energy of position 'potential energy'. It seemed that a body's kinetic energy plus its potential energy (its 'mechanical energy') remained nearly the same during the course of its movement, and this was called 'conservation of mechanical energy'. However, mechanical energy was not *perfectly* conserved. Some was lost to friction, to air resistance, and so on.

What Joule's experiments showed above all was that such conservation could be made exact when heat was taken into account, for, when mechanical energy was lost to friction or air resistance, it appeared as heat. Take that heat into account, and one can show, without qualification, that no new energy is created and no old energy destroyed. The first person actually to put this notion into words was Heinrich von Helmholtz. In 1847, von Helmholtz enunciated the 'law of conservation of energy', which states that energy can be converted from one form to another but cannot be created or destroyed. Whenever a certain amount of energy seems to disappear in one place, an equivalent amount must appear in another. This is also called 'the first law of thermodynamics'. It remains a foundation block of modern physics, undisturbed by either quantum theory or relativity.

Now, although any form of work can be converted entirely into heat, the reverse is not true. When heat is turned to work, some of it is unusable and is unavoidably wasted. In running a steam engine, the heat of the steam is converted into work only until the temperature of the steam is reduced to the temperature of the environment; after that, although there is much remaining heat in the cold water formed from the steam, no more of it can be converted to work. Even in the temperature range at which work can be extracted, some of the heat does not go into work but is used up in heating the engine and the air around it, in overcoming friction between the piston and the cylinder, and so on.

In any energy conversion – e.g., electric energy into light energy, or magnetic energy into energy of motion – some of the energy is wasted. It is not lost; that would be contrary to the first law. But it is converted to heat that is dissipated in the environment.

The capacity of any system to perform work is its 'free energy'. The portion of the energy that is unavoidably lost as non-useful heat is reflected in the measurement of the 'entropy' – a term first used in 1850 by the German physicist Rudolf Julius Emmanuel Clausius.

Clausius pointed out that in any process involving a flow of energy there is always some loss, so that the entropy of the universe is continually increasing. This continual increase of entropy is called the 'second law of thermodynamics'. It is sometimes referred to as the 'running-down of the universe' or the 'heat-death of the universe'. Fortunately, the quantity of usable energy (supplied almost entirely by the stars, which are, of course, 'running down' at a tremendous rate) is so vast that there is enough for all purposes for many thousands of millions of years.

A clear understanding of the nature of heat finally came with the understanding of the atomic nature of matter. It developed from the realization that the molecules composing a gas were in continual motion, bouncing off one another and off the walls of their container. The first investigator who attempted to explain the properties of gases from this standpoint was the Swiss mathematician Daniel Bernoulli, in 1738, but he was ahead of his time. In the mid nineteenth century, Maxwell and Boltzmann (see p. 220) worked out the mathematics adequately and established the 'kinetic theory of gases' ('kinetic' comes from a Greek word meaning 'motion'). The theory showed heat to be equivalent to the motion of molecules. Thus the caloric theory of heat received its death-blow. Heat was seen to be a vibrational phenomenon: the movement of molecules in gases and liquids or the jittery to-and-fro trembling of molecules in solids.

When a solid is heated to a point where the to-and-fro trembling is strong enough to break the bonds that hold neighbouring molecules together, the solid melts and becomes a liquid. The stronger the bond between neighbouring molecules in a solid, the more heat is needed to make it vibrate violently enough to break the bond. This means that the substance has a higher melting point.

In the liquid state, the molecules can move freely past one another. When the liquid is heated further, the movements of the molecules finally become sufficiently energetic to set them free of the body of the liquid altogether, and then the liquid boils. Again the boiling point is higher where the intermolecular forces are stronger.

In converting a solid to a liquid, all of the energy of heat goes into breaking the intermolecular bonds. This is why the heat absorbed by melting ice does not raise the ice's temperature. The same is true of a liquid being boiled.

Now we can distinguish between heat and temperature easily. Heat is the total energy contained in the molecular motions of a given quantity of matter. Temperature represents the average energy of motion per molecule in that matter. Thus a quart of water at 60°C contains twice as much heat as a pint of water at 60°C (twice as many molecules are vibrating), but the quart and pint have the same temperature, for the average energy of molecular motion is the same in each case.

There is energy in the very structure of a chemical compound – that is, in the bonding forces that hold an atom or ion or molecule to its neighbour. If these bonds are broken and rearranged into new bonds involving less energy, the excess of energy will make its appearance as heat or light or both. Sometimes the energy is released so quickly that an explosion is the result.

It is possible to calculate the chemical energy contained in any substance and show what the amount of heat released in any reaction must be. For instance, the burning of coal involves breaking the bonds between carbon atoms in the coal and the bonds between the oxygen molecules' atoms, with which the carbon recombines. Now the energy of the bonds in the new compound (carbon dioxide) is less than that of the bonds in the original substances that formed it. This difference, which can be measured, is released as heat and light.

In the 1870s, the American physicist Josiah Willard Gibbs worked out the theory of 'chemical thermodynamics' in such detail that this branch of science was brought from virtual nonexistence to complete maturity at one stroke.

The long paper in which Gibbs described his reasoning was far above the heads of others in America, and it was published in the *Transactions of the Connecticut Academy of Arts and Sciences* only after considerable hesitation. Even afterwards, its close-knit mathematical argument and the retiring nature of Gibbs himself combined to keep the subject under a bushel until the Russian-German physical chemist Wilhelm Ostwald discovered the work in 1883, translated the paper into German, and proclaimed the importance of Gibbs to the world.

As an example of the importance of Gibbs's work, his equations demonstrated the simple, but rigorous, rules governing the equilibrium between different substances existing simultaneously in more than one phase (i.e., in both solid form and in solution, in two immiscible liquids and a vapour, and so on). This 'phase rule' is the breath of life to metallurgy and to many other branches of chemistry.

Mass to Energy

With the discovery of radioactivity in 1896 (p. 240), a totally new question about energy arose at once. The radioactive substances uranium and thorium were giving off particles with astonishing energies. Moreover, Marie Curie found that radium was continually emitting heat in substantial quantities: an ounce of radium gave off 4,000 calories per hour, and this would go on hour after hour, week after week, decade after decade. The most energetic chemical reaction known could not produce a millionth of the energy liberated by radium. And, what was no less surprising, this production of energy, unlike chemical reactions, did not depend on temperature: it went on just as well at the very low temperature of liquid hydrogen as it did at ordinary temperatures!

Quite plainly an altogether new kind of energy, very different from chemical, was involved here. Fortunately physicists did not have to wait long for the answer. Once again, it was supplied by Einstein, in his Special Theory of Relativity.

Einstein's mathematical treatment of energy showed that mass could be considered a form of energy – a very concentrated form, for a very small quantity of mass would be converted into an immense quantity of energy.

Einstein's equation relating mass and energy is now one of the most famous equations in the world. It is:

$$e = mc^2$$

Here 'e' represents energy (in ergs), 'm' represents mass (in grams) and 'c' represents the speed of light (in centimetres per second).

Since light travels at 30,000 million centimetres per second, the value of c^2 is 900 trillion. This means that the conversion of one gram of mass energy will produce 900 trillion ergs. The erg is a small unit of energy not translatable into any common terms, but we can get an idea of what this number means when we are told that the energy in one gram of mass is sufficient to keep a 1,000-watt electric-light bulb running for 2,850 years. Or, to put it another way, the complete conversion of a gram of mass into energy would yield as much as the burning of 2,000 tons of petrol.

Einstein's equation destroyed one of the sacred conservation laws of science. Lavoisier's 'law of conservation of mass' had stated that matter could neither be created nor destroyed. Actually, every energy-releasing chemical reaction changes a small amount of mass into energy: the products, if they could be weighed with utter precision, would not quite equal the original matter. But the mass lost in ordinary chemical reactions is so small that no technique available to the chemists of the nineteenth century could conceivably have detected it. Physicists, however, were now dealing with a completely different phenomenon, the nuclear reaction of radioactivity rather than the chemical reaction of burning coal. Nuclear reactions released so much energy that the loss of mass was large enough to be measured.

By postulating the interchange of mass and energy, Einstein merged the laws of conservation of energy and of mass into one law – the conservation of mass–energy. The first law of thermodynamics not only still stood: it was more unassailable than ever.

The conversion of mass to energy was confirmed experimentally by Francis W. Aston through his mass spectrograph. This could measure the mass of atomic nuclei very precisely by the amount of their deflection by a magnetic field. What Aston did with an improved instrument in 1925 was to show that the various nuclei were not exact multiples of the masses of the neutrons and protons that composed them.

Let us consider the masses of these neutrons and protons for a moment. For a century, the masses of atoms and sub-atomic particles generally have been measured on the basis of allowing the atomic weight of oxygen to be exactly 16·00000 (see p. 232). In 1929, however, William Giauque had showed that oxygen consisted of three isotopes, oxygen 16, oxygen 17, and oxygen 18, and that the atomic weight of oxygen was the weighted average of the mass numbers of these three isotopes.

To be sure, oxygen 16 was by far the most common of the three, making up 99·759 per cent of all oxygen atoms. This meant that if oxygen had the over-all atomic weight of 16·00000, the oxygen-16 isotope had a mass number of *almost* 16. (The masses of the small quantities of oxygen 17 and oxygen 18 brought the value up to 16.) Chemists, for a generation after the discovery, did not let this disturb them, but kept the old basis for what came to be called 'chemical atomic weights'.

Physicists, however, reacted otherwise. They preferred to set the mass of the oxygen-16 isotope at exactly 16·00000 and determine all other masses on that basis. On this basis, the 'physical atomic weights' could be set up. On the oxygen 16 equals 16 standard, the atomic weight of oxygen itself, with its traces of heavier isotopes, was 16·0044. In general the physical atomic weights of all elements would be 0·027 per cent higher than their chemical atomic weight counterparts.

In 1961, physicists and chemists reached a compromise. It was agreed to determine atomic weights on the basis of allowing the carbon-12 isotope to have a mass of 12·0000. This based the atomic weights on a characteristic mass number and made them as fundamental as possible. In addition, this base made the atomic weights almost exactly what they were under the old system. Thus on the carbon 12 equals 12 standard, the atomic weight of oxygen is 15·9994.

Well then, let us start with a carbon-12 atom, with its mass equal to 12·00000. Its nucleus contains six protons and six neutrons. From mass-spectrographic measurements it becomes evident that, on the carbon 12 equals 12 standard, the mass of a proton is 1·007825 and that of a neutron is 1·008665. Six protons, then, should have a mass of 6·046950 and six neutrons 6·051990. Together the twelve nucleons should have a mass of 12·104940. But the mass of the carbon 12 is 12·00000. What has happened to the missing 0·104940?

This disappearing mass is the 'mass defect'. The mass defect divided by the mass number gives the mass defect per nucleon, or the 'packing fraction'. The mass has not really disappeared, of course. It has been converted into energy in accordance with Einstein's equation, so that the mass defect is also the 'binding energy' of the nucleus. To break a nucleus down into individual protons and neutrons would require the input of an amount of energy equal to the binding energy, since an amount of mass equivalent to that energy would have to be formed.

Aston determined the packing fraction of many nuclei, and he found that it increased rather quickly from hydrogen up to elements in the neighbourhood of iron and then decreased, rather slowly, for the rest of the periodic table. In other words, the binding energy per nucleon was highest in the middle of the periodic table. This meant that conversion of an element at either end of the table into one nearer the middle should release energy.

Take uranium 238 as an example. This nucleus breaks down by a series of decays to lead 206. In the process, it emits eight alpha particles. (It also gives off beta particles, but these are so light they can be ignored.) Now the mass of lead 206 is 205·9745 and that of eight alpha particles totals 32·0208. Altogether these products add up to a mass of 237·9953. But the mass of uranium 238, from which they came, is 238·0506. The difference, or loss of mass, is 0·0553. That loss of mass is just enough to account for the energy released when uranium breaks down.

When uranium breaks down to still smaller atoms, as it does in fission, a great deal more energy is released. And when hydrogen is converted to helium, as it is in stars, there is an even larger fractional loss of mass and a correspondingly richer development of energy.

Physicists began to look upon the mass–energy equivalence as a very reliable bookkeeping. For instance, when the positron was discovered in 1934, its mutual annihilation with an electron produced a pair of gamma rays whose energy was just equal to the mass of the two particles. Furthermore, as Blackett was first to point out, mass could be created out of appropriate amounts of energy. A gamma ray of the proper energy, under certain circumstances, would disappear and give rise to an 'electron–positron pair', created out of pure energy. Larger amounts of energy, supplied by cosmic particles or by particles fired out of proton synchrotrons (see p. 323), would bring about the creation of more massive particles, such as mesons and anti-protons.

It is no wonder that when the bookkeeping did not balance, as in the emission of beta particles of less than the expected energy, physicists invented the neutrino to balance the energy account rather than tamper with Einstein's equation (see p. 331).

If any further proof of the conversion of mass to energy was needed, the atomic bomb provided the final clincher.

Particles and Waves

In the 1920s, dualism reigned supreme in physics. Planck had shown that radiation was particle-like as well as wave-like. Einstein had shown that mass and energy were two sides of the same coin, and that space and time were inseparable. Physicists began to look for other dualisms.

In 1923, the French physicist Louis Victor de Broglie was able to show that, just as radiation had the characteristics of particles, so the particles of matter, such as electrons, should display the characteristics of waves. The waves associated with these particles, he predicted, would have a wavelength inversely related to the mass times the velocity (that is, the momentum) of the particle. The wavelength associated with electrons of moderate speed, de Broglie calculated, ought to be in the X-ray region.

In 1927, even this surprising prediction was borne out. Clinton

Joseph Davisson and Lester Halbert Germer of the Bell Telephone Laboratories were bombarding metallic nickel with electrons. As the result of a laboratory accident, which had made it necessary to heat the nickel for a long time, the metal was in the form of large crystals, which were ideal for diffraction purposes because the spacing between atoms in a crystal is comparable to the very short wavelengths of electrons. Sure enough, the electrons passing through those crystals behaved not as particles but as waves. The film behind the nickel showed interference patterns, alternate bands of fogging and clarity, just as it would have shown if X-rays rather than electrons had gone through the nickel.

Interference patterns were the very thing that Young had used more than a century earlier to prove the wave nature of light. Now they proved the wave nature of electrons. From the measurements of the interference bands, the wavelength associated with the electron could be calculated, and it turned out to be 1·65 ångström units, almost exactly what de Broglie had calculated it ought to be.

In the same year the British physicist George Paget Thomson, working independently and using different methods, also showed that electrons had wave properties.

De Broglie received the Nobel Prize in physics in 1929, and Davisson and Thomson shared the Nobel Prize in physics in 1937.

This entirely unlooked-for discovery of a new kind of dualism was put to use almost at once in microscopy. Ordinary optical microscopes, as I have mentioned, cease to be useful at a certain point because there is a limit to the size of objects that light waves can define sharply. As objects get smaller, they also get fuzzier, because the light waves begin to pass round them – something first pointed out by the German physicist Ernst Karl Abbe in 1878. (For the same reason, the long radio waves give a fuzzy picture even of large objects in the sky.) The cure, of course, is to try to find shorter wavelengths with which to resolve the smaller objects. Ordinary-light microscopes can distinguish two dots 1/5,000 millimetre apart, but ultra-violet microscopes can distinguish dots 1/10,000 millimetre apart. X-rays would be better still, but there are no lenses for X-rays. This problem can be solved, however, by using the waves associated with electrons, which have about the same wavelength as X-rays, but are easier to manipulate. For one thing, a magnetic field can bend the 'electron rays', because the waves are associated with a charged particle.

Just as the eye can see an expanded image of an object if the light rays involved are appropriately manipulated by lenses, so a photograph can register an expanded image of an object if electron waves are appropriately manipulated by magnetic fields. And, since the wavelengths associated with electrons are far smaller than those of ordinary light, the resolution obtainable with an 'electron microscope' at high magnification is much greater than that available to an ordinary microscope.

A crude electron microscope capable of magnifying 400 times was made in Germany in 1932 by Ernst Ruska and Max Knoll, but the first really usable one was built in 1937 at the University of Toronto by James Hillier and Albert F. Prebus. Their instrument could magnify an object 7,000 times, whereas the best optical microscopes reach their limit with a magnification of about 2,000. By 1939, electron microscopes were commercially available and, eventually, Hillier and others developed electron microscopes capable of magnifying up to 2,000,000 times.

A 'proton microscope', if one were developed, would magnify to a far greater extent than does an electron microscope, because the waves associated with protons are shorter. In a sense, the proton synchrotron is a kind of proton microscope, probing the interior of the nucleus with its speeding protons. The greater the speed of a proton, the greater its momentum and the shorter the wavelength associated with it. Protons with an energy of one Mev can 'see' the nucleus, while at 20 Mev they can begin to 'see' detail within the nucleus. This is another reason why physicists are eager to pile more and more electron volts into their atom smasher – so that they may 'see' the ultra-small more clearly.

It ought not be too surprising that this particle–wave dualism works in reverse and that phenomena ordinarily considered wave-

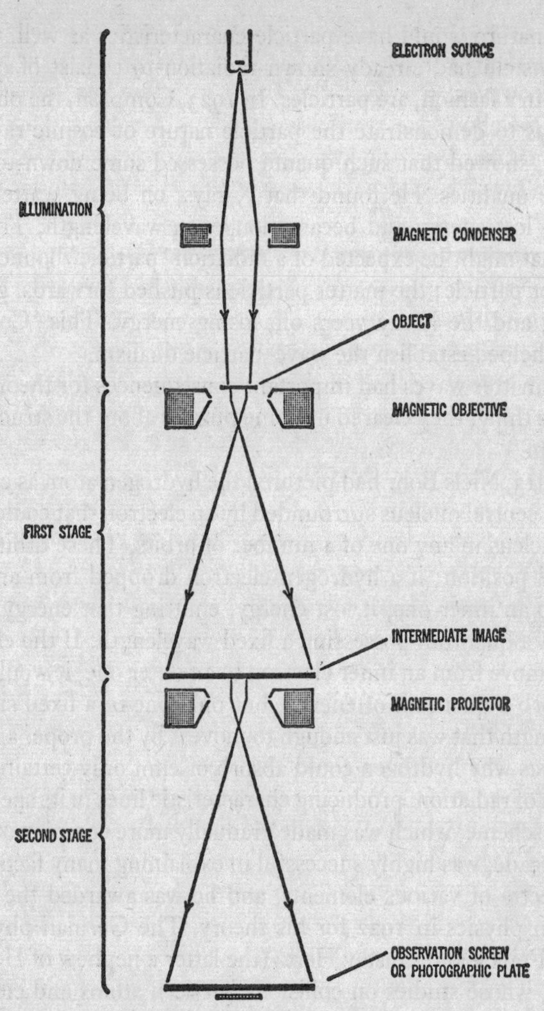

Diagram of electron microscope. The magnetic condenser directs the electrons in a parallel beam. The magnetic objective functions like a convex lens, producing an enlarged image, which is then further magnified by a magnetic projector. The image is projected on a fluorescent observation screen or a photographic plate.

like in nature would have particle characteristics as well. Planck and Einstein had already shown radiation to consist of quanta, which, in a fashion, are particles. In 1923, Compton, the physicist who was to demonstrate the particle nature of cosmic rays (see p. 311), showed that such quanta possessed some down-to-earth particle qualities. He found that X-rays, on being scattered by matter, lost energy and became longer in wavelength. This was just what might be expected of a radiation 'particle' bouncing off a matter particle; the matter particle is pushed forwards, gaining energy, and the X-ray veers off, losing energy. This 'Compton effect' helped establish the wave–particle dualism.

The matter waves had important consequences for theory, too. For one thing, they cleared up some puzzles about the structure of the atom.

In 1913, Niels Bohr had pictured the hydrogen atom as consisting of a central nucleus surrounded by an electron that could circle that nucleus in any one of a number of orbits. These orbits were in fixed position; if a hydrogen electron dropped from an outer orbit to an inner one, it lost energy, emitting that energy in the form of a quantum possessing a fixed wavelength. If the electron was to move from an inner electron to an outer one, it would have to absorb a quantum of energy, but only one of a fixed size and wavelength that was just enough to move it by the proper amount. That was why hydrogen could absorb or emit only certain wavelengths of radiation, producing characteristic lines in its spectrum. Bohr's scheme, which was made gradually more complex over the next decade, was highly successful in explaining many facts about the spectra of various elements, and he was awarded the Nobel Prize in physics in 1922 for his theory. The German physicists James Franck and Gustav Hertz (the latter a nephew of Heinrich Hertz), whose studies on collisions between atoms and electrons lent an experimental foundation to Bohr's theories, shared the Nobel Prize in physics in 1925.

Bohr had no explanation of why the orbits were fixed in the positions they held. He simply chose the orbits that would give the correct results, so far as absorption and emission of the actually observed wavelengths of light were concerned.

In 1926, the German physicist Erwin Schrödinger decided to take another look at the atom in the light of the de Broglie theory of the wave nature of particles. Considering the electron as a wave, he decided that the electron did not circle round the nucleus as a planet circles round the sun but constituted a wave that curved all round the nucleus, so that it was in all parts of its orbit at once, so to speak. It turned out that, on the basis of the wavelength predicted by de Broglie for an electron, a whole number of electron waves would exactly fit the orbits outlined by Bohr. Between the orbits, the waves would not fit in a whole number but would join up 'out of phase'; and such orbits could not be stable.

Schrödinger worked out a mathematical description of the atom called 'wave mechanics' or 'quantum mechanics', and this became a more satisfactory method of looking at the atom than the Bohr system had been. Schrödinger shared the Nobel Prize in 1933 with Dirac, the author of the theory of anti-particles (see p. 311), who also contributed to the development of this new picture of the atom. The German physicist Max Born, who contributed further to the mathematical development of quantum mechanics, shared in the Nobel Prize in physics in 1954.

By this time the electron had become a pretty vague 'particle'. And this vagueness was soon to grow worse. Werner Heisenberg of Germany proceeded to raise a profound question that projected particles, and physics itself, almost into a realm of the unknowable.

Heisenberg had presented his own model of the atom. He had abandoned all attempts to picture the atom as composed either of particles or of waves. He decided that any attempt to draw an analogy between atomic structure and the structure of the world about us was doomed to failure. Instead, he described the energy levels or orbits of electrons purely in terms of numbers, without a trace of picture. Since he used a mathematical device called a 'matrix' to manipulate his numbers, his system was called 'matrix mechanics'.

Heisenberg received the Nobel Prize in physics in 1932 for his contributions to quantum mechanics, but his 'matrix' system was less popular with physicists than Schrödinger's wave mechanics, since the latter seemed just as useful as Heisenberg's abstractions, and it is difficult for even a physicist to force himself to abandon all attempts to picture what he is talking about.

By 1944, physicists seemed to have done the correct thing, for the Hungarian-American mathematician John von Neumann presented a line of argument that seemed to show that matrix mechanics and wave mechanics were mathematically equivalent. Everything that was demonstrated by one could equally well be demonstrated by the other. Why not choose the less abstract version therefore? (Nevertheless, in 1964, Dirac raised the question as to whether the two are really equivalent. He thinks not and favours Heisenberg over Schrödinger; the matrices over the waves.)

After having introduced matrix mechanics (to jump back in time again), Heisenberg went on to consider the problem of describing the position of a particle. How could one determine where a particle was? The obvious answer is: Look at it. Well, let us imagine a microscope that could make an electron visible. We must shine a light or some appropriate kind of radiation on it to see it. But an electron is so small that a single photon of light striking it would move it and change its position. In the very act of measuring its position, we would have changed that position.

This is a phenomenon that occurs in ordinary life. When we measure the air pressure in a tyre with a gauge, we let a little air out of the tyre and change the pressure slightly in the act of measuring it. Likewise when we put a thermometer in a bath-tub of water to measure the temperature, the thermometer's absorption of heat changes the temperature slightly. A meter measuring electric current takes away a little current for moving the pointer on the dial. And so it goes in every measurement of any kind that we make.

However, in all ordinary measurements the change in the subject we are measuring is so small that we can ignore it. The situation is quite different when we come to look at the electron. Our measuring device now is at least as large as the thing we are measuring; there is no usable measuring agent smaller than the electron. Consequently our measurement must inevitably have, not a negligible, but a decisive, effect on the object measured. We could stop the electron and so determine its position at a given instant. But in that case we could not know what its motion or velocity was. On the other hand, we might record its velocity, but then we could not fix its position at any given moment.

Heisenberg showed that there is no way of devising a method of pin-pointing the position of a sub-atomic particle unless you are willing to be quite uncertain as to its exact motion. And, in reverse, there is no way of pinpointing a particle's exact motion unless you are willing to be quite uncertain as to its exact position. To calculate both exactly, at the same instant of time, is impossible.

If this is so, then even at absolute zero, there cannot be complete lack of energy. If energy reached zero and particles became completely motionless, then only position need be determined since velocity could be taken as zero. It would be expected, therefore, that some residual 'zero-point energy' must remain, even at absolute zero, to keep particles in motion and, so to speak, uncertain. It is this zero-point energy, which cannot be removed, that is sufficient to keep helium liquid even at absolute zero (see p. 272).

In 1930, Einstein showed that the uncertainty principle, which stated it was impossible to reduce the error in position without increasing the error in momentum, implied it was also impossible to reduce the error in measurement of energy without increasing the uncertainty of time during which the measurement could take place. He thought he could use this as a springboard for the disproof of the uncertainty principle, but Bohr proceeded to show that Einstein's attempted disproof was wrong.

Indeed, Einstein's version of uncertainty proved very useful, since it meant that in sub-atomic processes, the law of conservation of energy could be violated for very brief periods of time, provided all was brought back to the conservational state by the end of those periods – the greater the deviation from conservation, the briefer the time-interval allowed. Yukawa used this notion in working out his theory of pions (see p. 340). It was even possible to explain certain sub-atomic phenomena by assuming that particles were produced out of nothing in defiance of energy conservation, but ceased to exist before the time allotted for their detection, so that they were only 'virtual particles'. The theory of virtual particles was worked out in the late 1940s by three men: the American physicists Julian Schwinger and Richard Phillips Feynman, and the Japanese physicist Sin-itiro Tomonaga. The three were jointly awarded the 1964 Nobel Prize in physics in consequence.

The uncertainty principle has profoundly affected the thinking of physicists and philosophers. It had a direct bearing on the philosophical question of 'causality' (that is, the relationship of cause and effect). But its implications for science are not those that are commonly supposed. One often reads that the principle of indeterminacy removes all certainty from nature and shows that science after all does not and never can know what is really going on, that scientific knowledge is at the mercy of the unpredictable whims of a universe in which effect does not necessarily follow cause. Whether or not this interpretation is valid from the standpoint of philosophy, the principle of uncertainty has in no way shaken the attitude of scientists towards scientific investigation. If, for instance, the behaviour of the individual molecules in a gas cannot be predicted with certainty, nevertheless on the average the molecules do obey certain laws, and their behaviour can be predicted on a statistical basis, just as insurance companies can calculate reliable mortality tables even though it is impossible to predict when any particular individual will die.

In most scientific observations, indeed, the indeterminacy is so small compared with the scale of the measurements involved that it can be neglected for all practical purposes. One can determine simultaneously both the position and the motion of a star, of a planet, or a billiard ball, or even of a grain of sand, with completely satisfactory accuracy.

As for the uncertainty among the sub-atomic particles themselves this does not hinder but actually helps physicists. It has been used to explain facts about radioactivity and about the absorption of sub-atomic particles by nuclei, as well as many other sub-atomic events, more reasonably than would have been possible without the uncertainty principle.

The uncertainty principle means that the universe is more complex than was thought, but not that it is irrational.

8 The Machine

Fire and Steam

The first law of thermodynamics states that energy cannot be created out of nothing. But there is no law against turning one form of energy into another. The whole civilization of mankind has been built upon finding new sources of energy and harnessing it in ever more efficient and sophisticated ways. In fact, the greatest single discovery in man's history involved methods for converting the chemical energy of a fuel such as wood into heat and light.

It was perhaps half a million years ago that our man-like ancestors 'discovered' fire. No doubt they had encountered – and been put to flight by – lightning-ignited brush fires and forest fires before that. But the discovery of its virtues did not come until curiosity overcame fear. Some pre-man may have been attracted to the quietly burning remnants of such a fire and found amusement in playing with it, in feeding it sticks, and in watching the dancing flames. At night he would have appreciated the light and warmth of the fire, and the fact that it kept other animals away. Eventually he would learn to make a fire himself by rubbing dry sticks together, the more easily and surely to use it, to warm his camp or cave with it, to roast his game in order to make it easier to chew and better tasting, and so on.

Fire provided man with a practically limitless supply of energy, which is why it is considered the greatest single human discovery – the one that hastened man's rise above the state of an animal. Yet

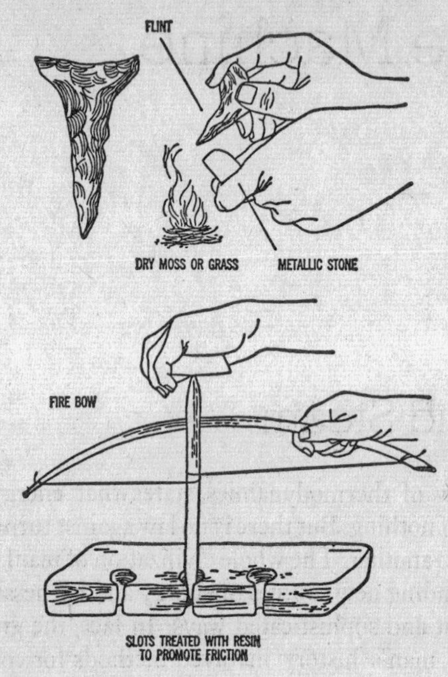

Early fire-making methods.

curiously enough, for many thousands of years – in fact, up to the Industrial Revolution – man realized only a small part of its possibilities. He used it to light and warm his home, to cook his food, to work metals and make pottery and glass – and that was about all.

Meanwhile he was discovering other sources of energy. And some of the most important of them were developed during the so-called 'Dark Ages'. It was in medieval times that man began to burn the black rock called coal in his metallurgical furnaces, to harness the wind with windmills, to use water mills for grinding grain, to employ magnetic energy in the compass, and to use explosives in warfare.

About A.D. 670, a Syrian alchemist, Callinicus, is believed to have invented 'Greek fire', a primitive incendiary bomb composed of sulphur and naphtha, which was credited with saving Constantinople from its first siege by the Moslems. Gunpowder arrived in

Europe in the thirteenth century. Roger Bacon described it about A.D. 1280, but it had been known in Asia for centuries before that and may have been introduced to Europe by the Mongol invasions beginning in A.D. 1240. In any case, artillery powered by gunpowder came into use in Europe in the fourteenth century, and cannons are supposed to have appeared first at the battle of Crécy in A.D. 1346.

The most important of all the medieval inventions is the one credited to Johann Gutenberg of Germany. About A.D. 1450, he cast the first movable type and thereby introduced printing as a powerful force in human affairs. He also devised printer's ink, in which carbon black was suspended in linseed oil, rather than, as hitherto, in water. Together with the replacement of parchment by paper (which had been invented by a Chinese eunuch, Ts'ai Lun – according to tradition – about A.D. 50 and which reached modern Europe, by way of the Arabs, in the thirteenth century), this made possible the large-scale production and distribution of books and other written material. No invention prior to modern times was adopted so rapidly. Within a generation of the discovery, 40,000 books were in print.

The recorded knowledge of mankind was no longer buried in royal collections of manuscripts but was made accessible in libraries available to all who could read. Pamphlets began to create and give expression to public opinion. (Printing was largely responsible for the success of Martin Luther's revolt against the Papacy, which might otherwise have been nothing more than a private quarrel.) And it was printing that created one of the prime instruments that gave rise to science as we know it. That indispensable instrument is the wide communication of ideas. Science had been a matter of personal communications among a few devotees; now it became a major field of activity, which enlisted more and more workers, elicited the prompt and critical testing of theories, and ceaselessly opened new frontiers.

The great turning point in man's harnessing of energy came at the end of the seventeenth century, although there had been a dim foreshadowing in ancient times. The Greek inventor, Hero of

Alexandria, sometime during the first centuries A.D. (his life cannot be pinned down even to a particular century), built a number of devices that ran on steam power. He used the expanding push of steam to open temple doors, whirl spheres, and so on. The ancient world, then deep in decline, could not follow up this premature advance.

Then, over fifteen centuries later, a new and vigorously expanding society had a second chance. It arose out of the increasingly acute necessity of pumping water out of mines that were being driven ever deeper. The old hand pump (see p. 167) made use of a vacuum to lift water; and as the seventeenth century proceeded, men came to appreciate, ever more keenly, just how great the power of a vacuum was (or, rather, the power of air pressure that was called into play by the existence of a vacuum).

In 1650, for instance, the German physicist (and mayor of the city of Magdeburg) Otto von Guericke invented an air pump worked by muscle power. He proceeded to put two flanged metal hemispheres together and to pump the air out from between them by means of a nozzle that one hemisphere possessed. As the air pressure within dropped lower and lower, the air pressure from without, no longer completely counterbalanced, pushed the hemispheres together more and more powerfully. At the end, two teams of horses straining in opposite directions could not pull the hemispheres apart, but when air was allowed to re-enter, they fell apart of themselves. This experiment was conducted before important people including, on one occasion, the German emperor himself, and it made a big splash.

Now it occurred to several inventors: Why not use steam instead of muscle power to create the vacuum? Suppose one filled a cylinder (or similar vessel) with water and heated the water to a boil. Steam, as it formed, would push out the water. If the vessel was cooled (e.g., by means of cold water played on the outside surface), the steam in the vessel would condense to a few drops of water and leave a virtual vacuum. The water that one wanted to raise (e.g., out of a flooded mine) could then rise through a valve into this evacuated vessel.

The first to translate this idea into a practical working device was an English military engineer named Thomas Savery. His 'steam engine' (the word 'engine' originally meant any ingenious device, and came from the same Greek root as 'ingenious') could be used to pump water out of a mine or a well or to drive a water wheel, so he called it 'The Miner's Friend'. But it was dangerous (because the high pressure of the steam might burst the vessels and pipes) and very inefficient (because the heat of the steam was lost each time the container was cooled). Seven years after Savery patented his engine in 1698, an English blacksmith named Thomas Newcomen built an improved engine that operated at low steam pressure; it had a piston in a cylinder and employed air pressure to push down the piston.

Newcomen's engine was not very efficient either, and the steam engine remained a minor gadget for more than sixty years until a Scottish instrument maker named James Watt found the way to make it effective. Hired by the University of Glasgow to fix a model of a Newcomen engine that was not working properly, Watt fell to thinking about the device's wasteful use of fuel. Why, after all, should the steam vessel have to be cooled off each time? Why not keep the steam chamber steam-hot at all times and lead the steam into a separate condensing chamber that could be kept cold? Watt went on to add a number of other improvements: employing steam pressure to help push the piston, devising a set of mechanical linkages that kept the piston moving in a straight line, hitching the back-and-forth motion of the piston to a shaft that turned a wheel, and so on. By 1782 his steam engine, which got at least three times as much work out of a ton of coal as Newcomen's, was ready to be put to work as a universal work horse.

In the times after Watt, steam-engine efficiency was continually increased, chiefly through the use of hotter and hotter steam at higher and higher pressure. Carnot's founding of thermodynamics (see Chapter 7) arose mainly out of the realization that the maximum efficiency with which any heat engine could be run was proportional to the difference in temperature between the hot reservoir (steam, in the usual case) and the cold.

The first application of the steam engine to some labour more dramatic than that of pumping water out of mines was the steamship. In 1787 the American inventor John Fitch built a steamboat that worked, but it failed as a financial venture and Fitch died unknown and unappreciated. Robert Fulton, a more able promoter than Fitch, launched his steamship, the *Clermont*, in 1807 with so much more fanfare and support that he came to be considered the inventor of the steamship, though actually he was no more the builder of the first such ship than Watt was the builder of the first steam engine.

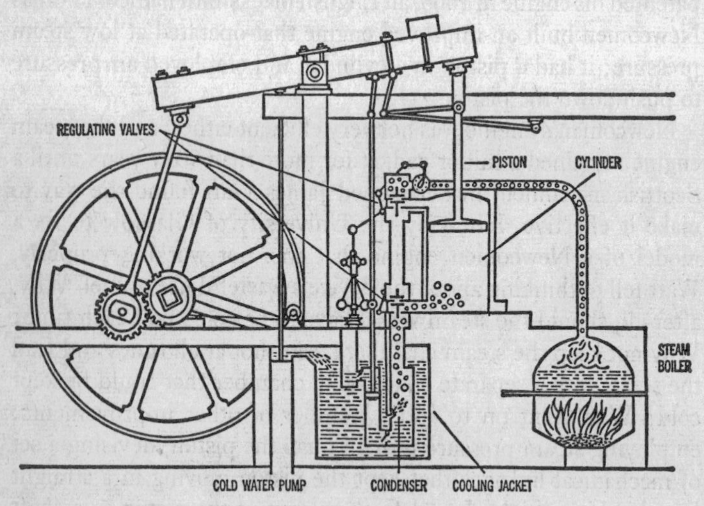

Watt's steam engine.

Fulton should, perhaps, better be remembered for his strenuous attempts to build underwater craft. His submarines were not practical, but they anticipated a number of modern developments. He built one called the *Nautilus*, which probably served as inspiration for Jules Verne's fictional submarine of the same name in *Twenty Thousand Leagues under the Sea*, published in 1870. That, in turn, was the inspiration for the naming of the first nuclear-powered submarine (see p. 482).

By the 1830s, steamships were crossing the Atlantic and were being driven by the screw propeller, a considerable improvement over the side paddle wheels. And by the 1850s, the speedy and beautiful Yankee Clippers had begun to furl their sails and to be replaced by steamers in the merchant fleets and navies of the world.

Meanwhile the steam engine had also begun to dominate land transportation. In 1814, the English inventor George Stephenson (owing a good deal to the prior work of an English engineer, Richard Trevithick) built the first practical steam locomotive. The in-and-out working of steam-driven pistons could turn metal wheels along steel rails as they could turn paddle wheels in the water. And in 1830, the American manufacturer Peter Cooper built the first steam locomotive in the Americas. For the first time in history land travel became as convenient as sea travel, and overland commerce could compete with seaborne trade. By 1840, the railway had reached the Mississippi River, and, by 1869, the full width of the United States was spanned by rail.

British inventors also led in introducing the steam engine into factories to run machinery. With this Industrial Revolution (a term introduced in 1837 by the French economist Jérôme Adolphe Blanqui), man completed his graduation from muscle power to mechanical power.

Electricity

In the nature of things, the steam engine is suitable only for large-scale, steady production of power. It cannot efficiently deliver energy in small packages or intermittently at the push of a button: a 'little' steam engine, in which the fires were damped down or started up on demand, would be an absurdity. But the same generation that saw the development of steam power also saw the discovery of a means of transforming energy into precisely the form I have mentioned – a ready store of energy which could be delivered anywhere, in small amounts or large, at the push of a button. This form, of course, is electricity.

The Greek philosopher Thales, about 600 B.C., noted that a fossil resin found on the Baltic shores, which we call amber and they called 'elektron', gained the ability to attract feathers, threads, or bits of fluff when it was rubbed with a piece of fur. It was William Gilbert of England, the investigator of magnetism (see p. 189), who first suggested that this attractive force be called 'electricity', from the Greek word 'elektron'. Gilbert found that, in addition to amber, some other materials, such as glass, gained electric properties on being rubbed.

In 1733, the French chemist Charles Francis de Cisternay Du Fay discovered that if two amber rods, or two glass rods, were electrified by rubbing, they repelled each other. However, an electrified glass rod attracted an electrified amber rod. If the two were allowed to touch, both lost their electricity. He felt this showed there were two kinds of electricity, 'vitreous' and 'resinous'.

The American scholar Benjamin Franklin, who became intensely interested in electricity, suggested that it was a single fluid. When glass was rubbed, electricity flowed into it, making it 'positively charged'; on the other hand, when amber was rubbed, electricity flowed out of it, and it therefore became 'negatively charged'. And when a negative rod made contact with a positive one, the electric fluid would flow from the positive to the negative until a neutral balance was achieved.

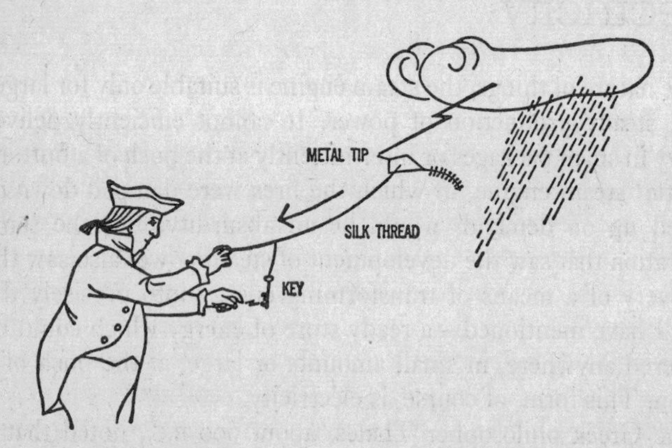

Franklin's experiment.

This was a remarkably shrewd speculation. If we substitute the word electrons for Franklin's 'fluid' and reverse the direction of flow (actually electrons flow from the amber to the glass), his guess was essentially correct.

A French inventor named John Théophile Desaguliers suggested in 1740 that substances through which the electric fluid travelled freely (e.g., metals) be termed 'conductors' and those through which it did not move freely (e.g., glass and amber) be called 'insulators'.

Experimenters found that a large electric charge could gradually be accumulated in a conductor if it was insulated from loss of electricity by glass or a layer of air. The most spectacular device of this kind was the 'Leyden jar'. It was first devised in 1745 by the German scholar E. Georg von Kleist, but it was first put to real use at the University of Leyden in Holland, where it was independently constructed a few months later by the Dutch scholar Peter van Musschenbroek. The Leyden jar is an example of what is today called a 'condenser', or 'capacitor', that is, two conducting surfaces, separated by a small thickness of insulator, within which one can store a quantity of electric charge.

In the case of the Leyden jar, the charge is built up on tinfoil coating a glass jar, via a brass chain stuck into the jar through a stopper. When you touch the charged jar, you get a startling electric shock. The Leyden jar can also produce a spark. Naturally, the greater the charge on a body, the greater its tendency to escape. The force driving the electrons away from the region of highest excess (the 'negative pole') towards the region of greatest deficiency (the 'positive pole') is the 'electromotive force' (EMF) or 'electric potential'. If the electric potential becomes high enough, the electrons will even jump an insulating gap between the negative and positive poles. Thus they will leap across an air gap, producing a bright spark and a crackling noise. The light of the spark is caused by the radiation resulting from the collisions of innumerable electrons with air molecules, and the noise arises from the expansion of the quickly heated air, followed by the clap of cooler air rushing into the partial vacuum momentarily produced.

Naturally one wondered whether lightning and thunder were the same phenomenon, on a vast scale, as the little trick performed by a Leyden jar. A British scholar, William Wall, had suggested just this in 1708. This thought was sufficient to prompt Benjamin Franklin's famous experiment in 1752. The kite he flew in a thunderstorm had a pointed wire, to which he attached a silk thread which could conduct electricity down from the thunder-clouds. When Franklin put his hand near a metal key tied to the silk thread, the key sparked. Franklin charged it again from the clouds, then used it to charge a Leyden jar, obtaining the same kind of charged Leyden jar in this fashion as in any other. Thus Franklin demonstrated that the thunderclouds were charged with electricity and that thunder and lightning were indeed the effect of a Leyden-jar-in-the-sky in which the clouds formed one pole and the earth another.

The luckiest thing about the experiment, from Franklin's personal standpoint, was that he survived. Some others who tried it were killed, because the induced charge on the kite's pointed wire accumulated to the point of producing a fatally intense discharge to the body of the man holding the kite.

Franklin at once followed up this advance in theory with a practical application. He devised the 'lightning rod', which was simply an iron rod attached to the highest point of a structure and connected to wires leading to the ground. The sharp point bled off electric charges from the clouds above, as Franklin showed by experiment, and, if lightning did strike, the charge was carried safely to the ground.

Lightning damage diminished drastically as the rods rose over structures all over Europe and the American colonies, no small accomplishment. Yet even today, two thousand million lightning flashes strike each year, killing (it is estimated) twenty people a day and hurting eighty more.

Franklin's experiment had two electrifying (please pardon the pun) effects. In the first place, the world at large suddenly became interested in electricity. Second, it put the American colonies on the map, culturally speaking. For the first time an American had actually displayed sufficient ability as a scientist to impress the cultivated Europeans of the Age of Reason. When, a quarter of a century later, Franklin represented the infant United States at Versailles and sought assistance, he won respect, not only as the simple envoy of a new republic, but also as a mental giant who had tamed the lightning and brought it humbly to earth. That flying kite contributed more than a little to the cause of American independence.

Following Franklin's work, electrical research advanced by leaps. Quantitative measurements of electrical attraction and repulsion were carried out in 1785 by the French physicist Charles Augustin de Coulomb. He showed that this attraction (or repulsion) between given charges varied inversely as the square of the distance. In this, electrical attraction resembled gravitational attraction. In honour of this finding, the 'coulomb' has been adopted as a name for a common unit of quantity of electricity.

Shortly thereafter, the study of electricity took a new, startling, and very fruitful turning. What we have been looking at above is, of course, 'static electricity'. This refers to an electric charge that is placed on an object and then stays there. The discovery of an electric charge that moved, of electric currents or 'dynamic electricity', began with the Italian anatomist Luigi Galvani. In 1791, he accidentally discovered that thigh muscles from dissected frogs would contract if simultaneously touched by two different metals (thus adding the verb 'to galvanize' to the English language).

The muscles behaved as though they had been stimulated by an electric spark from a Leyden jar, and so Galvani assumed that muscles contained something he called 'animal electricity'. Others, however, suspected that the origin of the electric charge might lie in the junction of the two metals rather than in muscle. In 1800, the Italian physicist Alessandro Volta studied combinations of dissimilar metals, connected not by muscle tissue but by simple solutions.

He began by using chains of dissimilar metals connected by bowls half-full of salt water. To avoid too much liquid too easily spilled, he prepared small disks of copper and of zinc, piling them alternately. He also made use of cardboard discs moistened with salt water so that his 'voltaic pile' consisted of silver, cardboard, zinc, silver, cardboard, zinc, silver, and so on. From such a set-up, electric current could be drawn off continuously.

Any series of similar items indefinitely repeated may be called a battery. Volta's instrument was the first 'electric battery'. It may also be called an 'electric cell'. It was to take a century before scientists would understand how chemical reactions involved electron transfers and how to interpret electric currents in terms of shifts and flows of electrons. Meanwhile, however, they made use of the current without understanding all its details.

Humphry Davy used an electric current to pull apart the atoms of tightly bound molecules and was able for the first time, in 1807 and 1808, to prepare such metals as sodium, potassium, magnesium, calcium, strontium, and barium. Faraday (Davy's assistant and protégé) went on to work out the general rules of such molecule-breaking 'electrolysis' and this, a half century later, was to guide Arrhenius in his working out of the hypothesis of ionic dissociation (see p. 186).

The manifold uses of dynamic electricity in the century and a half since Volta's battery seem to have placed static electricity in the shade and to have reduced it to a mere historical curiosity. Not so, for knowledge and ingenuity need never be static. By 1960, the American inventor Chester Carlson had perfected a practical device for copying material by attracting carbon-black to paper through localized electrostatic action. Such copying, involving no solutions or wet media, is called 'xerography' (from Greek words meaning 'dry writing') and has revolutionized office procedures.

The names of the early workers in electricity have been immortalized in the names of the units used for various types of measurement involving electricity. I have already mentioned 'coulomb' as a unit of quantity of electricity. Another unit is the 'faraday', for 96,500 coulombs is equal to one faraday. Faraday's name is used a second time, for a 'farad' is a unit of electrical capacity. Then, too, the unit of electrical intensity (the quantity of electric current passing through a circuit in a given time) is called the 'ampere', after the French physicist Ampère (see p. 191). One ampere is equal to one coulomb per second. The unit of electromotive force (the force that drives the current) is the 'volt', after Volta.

A given EMF did not always succeed in driving the same quantity of electricity through different circuits. It would drive a great deal of current through good conductors, little current through poor conductors, and virtually no current through non-conductors. In 1827, the German mathematician George Simon Ohm studied this 'resistance' to electrical flow and showed that it could be precisely related to the amperes of current flowing through a circuit under the push of a known EMF. The resistance could be determined by taking the ratio of volts to amperes. This is 'Ohm's law', and the unit of electrical resistance is the 'ohm', so that one ohm is equal to one volt divided by one ampere.

The conversion of chemical energy to electricity, as in Volta's battery and the numerous varieties of its descendants, has always been relatively expensive because the chemicals involved are not common or cheap. For this reason, although electricity could be used in the laboratory with great profit in the early nineteenth century, it could not be applied to large-scale uses in industry.

There have been sporadic attempts to make use of the chemical reactions involved in the burning of ordinary fuels as a source of electricity. Fuels such as hydrogen (or, better still, coal) are much cheaper than metals such as copper and zinc. As long ago as 1839, the English scientist William Grove devised an electric cell running on the combination of hydrogen and oxygen. It was interesting, but not practical. In recent years, physicists have been working hard to prepare practical varieties of such 'fuel cells'. The theory is all set; it is only the practical problems that must be ironed out, and these are proving most refractory.

When the large-scale use of electricity came into being in the latter half of the nineteenth century, it is not surprising, then, that it did not arrive by way of the electric cell. As early as the 1830s Faraday had produced electricity by means of the mechanical motion of a conductor across the lines of force of a magnet (see pp. 191–2). In such an 'electric generator' or 'dynamo' (from a Greek word for 'power'), the kinetic energy of motion could be turned into electricity. Such motion could be kept in being by steam power, which in turn could be generated by burning fuel. Thus, much more indirectly than in a fuel cell, the energy of burning coal or oil (or even wood) could be converted into electricity. By 1844, large, clumsy versions of such generators were being used to power machinery.

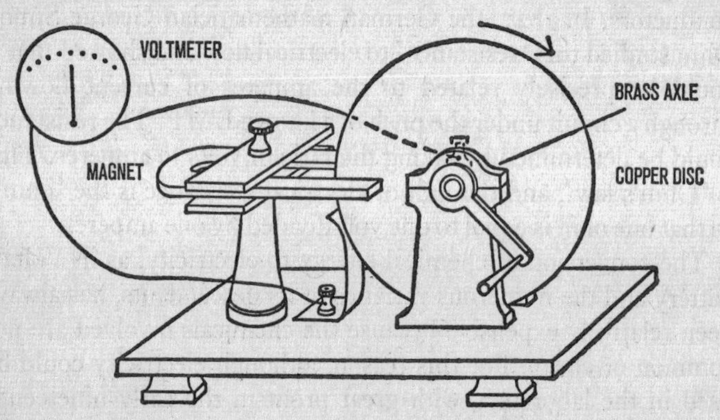

Faraday's 'dynamo'. The rotating copper disk cuts the magnet's lines of force, inducing a current on the voltmeter.

What was needed were ever stronger magnets, so that motion across the intensified lines of force could produce larger floods of electricity. These stronger magnets were obtained, in turn, by the use of electric currents. In 1823, the English electrical experimenter, William Sturgeon, wrapped eighteen turns of bare copper wire about a U-shaped iron bar and produced an 'electromagnet'. When the current was on, the magnetic field it produced was concentrated in the iron bar which could then lift twenty times its own weight of iron. With the current off, it was no longer a magnet and would lift nothing.

In 1829, the American physicist Joseph Henry improved this gadget vastly by using insulated wire. Once the wire was insulated it was possible to wind it in close loops over and over without fear of short circuits. Each loop increased the intensity of the magnetic field and the power of the electromagnet. By 1831, Henry had produced an electromagnet, of no great size, that could lift over a ton of iron.

The electromagnet was clearly the answer to better electrical generators. In 1845, the English physicist Charles Wheatstone made use of such an electromagnet for this purpose. Better understanding of the theory behind lines of force came about with

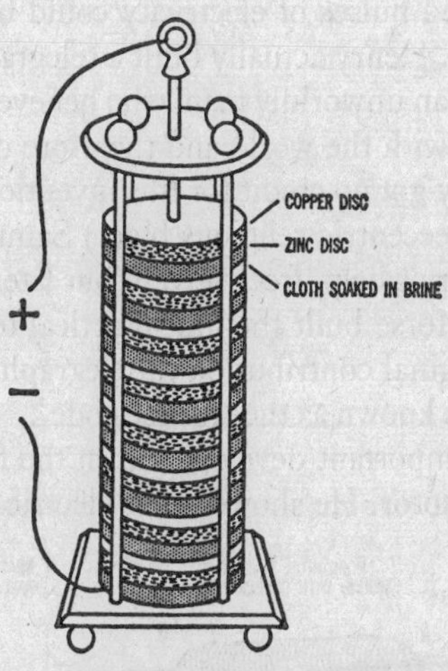

Volta's battery. The two different metals in contact give rise to a flow of electrons, which are conducted from one 'cell' to the next by the salt-soaked cloth. The familiar 'dry battery' or 'flashlight battery' of today, involving carbon and zinc, was first devised by Bunsen (of spectroscopy fame) in 1841.

Maxwell's mathematical interpretation of Faraday's work (see pp. 192–3) in the 1860s, and, in 1872, the German electrical engineer Friedrich von Hefner-Alteneck designed the first really efficient generator. At last electricity could be produced cheaply and in floods, and it could be done, not only from burning fuel, but from falling water.

For the work that led to the early application of electricity to technology, the lion's share of the credit must fall to Joseph Henry. Henry's first application of electricity was the invention of telegraphy. He devised a system of relays which made it possible to transmit an electric current over miles of wire. The strength of a current declines fairly rapidly as it travels at constant voltage across longer and longer stretches of resisting wire; what Henry's relays did was to use the dying signal to activate a small electromagnet that operated a switch that turned on a boost in power from stations placed at appropriate intervals. Thus a message consisting of coded pulses of electricity could be sent for a considerable distance. Henry actually built a telegraph that worked.

Because he was an unworldly man, who believed that knowledge should be shared with the world and therefore did not patent his discoveries, Henry got no credit for this invention. The credit fell to the artist (and eccentric religious bigot) Samuel Finley Breese Morse. With Henry's help, freely given (but later only grudgingly acknowledged), Morse built the first practical telegraph in 1844. Morse's main original contribution to telegraphy was the system of dots and dashes known as the 'Morse code'.

Henry's most important development in the field of electricity was the electric motor. He showed that electric current could be

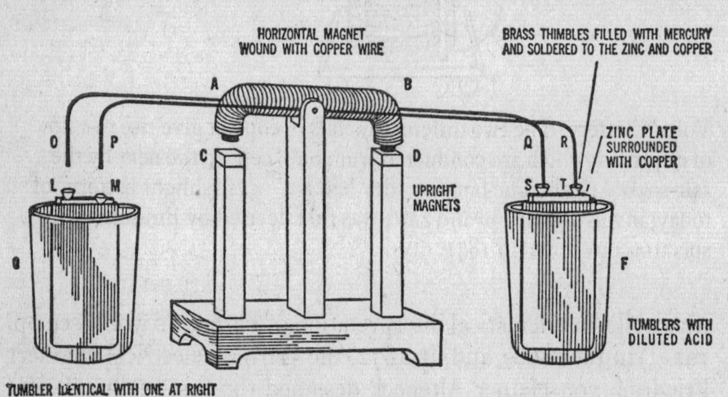

Henry's motor. The upright bar magnet *D* attracts the wire-wound magnet *B*, pulling the long metal probes *Q* and *R* into the brass thimbles *S* and *T*, which act as terminals for the wet cell *F*. Current flows into the horizontal magnet, producing an electromagnetic field that pulls *A* and *C* together. The whole process is then repeated on the opposite side. Thus the horizontal bars oscillate up and down.

used to turn a wheel, just as the turning of a wheel can generate current in the first place. And an electrically driven wheel (or motor) could be used to run machinery. The motor could be carried anywhere; it could be turned on or off at will (without waiting to build up a head of steam); and it could be made as small as one wished.

The catch was that electricity had to be transported from the generating station to the place where the motor was to be used. Some way had to be found to cut down the loss of electrical energy (taking the form of dissipated heat) as it travelled over wires.

One answer was the 'transformer'. The experimenters with currents found that electricity suffered far less loss if it was transmitted at a low rate of flow. So the output from the generator was stepped up to a high voltage by means of a transformer that, while multiplying the voltage, say, three times, reduces the current (rate of flow) to one third. At the receiving station, the voltage can be stepped down again so that the current is correspondingly increased for use in motors.

The transformer works by using the 'primary' current to induce a current at high voltage in a secondary coil. This induction requires varying the magnetic field through the second coil. Since a steady current will not do this, the current used is a continually changing one that builds up to a maximum and then drops to zero and starts building in the opposite direction – in other words, an 'alternating current'.

Alternating current (a.c.) did not win out over direct current (d.c.) without a struggle. Thomas Alva Edison, the greatest name in electricity in the final decades of the nineteenth century, championed d.c. and established the first d.c. generating station in New York in 1882 to supply current for the electric light he had invented. He fought a.c. on the ground that it was more dangerous (pointing out, for instance, that it was used in electric chairs). He was bitterly opposed by Nikola Tesla, an engineer who had worked for Edison and been shabbily treated. Tesla developed a successful system of a.c. in 1888. In 1893, George Westinghouse, also a believer in a.c., won a crucial victory over Edison by obtaining for his electric company the contract to develop the Niagara Falls power plants on an a.c. basis. In the following decades, Steinmetz established the theory of alternating currents on a firm mathematical basis. Today alternating current is all but universal in systems of power distribution. (In 1966, to be sure, engineers at General Electric devised a direct-current transformer – something long held to be impossible – but it involves liquid-helium temperatures and a low efficiency. It is fascinating, theoretically, but of no likely commercial use at the moment.)

Electrical Gadgets

The steam engine is a 'prime mover'. It takes energy already existing in nature (the chemical energy of wood, oil, or coal) and turns it into work. The electric motor is not; it converts electricity into work, but the electricity must itself be formed from the energy of burning fuel or falling water. For this reason, electricity is more expensive than steam for heavy jobs. Nevertheless, it can be used for the purpose. At the Berlin Exhibition of 1879, an electric-powered locomotive (using a third rail as its source of current) successfully pulled a train of coaches. Electrified trains are common now, especially for rapid transit within cities, for the added expense is more than made up for by increased cleanliness and smoothness of operation.

Where electricity really comes into its own, however, is where it performs tasks that steam cannot. There is, for instance, the telephone, patented by the Scottish-born inventor Alexander Graham Bell in 1876. In the telephone mouthpiece, the speaker's sound waves strike a thin steel diaphragm and make it vibrate in accordance with the pattern of the waves. The vibrations of the diaphragm in turn set up an analogous pattern in an electric current, by way of carbon powder. When the diaphragm presses on the carbon powder, the powder conducts more current; when the diaphragm moves away, it conducts less. Thus the electric current strengthens and weakens in exact mimicry of the sound waves. At the telephone receiver the fluctuations in the strength of the current actuate an electromagnet that makes a diaphragm vibrate and reproduce the sound waves.

In 1877, a year after the invention of the telephone, Edison patented his 'phonograph'. The first records had the grooves scored on tinfoil wrapped around a rotating cylinder. The American inventor Charles Sumner Tainter substituted wax cylinders in 1885 and then Emile Berliner introduced wax-coated discs in 1887. In 1925, recordings began to be made by means of electricity through the use of a 'microphone', which translated sound into a mimicking electric current via a piezoelectric crystal instead of a metal diaphragm – the crystal allowing a better quality of reproduction of the sound. In the 1930s, the use of radio tubes for amplification was introduced. Then, in the post-Second World War era, came the long-playing record, the 'hi-fi', and 'stereophonic' sound, which have had the effect, so far as the sound itself is concerned, of practically removing all mechanical barriers between the orchestra or singer and the listener!

'Tape-recording' of sound was invented in 1898 by a Danish electrical engineer named Valdemar Poulsen, but had to await certain technical advances to become practical. An electromagnet, responding to an electric current carrying the sound pattern, magnetizes a powder coating on a tape or wire moving past it, and the playback is accomplished through an electromagnet which picks up this pattern of magnetism and translates it again into a current that will reproduce the sound.

Of all the tricks performed by electricity, certainly the most popular was its turning night into day. Mankind had fought off the daily crippling darkness-after-sundown with the camp-fire, the torch, and the candle; for a hundred thousand years or so, the level of artificial light remained dim and flickering. Then, in the nineteenth century came whale oil, kerosene, and gas, and man-made light became somewhat stronger. Now electricity brought to pass a far better kind of lighting – safer, more convenient, and as brilliant as one could wish.

The problem was to heat a filament by electricity to an incandescent glow. It seemed simple, but many tried and failed to produce a durable lamp. Naturally, the filament had to glow in the absence of oxygen or be oxidized to destruction almost at once. The first attempts to remove oxygen involved the straightforward route of removing air. By 1875, Crookes (in connection with his work on cathode rays; see p. 249) had devised methods for producing a good enough vacuum for this purpose, and with sufficient speed and economy. Nevertheless, the filaments used remained unsatisfactory. They broke too easily. In 1878, Thomas Edison, fresh from his triumph in creating the phonograph, announced that he would tackle the problem. He was only thirty-one, but such was his reputation as an inventor that his announcement caused the stocks of gas companies to tumble on the New York and London stock exchanges.

After hundreds of experiments and fabulous frustrations, Edison finally found a material that would serve as the filament – a scorched cotton thread. On 21 October 1879, he lit his bulb. It burned for forty continuous hours. On the following New Year's Eve, Edison put his lamps on triumphant public display by lighting up the main street of Menlo Park, New Jersey, where his laboratory was located. He quickly patented his lamp and began to produce it in quantity.

Yet Edison was not the sole inventor of the incandescent lamp. At least one other inventor had about an equal claim – Joseph Swan of England, who exhibited a carbon-filament lamp at a meeting of the Newcastle-upon-Tyne Chemical Society on 18 December 1878, but did not get his lamp into production until 1881.

Edison proceeded to work on the problem of providing homes with a steady and sufficient supply of electricity for his lamps – a task which took as much ingenuity as the invention of the lamp itself. Two major improvements were later made in the lamp. In 1910, William David Coolidge of the General Electric Company adopted the heat-resisting metal tungsten as the material for the filament, and, in 1913, Irving Langmuir introduced the inert gas nitrogen in the lamp to prevent the evaporation and breaking of the filament that occurred in a vacuum.

Argon (the use of which was introduced in 1920) serves the

purpose even better than nitrogen, for argon is completely inert. Krypton, another inert gas, is still more efficient, allowing a lamp filament to reach higher temperatures and burn more brightly without loss of life expectancy.

Other kinds of electric lamp have, of course, been developed. The so-called 'neon lights' (introduced by the French chemist

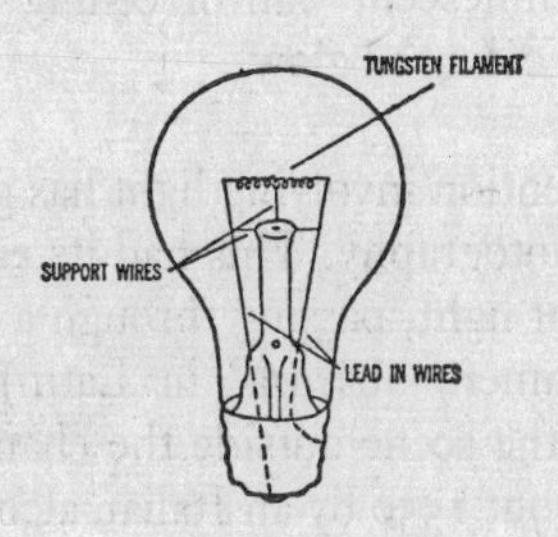

Incandescent lamp.

Georges Claude in 1910) are tubes in which an electric discharge excites atoms of neon gas to emit a bright, red glow. The 'sun lamp' contains mercury vapour, which when excited by a discharge yields radiation rich in ultra-violet light; this can be used not only to produce a tan but also to kill bacteria or generate fluorescence. And the latter in turn leads to fluorescent lighting, introduced in its contemporary form in 1939 at the New York World's Fair. Here the ultra-violet light from mercury vapour excites fluorescence in a 'phosphor' coating the inside of the tube. Since this cool light wastes little energy in heat, it consumes less electric power.

A 40-watt fluorescent tube supplies as much light, and far less heat, than a 150-watt incandescent light. Since the Second World War, therefore, there has been a massive swing towards the fluorescent. The first fluorescent tubes made use of beryllium salts as phosphors. This resulted in cases of serious poisoning ('berylliosis') induced by breathing dusts containing these salts or by introducing the substance through cuts caused by broken tubes. After 1949, other far less dangerous phosphors were used.

The latest promising development is a method that converts

electricity directly into light without the prior formation of ultra-violet light. In 1936, the French physicist Georges Destriau discovered that an intense alternating current could make a phosphor, such as zinc sulphide, glow. Electrical engineers are now distributing the phosphor through plastic or glass and are using this phenomenon, called 'electroluminescence', to develop glowing panels. Thus a luminescent wall or ceiling could light a room, bathing it in a soft, coloured glow.

Probably no invention involving light has given mankind more enjoyment than photography. This had its earliest beginnings in the observation that light, passing through a pinhole into a small dark chamber ('camera obscura' in Latin), will form a dim, inverted image of the scene outside the chamber. Such a device was constructed about 1550 by an Italian alchemist, Giambattista della Porta. This is the 'pinhole camera'.

In a pinhole camera, the amount of light entering is very small.

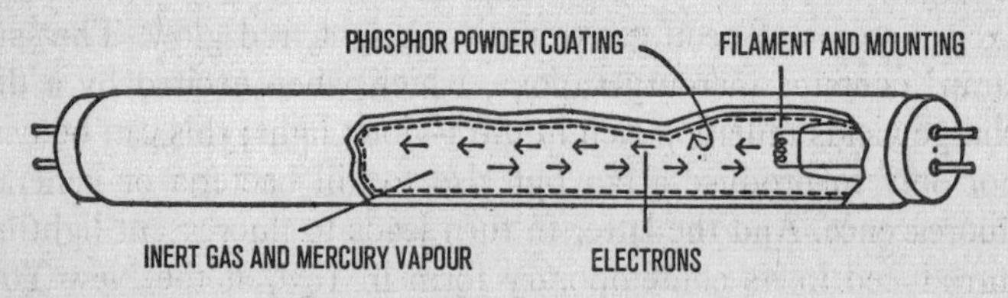

Fluorescent lamp. A discharge of electrons from the filament excites the mercury vapour in the tube, producing ultra-violet radiation. The ultra-violet makes the phosphor glow.

If, however, a lens is substituted for the pinhole, a considerable quantity of light can be brought to a focus, and the image is then much brighter. With that accomplished it is necessary to find some chemical reaction that will respond to light. A number of men laboured in this cause, including, most notably, the Frenchmen Joseph Nicéphore Niepce and Louis Jacques Mande Daguerre, and the Englishman William Henry Fox Talbot. By the mid nineteenth century, permanent images painted in chemicals could be produced.

The image is focused on an emulsion of a silver compound smeared (at first) on a glass plate. The light produces a chemical change in the compound, the amount of change being proportional to the intensity of the light at any given point. In the developing process, the chemical developer converts those parts changed by the light into metallic silver, again to an extent proportional to the intensity of light. The unaffected silver compound is then dissolved away, leaving a 'negative' on which the image appears as a pattern of blackening in various degrees. Light projected through the negative reverses the light and dark spots and forms the positive image. Photography went on to prove its value in human documentation almost at once when, in the 1850s, the British photographed Crimean war scenes and when, in the next decade, the American photographer Matthew Brady, with what we would now consider impossibly primitive equipment, took classic photographs of the American Civil War in action.

Throughout the nineteenth century, the process was gradually made faster and simpler. The American inventor George Eastman developed dry plates (in place of the original moist emulsion) and then adopted plastic film as the backing for the emulsion. More sensitive emulsions were created, so that faster shots could be made and the subject did not need to 'pose'.

Since the Second World War picture-taking has been further simplified by means of the 'Land camera', invented by Edwin Herbert Land of the Polaroid Corporation. It uses two films on which the negative and positive are developed automatically by chemicals incorporated in the film.

In the early twentieth century, a process of colour photography was developed by the Luxembourg-born French physicist Gabriel Lippmann, which won him the Nobel Prize for physics in 1908. That proved a false alarm, however, and practical colour photography was not developed until 1936. This second, and successful, try was based on the observation, in 1855, by Maxwell and von Helmholtz that any colour in the spectrum could be produced by combining red, green, and blue light. On this principle, the colour film is composed of emulsions in three layers, one sensitive to the red, one to the green, and one to the blue components of the image.

Three separate but superimposed pictures are formed, each reproducing the intensity of light in its part of the spectrum as a pattern of black-and-white shading. The film is then developed in three successive stages, using red, blue, and green dyes to deposit the appropriate colours on the negative. Each spot in the picture is a specific combination of red, green, and blue, and the brain interprets these combinations to reconstitute the full range of colour.

In 1959, Land presented a new theory of colour vision. The brain, he maintained, does not require a combination of three colours to create the impression of full colour. All it needs is two different wavelengths, or sets of wavelengths, one longer than the other by a certain minimum amount. For instance, one of the sets of wavelengths may be an entire spectrum, or white light. Because the average wavelength of white light is in the yellow-green region, it can serve as the 'short' wavelength. Now a picture reproduced through a combination of white light and red light (serving as the long wavelength) comes out in full colour. Land has also made pictures in full colour with filtered green light and red light and with other appropriate dual combinations.

The invention of motion pictures came from an observation first made by the English physician Peter Mark Roget in 1824. He noted that the eye forms a persistent image, which lasts for an appreciable fraction of a second. After the inauguration of photography, many experimenters, particularly in France, made use of this fact to create the illusion of motion by showing a series of pictures in rapid succession. Everyone is familiar with the parlour gadget consisting of a series of picture cards which, when riffled rapidly, make a figure seem to move and perform acrobatics. If a series of pictures, each slightly different from the one before, is flashed on a screen at intervals of about one sixteenth of a second apart, the persistence of the successive images in the eye will cause them to blend together and so give the impression of continuous motion.

It was Edison who produced the first 'movie'. He photographed a series of pictures on a strip of film and then ran the film through a projector, which showed each in succession with a burst of light.

The first motion picture was put on display for public amusement in 1894, and, in 1914, theatres showed the full-length motion picture, *The Birth of a Nation.*

To the silent movies, a sound track was added in 1927. The 'sound track' also takes the form of light: the wave pattern of music and the actor's speech is converted into a varying current of electricity by a microphone, and this current lights a lamp which is photographed along with the action of the motion picture. When the film, with this track of light at one side, is projected on the screen, the brightening and dimming of the lamp in the pattern of the sound waves is converted back to an electric current by means of a 'phototube', using the photoelectric effect, and the current in turn is reconverted to sound.

Within two years after the first 'talking picture', *The Jazz Singer*, silent movies were a thing of the past, and so, almost, was vaudeville. By the late 1930s, the 'talkies' had added colour. In addition, the 1950s saw the development of wide-screen techniques and even a short-lived fad for three-dimensional (3D) effects, involving two pictures thrown on the same screen. By wearing polarized spectacles, an observer saw a separate picture with each eye, thus producing a stereoscopic effect.

Internal-Combustion Engines

While petroleum gave way to electricity in the field of artificial illumination, it became indispensable for another technical development that revolutionized modern life as deeply, in its way, as did the introduction of electrical gadgetry. This development was the internal-combustion engine, so-called because, in such an engine, fuel is burned within the cylinder so that the gases formed push the piston directly. Ordinary steam engines are 'external-combusion engines', the fuel being burned outside and the steam being then led, ready-formed, into the cylinder.

This compact device, with small explosions set off within the cylinder, made it possible to apply motive power to small vehicles in ways for which the bulky steam engine was not well-suited. To be sure, steam-driven 'horseless carriages' were devised as long ago as 1786, when William Murdock, a partner of James Watt, built one. A century later, the American inventor Francis Edgar Stanley invented the famous 'Stanley Steamer', which for a while competed with the early cars equipped with internal-combustion machines. The future, however, lay with the latter.

Actually, some internal-combustion engines were built at the beginning of the nineteenth century, before petroleum came into common use. They burned turpentine vapours or hydrogen as fuel. But it was only with petrol, the one vapour-producing liquid that is both combustible and obtainable in large quantities, that such an engine could become more than a curiosity.

The first practical internal-combustion engine was built in 1860 by the French inventor Étienne Lenoir; in 1876, the German technician Nikolaus August Otto built a 'four-cycle' engine. First a piston fitting tightly in a cylinder is pushed outwards, so that a mixture of petrol and air is sucked into the vacated cylinder. Then the piston is pushed in again to compress the vapour. At the point of maximum compression the vapour is ignited and explodes. The explosion drives the piston outwards, and it is this powered motion that drives the engine. It turns a wheel which pushes the piston in again to expel the burned residue or 'exhaust' – the fourth and final step in the cycle. Now the wheel moves the piston outwards to start the cycle over again.

A Scottish engineer named Dugald Clerk almost immediately added an improvement. He hooked up a second cylinder, so that its piston was being driven while the other was in the recovery stage: this made the power output steadier. Later the addition of more cylinders (eight is now a common number) increased the smoothness and power of this 'reciprocating engine'.

The ignition of the petrol–air mixture at just the right moment presented a problem. All sorts of ingenious devices were used, but by 1923 it became common to depend on electricity. The supply comes from a 'storage battery'. This is a battery that, like any other, delivers electricity as the result of a chemical reaction. But it can be recharged by sending an electric current through it in the

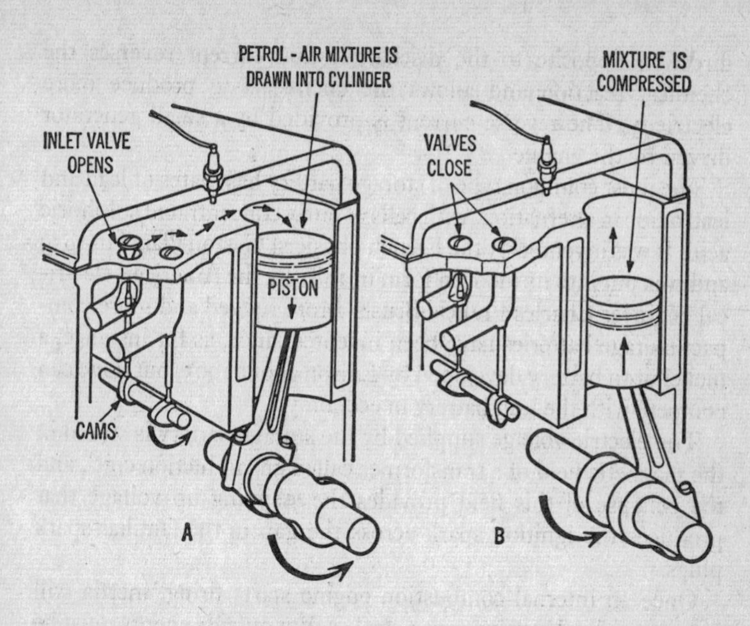

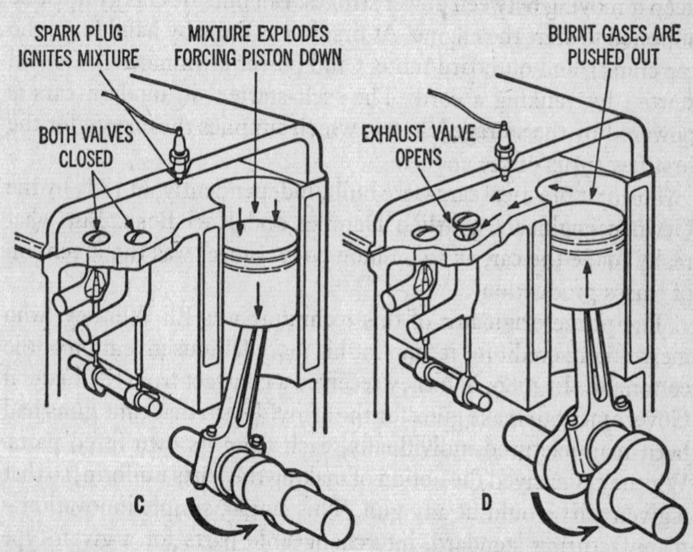

Nikolaus Otto's 'four-cycle' engine, built in 1876.

direction opposite to the discharge; this current reverses the chemical reaction and allows the chemicals to produce more electricity. The reverse current is provided by a small generator driven by the engine.

The most common type of storage battery has plates of lead and lead oxide in alternation, with cells of fairly concentrated sulphuric acid. It was invented by the French physicist Gaston Planté in 1859 and was put into its modern form in 1881 by the American electrical engineer Charles Francis Brush. More rugged and more compact storage batteries have been invented since, as for instance, a nickel–iron battery developed by Edison about 1905, but none can compete with the lead battery in economy.

The electric voltage supplied by the storage battery is stored in the magnetic field of a transformer called an 'induction coil', and the collapse of this field provides the stepping-up voltage that produces the ignition spark across the gap in the familiar spark plugs.

Once an internal-combustion engine starts firing, inertia will keep it moving between power strokes. But outside energy must be supplied to start the engine. At first it was done by hand (e.g., the car crank), and outboard motors and power lawn mowers are still started by yanking a cord. The 'self-starter' in modern cars is powered by the storage battery, which supplies the energy for the first few turns of the engine.

The first practical cars were built, independently, in 1885 by the German engineers Gottlieb Daimler and Karl Benz. But what really made the car, as a common conveyance, was the invention of 'mass production'.

The prime originator of this technique was Eli Whitney, who merits more credit for it than for his more famous invention of the cotton gin. In 1789, Whitney received a contract from the Federal Government to make guns for the army. Up to that time guns had been manufactured individually, each from its own fitted parts. Whitney conceived the notion of making the parts uniform, so that a given part would fit any gun. This single, simple innovation – manufacturing standard, interchangeable parts for a given type of article – was perhaps as responsible as any other factor for the creation of modern mass-production industry. When power tools came in, they made it possible to stamp out standard parts in practically unlimited numbers.

It was the American engineer Henry Ford who first exploited the concept to the full. He had built his first motor car (a two-cylinder job) in 1892, then had gone to work for the Detroit Automobile Company in 1899 as chief engineer. The company wanted to produce custom-made cars, but Ford had another notion. He resigned in 1902 to produce cars on his own – in quantity. In 1909, he began to turn out the Model T Ford and by 1913 he began to manufacture it on the Whitney plan – car for car, each just like the one before and all made with the same parts.

Ford saw that he could speed up production by using human workers as one used machines, performing the same small job over and over with uninterrupted regularity. The American inventor Samuel Colt (who had invented the revolver or 'six-shooter') had taken the first steps in this direction in 1847, and the car manufacturer Ransom E. Olds had applied the system to the motor car in 1900. Olds lost his financial backing, however, and it fell to Ford to carry this movement to its fruition. Ford set up the 'assembly line', with workers adding parts to the construction as it passed them on moving belts until the finished car rolled off at the end of the line. Two economic advances were achieved by this system: high wages for the workers and cars that could be sold at amazingly low prices.

By 1913, Ford was manufacturing 1,000 Model Ts a day. Before the line was discontinued in 1927, 15 million had been turned out, and the price had dropped to 290 dollars. The passion for yearly change then won out, and Ford was forced to join the parade of variety and superficial novelty that has raised the price of cars tenfold and lost Americans much of the advantage of mass production.

In 1892, the German mechanical engineer Rudolf Diesel introduced a modification of the internal-combustion engine that was simpler and more economical of fuel. He put the fuel–air mixture under high pressure, so that the heat of compression alone was enough to ignite it. The 'diesel engine' made it possible to use higher-boiling fractions of petroleum, which do not knock. Because of the higher compression used, the engine must be more solidly constructed and is therefore considerably heavier than the petrol engine. Once an adequate fuel-injection system was developed in the 1920s it began to gain favour for trucks, tractors, buses, ships, and locomotives, and is now undisputed king of heavy transportation.

Improvements in petrol itself further enhanced the efficiency of the internal-combustion engine. Petrol is a complex mixture of molecules made up of carbon and hydrogen atoms ('hydrocarbons'), some of which burn more quickly than others. Too quick a burning rate is undesirable, for then the petrol–air mixture explodes in too many places at once, producing 'engine knock'. A slow rate of burning produces an even expansion of vapour that pushes the piston smoothly and effectively.

The amount of knock produced by a given petrol is measured as its 'octane rating', by comparing it to the knock produced by a hydrocarbon called 'iso-octane', which is particularly low in knock production, mixed with 'normal heptane', which is particularly high in knock production. One of the prime functions of petrol refining is, among many other things, to produce a hydrocarbon mixture with a high octane rating.

Car engines have been designed through the years with a higher and higher 'compression ratio'; that is, the petrol–air mixture is compressed to greater and greater density before ignition. This milks the petrol of more power, but it also encourages knock, so that petrol of continually higher octane rating has had to be developed.

The task has been made easier by the use of chemicals that, when added in small quantities to the petrol, reduce knock. The most efficient of these 'anti-knock compounds' is 'tetraethyl lead', a lead compound first introduced for the purpose in 1925. Petrol containing it is 'leaded petrol' or 'ethyl petrol'. If tetraethyl lead were present alone, the lead oxides formed during petrol combustion would foul and ruin the engine. For this reason, ethylene bromide is also added. The lead atom of tetraethyl lead combines with the bromide atom of ethylene bromide to form lead bromide, which, at the temperature of the burning petrol, is vaporized and expelled with the exhaust.

Diesel fuels are tested for ignition delay after compression (too great a delay is undesirable) by comparison with a hydrocarbon called 'cetane', which contains sixteen carbon atoms in its molecule as compared with eight for 'iso-octane'. For diesel fuels, therefore, one speaks of a 'cetane number'.

The greatest triumph of the internal-combustion engine came, of course, in the air. By the 1890s, man had achieved the age-old dream – older than Daedalus and Icarus – of flying on wings. Gliding had become an avid sport of the aficionados. The first man-carrying glider was built in 1853 by the English inventor George Cayley. The 'man' it carried, however, was only a boy. The first important practitioner of the sport, the German engineer Otto Lilienthal, was killed in 1896 during a glider flight. Meanwhile, a violent urge to take off in powered flight had begun.

The American physicist and astronomer Samuel Pierpont Langley tried in 1902 and 1903 to fly a glider powered by an internal-combustion engine and came within an ace of succeeding. Had his money not given out, he might have got into the air on the next try. As it was, the honour was reserved for the brothers Orville and Wilbur Wright, bicycle manufacturers who had taken up gliders as a hobby.

On 17 December 1903, at Kitty Hawk, North Carolina, the Wright brothers got off the ground in a propeller-driven glider and stayed in the air for fifty-nine seconds, flying 852 feet. It was the first aeroplane flight in history, and it went almost completely unnoticed by the world at large.

There was considerably more public excitement after the Wrights had achieved flights of twenty-five miles and more and when, in 1909, the French engineer Louis Blériot crossed the English Channel in an aeroplane. The air battles and exploits of the First World War further stimulated the imagination and the biplanes of that day, with their two wings held precariously together by struts and wires, were familiar to a generation of post-First World War movie-goers. The German engineer Hugo Junkers designed a successful monoplane just after the war and the thick single wing, without struts, took over completely. (In 1939, the Russian-American engineer Igor Ivan Sikorsky built a multi-engined plane and designed the first helicopter, a plane with upper vanes that made vertical take-offs and landings and even hovering practical.)

But through the early 1920s the aeroplane remained more or less a curiosity – merely a new and more horrible instrument of war and a plaything of stunt flyers and thrill-seekers. Aviation did not come into its own until Charles Augustus Lindbergh in 1927 flew non-stop from New York to Paris. The world went wild over the feat, and the development of bigger and safer aeroplanes began.

Two major innovations have been effected in the aeroplane engine since it was established as a means of transportation. The first was the adoption of the gas-turbine engine. In this engine the hot, expanding gases of the fuel drive a wheel by their pressure against its blades, instead of driving pistons in cylinders. The engine is simple, cheaper to run, and less vulnerable to trouble, and it needed only the development of alloys that could withstand the high temperatures of the gases to become a practicable affair. Such alloys were devised by 1939. Since then 'turboprop' planes, using a turbine engine to drive the propellers, have become increasingly popular.

But they are now being superseded, at least for long flights, by the second major development – the jet plane. In principle the driving force here is the same as the one that makes a toy balloon dart forward when its mouth is opened and the air escapes. This is action-and-reaction: the motion of the expanding, escaping air in one direction results in equal motion, or thrust, in the opposite direction, just as the forward movement of the bullet in a gun barrel makes the gun kick backwards in recoil. In the jet engine, the burning of the fuel produces hot, high-pressure gases that drive the plane forwards with great force as they stream backwards through the exhaust. A rocket is driven by exactly the same means, except that it carries its own supply of oxygen to burn the fuel.

Patents for 'jet propulsion' were taken out by a French engineer, René Lorin, as early as 1913, but at the time it was a completely impractical scheme for aeroplanes. Jet propulsion is economical only at speeds of more than 400 miles an hour. In 1939, an Englishman, Frank Whittle, flew a reasonably practical jet plane, and, in January 1944, jet planes were put into war use by Great Britain and the United States against the 'buzz-bombs', Germany's V-1 weapon, a pilotless robot plane carrying explosives in its nose.

After the Second World War, military jets were developed that approached the speed of sound. The speed of sound depends on the natural elasticity of air molecules, their ability to snap back and forth. When the plane approaches that speed, the air molecules cannot get out of the way, so to speak, and are compressed ahead of the plane, which then undergoes a variety of stresses and strains. There was talk of the 'sound barrier' as though it were something physical that could not be approached without destruction. However, tests in wind tunnels led the way to more efficient streamlining, and, on 14 October 1947, an American X-1 rocket plane, piloted by Charles E. Yeager, 'broke the sound barrier'; for the first time in history, man surpassed the speed of sound. The air battles of the Korean War in the early 1950s were fought by jet planes moving at such velocities that comparatively few planes were shot down.

The ratio of the velocity of an object to the velocity of sound (which is 740 miles per hour at 0°C) in the medium through which the object is moving is called the 'Mach number' after the Austrian physicist Ernst Mach, who first investigated, theoretically, the consequences of motion at such velocities in the mid nineteenth century. By the 1960s, aeroplane velocities surpassed Mach 5. This was done by the experimental rocket plane X-15, the rockets of which pushed it high enough, for short periods of time, to allow its pilots to qualify as astronauts. Military planes

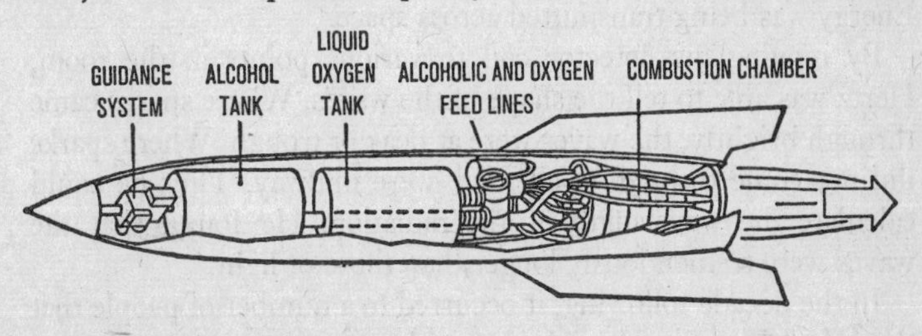

A simple liquid-fuelled rocket.

travel at lower velocities and commercial planes at lower velocities still.

A plane travelling at 'supersonic velocities' (over Mach 1) carries its sound waves ahead of it since it travels more quickly than the sound waves alone would. If close enough to the ground to begin with, the cone of compressed sound waves may intersect the ground with a loud 'sonic boom'. (The crack of a bullwhip is a miniature sonic boom, since, properly manipulated, the tip of such a whip can be made to travel at supersonic velocities.)

Radio

In 1888, Heinrich Hertz conducted the famous experiments that detected radio waves, predicted twenty years earlier by James Clerk Maxwell (see p. 367). What he did was to set up a high-voltage alternating current that surged into first one, then another of two metal balls separated by a small air gap. Each time the potential reached a peak in one direction or the other, it sent a spark across the gap. Under these circumstances, Maxwell's equations predicted, electromagnetic radiation should be generated. Hertz used a receiver consisting of a simple loop of wire with a small air gap at one point to detect that energy. Just as the current gave rise to radiation in the first coil, so the radiation ought to give rise to a current in the second coil. Sure enough, Hertz was able to detect small sparks jumping across the gap to his detector coil, placed across the room from the radiating coil. Energy was being transmitted across space.

By moving his detector coil to various points in the room, Hertz was able to tell the shape of the waves. Where sparks came through brightly, the waves were at peak or trough. Where sparks did not come through at all, they were midway. Thus he could calculate the wavelength of the radiation. He found that the waves were tremendously longer than those of light.

In the decade following, it occurred to a number of people that the 'Hertzian waves' might be used to transmit messages from one place to another, for the waves were long enough to go round obstacles. In 1890, the French physicist Édouard Branly made an improved receiver by replacing the wire loop with a glass tube filled with metal filings to which wires and a battery were attached.

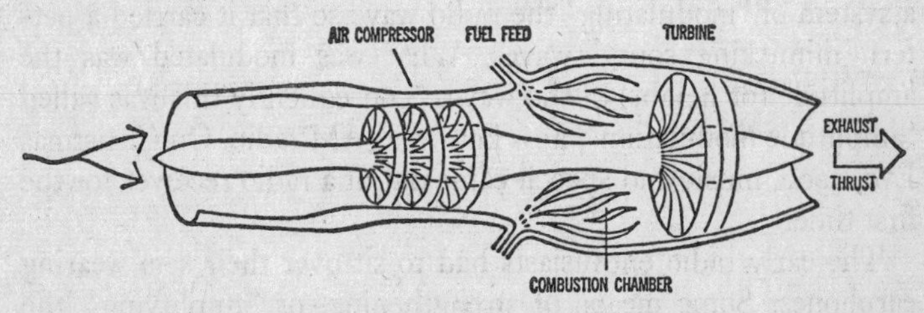

A turbojet engine. Air is drawn in, compressed, and mixed with fuel, which is ignited in the combustion chamber. The expanding gases power a turbine and produce thrust.

The filings would not carry the battery's current unless a high-voltage alternating current was induced in the filings, as Hertzian waves would do. With this receiver he was able to detect Hertzian waves at a distance of 150 yards. Then the English physicist Oliver Joseph Lodge (who later gained a dubious kind of fame as a champion of spiritualism) modified this device and succeeded in detecting signals at a distance of half a mile and in sending messages in Morse code.

The Italian inventor Guglielmo Marconi discovered that he could improve matters by connecting one side of the generator and receiver to the ground and the other to a wire, later called an 'antenna' (because it resembled, I suppose, an insect's feeler). By using powerful generators, Marconi was able to send signals over a distance of nine miles in 1896, across the English Channel in 1898, and across the Atlantic in 1901. Thus was born what the British still call 'wireless telegraphy' and the Americans named 'radiotelegraphy', or 'radio' for short.

Marconi worked out a system for excluding 'static' from other sources and tuning in only on the wavelength generated by the transmitter. For his inventions, Marconi shared the Nobel Prize in physics in 1909 with the German physicist Karl Ferdinand Braun, who also contributed to the development of radio.

The American physicist Reginald Aubrey Fessenden proceeded to develop a special generator of high-frequency alternating currents (doing away with the spark-gap device) and to devise a system of 'modulating' the radio wave so that it carried a pattern mimicking sound waves. What was modulated was the amplitude (or height) of the waves; consequently this was called 'amplitude modulation', now known as AM radio. On Christmas Eve, 1906, music and speech came out of a radio receiver for the first time.

The early radio enthusiasts had to sit over their sets wearing earphones. Some means of strengthening, or 'amplifying', the signal was needed, and the answer was found in a discovery that Edison had made – his only discovery in 'pure' science.

In one of his experiments, looking towards improving the electric lamp, Edison, in 1883, sealed a metal wire into a light bulb near the hot filament. To his surprise, electricity flowed from the hot filament to the metal wire across the air gap between them. Because this phenomenon had no utility for his purposes, Edison, a practical man, merely wrote it up in his notebooks and forgot it. But the 'Edison effect' became very important indeed when the electron was discovered and it became clear that current across a gap meant a flow of electrons. The British physicist Owen Willans Richardson showed, in experiments conducted between 1900 and 1903, that electrons 'boiled' out of metal filaments heated in vacuum. For this, he eventually received the Nobel Prize for physics in 1928.

In 1904, the English electrical engineer John Ambrose Fleming put the Edison effect to brilliant use. He surrounded the filament in a bulb with a cylindrical piece of metal (called a 'plate'). Now this plate could act in either of two ways. If it was positively charged, it would attract the electrons boiling off the heated filament and so would create a circuit that carried electric current. But if the plate was negatively charged, it would repel the electrons and thus prevent the flow of current. Suppose, then, that the plate was hooked up to a source of alternating current.

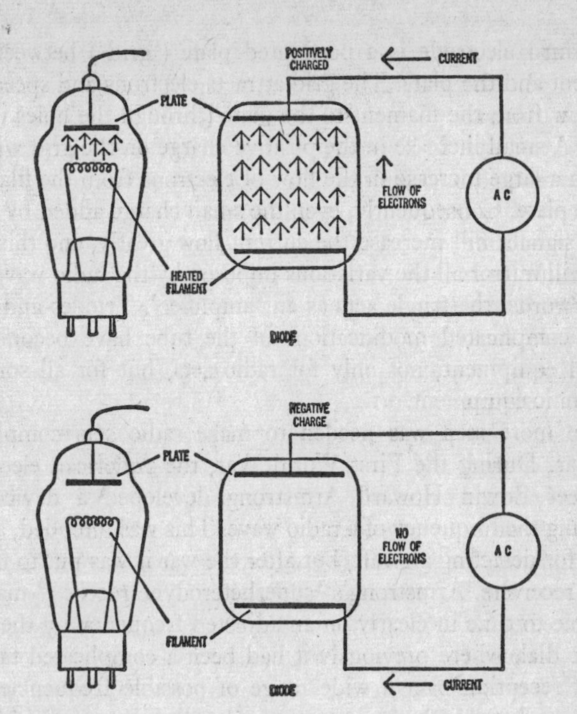

Principle of the vacuum-tube diode.

When the current flowed in one direction, the plate would get a positive charge and pass current in the tube; when the alternating current changed direction, the plate would acquire a negative charge and no current would flow in the tube. Thus the plate would pass current in only one direction; in effect, it would convert alternating to direct current. Because such a tube acts as a valve for the flow of current, the British logically call it a 'valve'. In the United States, it is vaguely called a 'tube'. Scientists took to calling it a 'diode', because it has two electrodes – the filament and the plate.

The diode serves in a radio set as a 'rectifier', changing alternating current to direct where necessary. In 1907, the American inventor Lee De Forest went a step further. He inserted a third electrode in the tube, making a 'triode' out of it.

The third electrode is a perforated plate ('grid') between the filament and the plate. The grid attracts electrons and speeds up the flow from the filament to the plate (through the holes in the grid). A small increase in the positive charge on the grid will result in a large increase in the flow of electrons from the filament to the plate. Consequently, even the small charge added by weak radio signals will increase the current flow greatly, and this current will mirror all the variations imposed by the radio waves. In other words, the triode acts as an 'amplifier'. Triodes and even more complicated modifications of the tube have become essential equipment, not only for radio sets, but for all sorts of electronic equipment.

One more step was needed to make radio sets completely popular. During the First World War, the American electrical engineer Edwin Howard Armstrong developed a device for lowering the frequency of a radio wave. This was intended, at the time, for detecting aircraft, but after the war it was put to use in radio receivers. Armstrong's 'superheterodyne receiver' made it possible to tune in clearly on an adjusted frequency by the turn of one dial, where previously it had been a complicated task to adjust reception over a wide range of possible frequencies. In 1921, regular radio programmes were begun by a station in Pittsburgh. Other stations were set up in rapid succession, and with the control of sound level and station tuning reduced to the turn of a dial, radio sets became hugely popular. By 1927, telephone conversations could be carried on across oceans, with the help of radio, and 'wireless telephony' was a fact.

There remained the problem of static. The systems of tuning introduced by Marconi and his successors minimized 'noise' from thunderstorms and other electrical sources, but did not eliminate it. Again it was Armstrong who found an answer. In place of amplitude modulation, which was subject to interference from the random amplitude modulations of the noise sources, he substituted frequency modulation. That is, he kept the amplitude of the radio carrier wave constant and superimposed a variation in frequency on it. Where the sound wave was large in amplitude, the carrier wave was made low in frequency, and vice versa. Frequency modulation (FM) virtually eliminated static, and FM radio came into popularity after the Second World War for programmes of serious music.

Television was an inevitable sequel to radio, just as talking movies were to the silents. The technical forerunner of television was the transmission of pictures by wire. This entailed translating

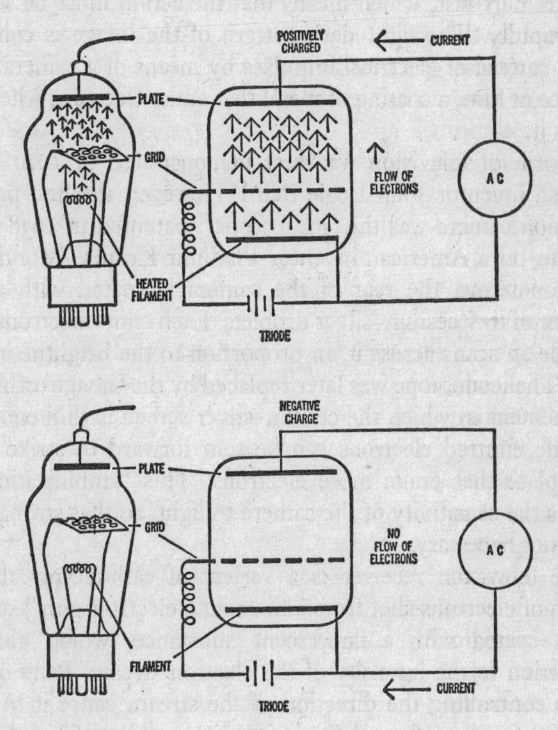

Principle of the triode.

a picture into an electric current. A narrow beam of light passed through the picture on a photographic film to a phototube behind. Where the film was comparatively opaque, a weak current was generated in the phototube; where it was clearer, a large current was formed. The beam of light swiftly 'scanned' the picture from left to right, line by line, and produced a varying current representing the entire picture. The current was sent over wires and at the destination reproduced the picture on film by a reverse process. Such 'wire-photos' were transmitted between London and Paris as early as 1907.

Television is the transmission of a 'movie' instead of still photographs – either 'live' or from a film. The transmission must be extremely fast, which means that the action must be scanned very rapidly. The light–dark pattern of the image is converted into a pattern of electrical impulses by means of a camera using, in place of film, a coating of metal that emits electrons when light strikes it.

A form of television was first demonstrated in 1926 by the Scottish inventor John Logie Baird. However, the first practical television camera was the 'iconoscope', patented in 1938 by the Russian-born American inventor Vladimir Kosma Zworykin. In the iconoscope, the rear of the camera is coated with a large number of tiny cesium–silver droplets. Each emits electrons as the light beam scans across it, in proportion to the brightness of the light. The iconoscope was later replaced by the 'image orthicon' – a refinement in which the cesium–silver screen is thin enough so that the emitted electrons can be sent forward to strike a thin glass plate that emits more electrons. This 'amplification' increases the sensitivity of the camera to light, so that strong lighting is not necessary.

The television receiver is a variety of cathode-ray tube. A stream of electrons shot from a filament ('electron-gun') strikes a screen coated with a fluorescent substance, which glows in proportion to the intensity of the electron stream. Pairs of electrodes controlling the direction of the stream cause it to sweep across the screen from left to right in a series of hundreds of horizontal lines, each slightly below the one before, and the entire 'painting' of a picture on the screen in this fashion is completed in a thirtieth of a second. The beam goes on painting successive pictures at the rate of thirty per second. At no instant of time is there more than one dot on the screen (bright or dark, as the case may be); yet, thanks to the persistence of vision, we see not only complete pictures but an uninterrupted sequence of movement and action.

Experimental television was broadcast in the 1920s, but television did not become practical in the commercial sense until 1947. Since then it has virtually taken over the field of entertainment.

In the mid 1950s two refinements were added. By the use of three types of fluorescent material on the television screen, designed to react to the beam in red, blue, and green colours, colour television was introduced. And 'video tape', a type of recording with certain similarities to the sound track on a movie film, made it possible to reproduce recorded programmes or events with better quality than could be obtained from motion-picture film.

The vacuum tube, the heart of all the electronic devices, eventually became a limiting factor. Usually the components of a device are steadily improved in efficiency as time goes on – which means that they are stepped up in power and flexibility and reduced in size and mass. (This is sometimes called 'miniaturization'.) But the vacuum tube became a bottleneck in the road to miniaturization. And then, quite by accident, an unexpected solution turned up.

In the 1940s, several scientists at the Bell Telephone Laboratories grew interested in the substances known as 'semi-conductors'. These substances, such as silicon and germanium, conduct electricity only moderately well, and the problem was to find out why that should be. The Bell Laboratories investigators discovered that such conductivity as they possessed was enhanced by traces of impurities mixed with the element in question.

Let us consider a crystal of pure germanium. Each atom has four electrons in its outermost shell, and in the regular array of atoms in the crystal each of the four electrons pairs up with an electron of a neighbouring germanium atom, so that all the electrons are paired in stable bonds. Because this arrangement is similar to that in diamond, germanium, silicon, and other such substances are called 'adamantine substances', from an old word for 'diamond'.

If, now, a little bit of arsenic is introduced into this contented adamantine arrangement, the picture grows complicated. Arsenic has five electrons in its outermost shell. An arsenic atom taking the place of a germanium atom in the crystal will be able to pair four of its five electrons with the neighbouring atoms, but the fifth can find no electron to pair with. It is left 'on the loose'. Now if an electric voltage is applied to this crystal, the loose electron will wander in the direction of the positive electrode. It will not move as freely as would electrons in a conducting metal, but the crystal will conduct electricity better than a non-conductor such as sulphur or glass.

This is not very startling, but now we come to a case which is somewhat more odd. Let us add a bit of boron, instead of arsenic, to the germanium. The boron atom has only three electrons in its outermost shell. These three can pair up with the electrons of three neighbouring germanium atoms. But what happens to the electron of the boron atom's fourth germanium neighbour? That electron is paired with a 'hole'! The word 'hole' is used advisedly, because this site, where the electron would find a partner in a pure germanium crystal, does in fact behave like a vacancy. If a voltage is applied to the boron-contaminated crystal, the next neighbouring electron, attracted towards the positive electrode, will move into the hole. In doing so, it leaves a hole where it was, and the electron next farther away from the positive electrode moves into *that* hole. And so the hole, in effect, travels steadily towards the negative electrode, moving exactly like an electron, but in the opposite direction. In short, it has become a conveyor of electric current.

To work well, the crystal must be almost perfectly pure with just the right amount of the specified impurity (i.e., arsenic or boron). The germanium–arsenic semi-conductor, with a wandering electron, is said to be 'n-type' – n for 'negative'. The germanium–boron semi-conductor, with a wandering hole that acts as if it were positively charged, is 'p-type' – p for 'positive'.

Unlike ordinary conductors, the electrical resistance of semi-conductors drops as the temperature rises. This is because higher temperatures weaken the hold of atoms on electrons and allow them to drift more freely. (In metallic conductors, the electrons are already free enough at ordinary temperatures. Raising the temperature introduces more random movement and impedes their flow in response to the electric field.) By determining the resistance of a semi-conductor, temperatures can be measured that are too high to be conveniently measured in other fashions. Such temperature-measuring semi-conductors are called 'thermistors'.

But semi-conductors in combination can do much more. Suppose we make a germanium crystal with one half p-type and the other half n-type. If we connect the n-type side to a negative electrode and the p-type side to a positive electrode, the electrons on the n-type side will move across the crystal towards the positive electrode, while the holes on the p-type side will travel in the opposite direction towards the negative electrode. Thus a current flows through the crystal. Now let us reverse the situation – that is, connect the n-type side to the positive electrode and the p-type to the negative electrode. This time the electrons of the n-side travel towards the positive electrode – which is to say, away from the p-side – and the holes of the p-side similarly move in the direction away from the n-side. As a result, the border regions at the junction between the n- and p-sides lose their free electrons and holes. This amounts to a break in the circuit, and no current flows.

In short, we now have a set-up that can act as a rectifier. If we hook up alternating current to this dual crystal, the crystal will pass the current in one direction, but not in the other. Therefore alternating current will be converted to direct current. The crystal serves as a diode, just as a vacuum tube (or 'valve') does.

With this device, electronics returned full circle to the first type of rectifier used for radio – namely, the crystal with a 'cat's whisker'. But the new type of crystal was far more effective and versatile. And it had impressive advantages over the vacuum tube. It was lighter, much less bulky, stronger, invulnerable to shocks, and it did not heat up – all of which gave it a much longer life than the tube. The new device was named, at the suggestion of John Robinson Pierce of the Bell Laboratories, the 'transistor', because it *trans*ferred a signal across a re*sistor*.

In 1948, William Shockley, Walter H. Brattain, and John

Bardeen at the Bell Laboratories went on to produce a transistor which could act as an amplifier. This was a germanium crystal with a thin p-type section sandwiched between two n-type ends. It was in effect a triode with the equivalent of a grid between the filament and the plate. By controlling the positive charge in the p-type centre, holes could be sent across the junctions in such a manner as to control the electron flow. Furthermore a small variation in the current of the p-type centre would cause a large variation in the current across the semi-conductor system. The semi-conductor triode could thus serve as an amplifier, just as a vacuum tube triode did. Shockley and his co-workers Brattain and Bardeen received the Nobel Prize in physics in 1956.

However well transistors might work in theory, their use in practice required certain comcomitant advances in technology. (This is invariably true in applied science.) Efficiency in transistors depended very strongly on the use of materials of extremely high purity, so that the nature and concentration of deliberately added impurities could be carefully controlled.

Fortunately, William G. Pfann introduced the technique of zone-refining in 1952. A rod of, let us say, germanium, is placed in the hollow of a circular heating element, which softens and begins to melt a section of the rod. The rod is drawn through the hollow so that the molten zone moves along it. The impurities in the rod tend to remain in the molten zone and are therefore literally washed to the ends of the rod. After a few passes of this sort, the main body of the germanium rod is unprecedentedly pure.

By 1953, tiny transistors were being used in hearing aids, making them so small that they could be fitted inside the ear. In short order, the transistor – steadily developed so that it could handle higher frequencies, withstand higher temperatures, and be reduced to almost microscopic size – took over many functions of the vacuum tube. Perhaps the most notable example is its use in electronic computers, which have been greatly reduced in size and improved in reliability. In the process, new substances have been developed with useful semi-conductor properties. Indium phosphide and gallium arsenide, for instance, have been

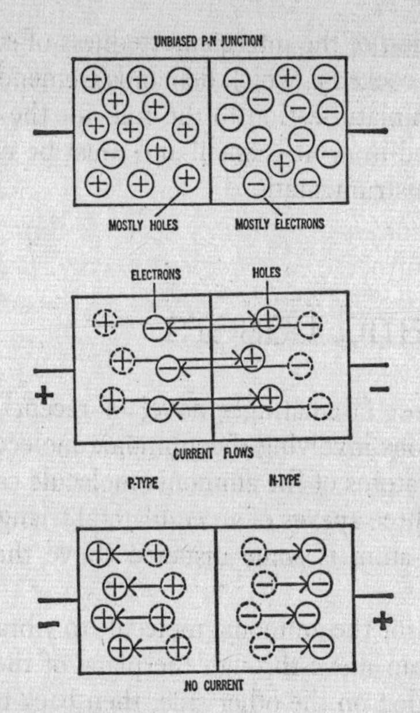

Principle of the junction transistor.

developed for use in transistors designed to work at high temperatures.

Nor do transistors represent the ultimate in miniaturization. In 1953, a simple two-wire mechanism, operating at liquid helium temperatures, was developed. It could act as a switch by the setting-up and breaking-down of super-conductivity in one wire by changes in the magnetic field of the other wire. Such switches are called 'cryotrons'.

In addition, there are tiny devices in which two thin films of metals such as aluminium and lead are separated by a thin film of insulator such as aluminium oxide. At temperatures in the super-conductive range, current will flow, tunnelling through the insulator if the voltage is high enough. The amount of current can be delicately controlled by voltage, temperature, and magnetic field intensity. Such 'tunnel sandwiches' offer another route to miniaturization.

As an indication of the interconnectedness of science, the new development of rocketry, which demands tremendous structures, also demands miniaturization to the full, for the payload that is eventually placed in orbit is small and must be crammed as full as can be with instrumentation.

Masers and Lasers

Perhaps the most fascinatingly novel of recent devices begins with investigations involving the ammonia molecule (NH_3). The three hydrogen atoms of the ammonia molecule can be viewed as occupying the three apexes of an equilateral triangle, whereas the single nitrogen atom is some distance above the centre of the triangle.

It is possible for the ammonia molecule to vibrate. That is, the nitrogen atom can move through the plane of the triangle to an equivalent position on the other side, then back to the first side, and so on, over and over. The ammonia molecule can, in fact, be made to vibrate back and forth with a natural frequency of 24,000 million times a second.

This vibration period is extremely constant, much more so than the period of any man-made vibrating device; much more constant, even, than the movement of astronomical bodies. Such vibrating molecules can be made to control electric currents, which will in turn control time-measuring devices with unprecedented precision, something that was first demonstrated in 1949 by the American physicist Harold Lyons. By the mid-1950s such 'atomic clocks' were surpassing all ordinary chronometers. Accuracies in time-measurement of one second in 100,000 years were reached in 1964 with a maser making use of hydrogen atoms.

The ammonia molecule in the course of these vibrations liberates a beam of electromagnetic radiation with a frequency of 24,000 million cycles per second. This radiation has a wavelength of 1·25 centimetres and is in the microwave region. Another way of looking at this fact is to imagine the ammonia molecule as capable of occupying one of two energy levels, with the energy difference equal to that of a photon representing a 1·25-centimetre radiation. If the ammonia molecule drops from the higher energy level to the lower, it emits a photon of this size. If a molecule in the lower energy level absorbs a photon of this size, it rises to the higher energy level.

But what if an ammonia molecule is already in the higher energy level and is exposed to such photons? As early as 1917, Einstein had pointed out that if a photon of just the right size struck such an upper-level molecule, the molecule would be nudged back down to the lower level and would emit a photon of exactly the size and moving in exactly the direction of the entering photon. There would be two identical photons where only one had existed before. This was confirmed experimentally in 1924.

Ammonia exposed to microwave radiation could, therefore, undergo two possible changes: molecules could be pumped up from lower level to higher or be nudged down from higher level to lower. Under ordinary conditions, the former process would predominate, for only a very small percentage of the ammonia molecules would, at any one instant, be at the higher energy level.

Suppose, though, that some method were found to place all or almost all the molecules in the upper energy level. Then it would be the movement from higher level to lower that would predominate. Indeed, something quite interesting would happen. The incoming beam of microwave radiation would supply a photon that would nudge one molecule downwards. A second photon would be released, and the two would speed on, striking two molecules, so that two more were released. All four would bring about the release of four more, and so on. The initial photon would let loose a whole avalanche of photons, all of exactly the same size and moving in exactly the same direction.

In 1953, the American physicist Charles Hard Townes devised a method for isolating ammonia molecules in the high-energy level and subjected them to stimulation by microwave photons of the correct size. A few photons entered, and a flood of such photons left. The incoming radiation was thus greatly amplified.

The process was described as '*m*icrowave *a*mplification by *s*timulated *e*mission of *r*adiation', and, from the initials of this phrase, the instrument came to be called a 'maser'.

Solid masers were soon developed; solids in which electrons could be made to take up one of two energy levels. The first masers, both gaseous and solid, were intermittent. That is, they had to be pumped up to the higher energy level first, then stimulated. After a quick burst of radiation, nothing more could be obtained until the pumping process had been repeated.

To circumvent this, it occurred to the Dutch-American physicist Nicolaas Bloembergen to make use of a three-level system. If the material chosen for the core of the maser can have electrons in any of three energy levels – a lower, a middle, and an upper – then pumping and emission can go on simultaneously. Electrons are pumped up from the lowest energy level to the highest. Once at the highest, proper stimulation will cause them to drop down: first to the middle level, then to the lower. Photons of different size are required for pumping and for stimulated emission, and the two processes will not interfere with each other. Thus, we end with a continuous maser.

As microwave amplifiers, masers can be used as very sensitive detectors in radio astronomy, where exceedingly feeble microwave beams received from outer space will be greatly intensified with great fidelity to the original radiation characteristics. (Reproduction without loss of original characteristics is to reproduce with little 'noise'. Masers are extraordinarily 'noiseless' in this sense of the word.) They have carried their usefulness into outer space, too. A maser was carried on board the Soviet satellite Cosmos 97, launched 30 November 1965, and did its work well.

For his work, Townes received the 1964 Nobel Prize for physics, sharing it with two Soviet physicists, Nicolai Gennediyevich Basov and Aleksandr Mikhailovich Prochorov, who had worked independently on maser theory.

In principle, the maser technique could be applied to electromagnetic waves of any wavelength, notably to those of visible light. Townes pointed out the possible route of such applications to light wavelengths in 1958. Such a light-producing maser might be called an 'optical maser'. Or else, this particular process might be called '*l*ight *a*mplification by *s*timulated *e*mission of *r*adiation', and the new set of initials 'laser' might be used. It is the latter word that has grown popular.

The first successful laser was constructed in 1960 by the American physicist Theodore Harold Maiman. He used a bar of synthetic ruby for the purpose, this being, essentially, aluminium oxide with a bit of chromium oxide added. If the ruby bar is exposed to light, the electrons of the chromium atoms are pumped to higher levels and, after a short while, begin to fall back. The first few photons of light emitted (with a wavelength of 694·3 nanometres) stimulate the production of other such photons, and the bar suddenly emits a beam of deep red light four times as intense as light at the sun's surface. Before 1960 was over, continuous lasers were prepared by an Iranian physicist, Ali Javan, working at Bell Laboratories. He used a gas mixture (neon and helium) as the light source.

The laser made possible light in a completely new form. The light was the most intense that had ever been produced, and the most narrowly monochromatic (single wavelength), but it was even more than that.

Ordinary light, produced in any other fashion, from a wood fire to the sun or to a firefly, consists of relatively short wave packets. They can be pictured as short bits of waves pointing in various directions. Ordinary light is made up of countless numbers of these.

The light produced by a stimulated laser, however, consists of photons of the same size and moving in the same direction. This means that the wave packets are all of the same frequency, and, since they are lined up precisely end to end, so to speak, they melt together. The light appears to be made up of long stretches of waves of even amplitude (height) and frequency (width). This is 'coherent light', because the wave packets seem to stick together. Physicists had learned to prepare coherent radiation for long wavelengths. It had never been done for light, though, until 1960.

The laser was so designed, moreover, that the natural tendency of the photons to move in the same direction was accentuated. The two ends of the ruby tube were accurately machined and silvered so as to serve as plane mirrors. The emitted photons flashed back and forth along the rod, knocking out more photons at each pass, until they had built up sufficient intensity to burst through the end that was more lightly silvered. Those that did come through were precisely those that had been emitted in a direction exactly parallel to the long axis of the rod, for those would move back and forth, striking the mirrored ends over and over. If any photon of proper energy happened to enter the rod in a different direction (even a very slightly different direction) and started a train of stimulated photons in that different direction, these would quickly pass out of the sides of the rod after only a few reflections at most.

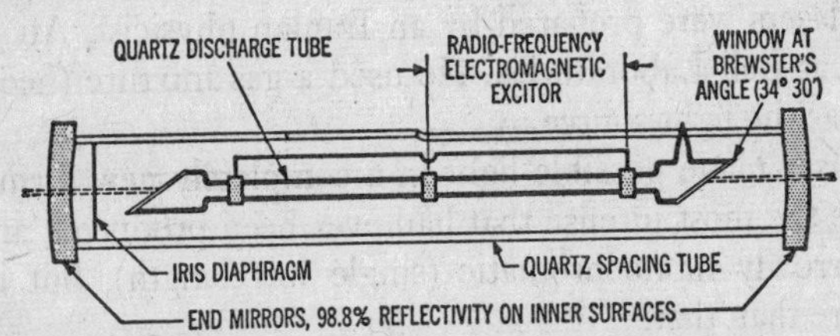

Continuous-wave laser with concave mirrors and Brewster angle windows on discharge tube. The tube is filled with a gas whose atoms are raised to high-energy states by electromagnetic excitation. These atoms are then stimulated to emit energy of a certain wavelength by the introduction of a light beam. Acting like a pipe organ, the resonant cavity builds up a train of coherent waves between the end mirrors. The thin beam that escapes is the laser ray. (After a drawing in *Science*, 9 October 1964.)

A beam of laser light is made up of coherent waves so parallel that it can travel through long distances without diverging to uselessness. It could be focused finely enough to heat a pot of coffee a thousand miles away. Laser beams even reached to the moon, in 1962, spreading out to a diameter of only two miles after having crossed nearly a quarter of a million miles of space!

Once the laser was devised, interest in its further development was nothing short of explosive. Within a few years individual lasers capable of producing coherent light in hundreds of different wavelengths, from the near ultra-violet to the far infra-red, were developed.

Laser action was obtained from a wide variety of solids, from metallic oxides, fluorides, tungstates, from semi-conductors, from liquids, from columns of gas. Each variety had its advantages and disadvantages.

In 1964, the first 'chemical laser' was developed by the American physicist Jerome V. V. Kasper. In such a laser, the source of energy is a chemical reaction (in the case of the first, the dissociation of CF_3I by a pulse of light). The advantage of the chemical laser over the ordinary variety is that the energy-yielding chemical reaction can be incorporated with the laser itself, and no outside energy-source is needed. This is analogous to a battery-powered device as compared with one that must be plugged into a wall socket. There is an obvious gain in portability here, to say nothing of the fact that chemical lasers seem to be considerably more efficient than the ordinary variety (12 per cent or more, as compared with 2 per cent or less).

Organic lasers – those in which a complex organic dye is used as the source of coherent light – were first developed in 1966 by John R. Lankard and Peter Sorokin. The complexity of the molecule makes it possible to produce light by a variety of electronic reactions and therefore in a variety of wavelengths. A single organic laser can be 'tuned' to deliver any wavelength within a range, rather than find itself confined to a single wavelength as is true of the others.

The narrowness of the beam of laser light means that a great deal of energy can be focused into an exceedingly small area; in that area, the temperature reaches extreme levels. The laser can vaporize metal for quick spectral investigation and analysis and can weld, cut, or punch holes of any desired shape through high-melting substances. By shining laser beams into the eye, surgeons have succeeded in welding loosened retinas so rapidly that surrounding tissues have no time to be affected by heat and, in similar fashion, to destroy tumours.

To show the vast range of laser applications, Arthur L. Shawlow has developed the trivial (but impressive) laser-eraser, which in an intensely brief flash evaporates the typewriter ink of the formed letters without so much as scorching the paper beneath; at the other extreme, laser interferometers can make unprecedentedly refined measurements. When earth-strains intensify, this can be detected by separated lasers, where shifts in the interference fringes of their light will detect tiny earth-movements with a delicacy of one part in a billion. Then, too, the first men on the moon left a reflector system designed to bounce back laser beams to earth. By such a method, the distance to the moon may be determined with greater accuracy than the distance, in general, from point to point on earth's surface.

One possible application that created excitement from the beginning has been the use of laser beams as carrier beams in communications. The high frequency of coherent light, as compared with that of the coherent radio waves used in radio and television today, holds forth the promise of being able to crowd many thousands of channels into the space that now holds one channel. The prospect arises that every human being on earth may have his own personal wavelength. Naturally, the laser light must be modulated. Varying electric currents produced by sound must be translated into varying laser light (either through changes in its amplitude on its frequency, or perhaps just by turning it on and off), which can in turn be used to produce varying electric current elsewhere. Such systems are being developed.

It may be that since light is so much more subject than radio waves to interference by clouds, mist, fog, and dust, that it will be necessary to conduct laser light through pipes containing lenses (to reconcentrate the beam at intervals) and mirrors (to reflect it around corners). However, a carbon-dioxide laser has been developed that produces continuous laser beams of unprecedented power that are far enough in the infra-red to be little affected by the atmosphere. Atmospheric communication may also be possible then.

A still more fascinating application of laser beams that is very much here-and-now involves a new kind of photography. In ordinary photography a beam of ordinary light reflected from an object falls on a photographic film. What is recorded is the cross-section of the light, which is by no means all the information it can potentially contain.

Suppose instead that a beam of light is split into two. One part strikes an object and is reflected with all the irregularities that this object would impose on it. The second part is reflected from a mirror with no irregularities. The two parts meet at the photographic film, and the interference of the various wavelengths is recorded. In theory, the recording of this interference would include all the data concerning each light beam. The photograph that records this interference pattern seems to be blank when developed, but if light is shone upon the film and passes through and takes on the interference characteristics, it produces an image containing the complete information. The image is as three-dimensional as was the surface from which light was reflected, and an ordinary photograph can be taken of the image from various angles that show the change in perspective.

This notion was first worked out by the Hungarian-British physicist Dennis Gabor in 1947, when he was trying to work out methods for the sharpening of images produced by electron microscopes. He called it 'holography', from a Latin word meaning 'the whole writing'.

While Gabor's idea was theoretically sound, it could not be implemented, because ordinary light would not do. With wavelengths of all sizes moving in all directions, the interference-fringes produced by the two beams of light would be so chaotic as to yield no information at all. It would be like producing a million dim images all superimposed in slightly different positions.

The introduction of laser light changed everything. In 1965, Emmet N. Leith and Juris Upatnieks, at the University of Michigan, were able to produce the first holograms. Since then, the technique has been sharpened to the point where holography in colour has become possible, and where the photographed interference-fringes can successfully be viewed with ordinary light. Micro-holography promises to add a new dimension (literally) to biological investigations, and where it will end, none can predict.

9 The Reactor

Nuclear Fission

The rapid advances in technology in the twentieth century have been bought at the expense of a stupendous increase in our consumption of the earth's energy resources. As the underdeveloped nations, with their thousands of millions of people, join the already industrialized countries in high living, the rate of consumption of fuel will jump even more spectacularly. Where will mankind find the energy supplies needed to support his civilization?

We have already seen a large part of the earth's timber disappear. Wood was man's first fuel. By the beginning of the Christian era, much of Greece, northern Africa, and the Near East had been ruthlessly deforested, partly for fuel, partly to clear the land for animal herding and agriculture. The uncontrolled felling of the forests was a double-barrelled disaster. Not only did it destroy the wood supply, but the drastic uncovering of the land meant a more or less permanent destruction of fertility. Most of these ancient regions, which once supported man's most advanced cultures, are sterile and unproductive now, populated by a ground-down and backward people.

The Middle Ages saw the gradual deforestation of western Europe, and modern times have seen the much more rapid deforestation of the North American continent. Almost no great stands of virgin timber remain in the world's temperate zones except in Canada and Siberia.

It seems unlikely that man will ever be able to get along without wood. Building timber and paper will always be necessities.

As for fuel, coal and oil have taken wood's place. Coal was mentioned by the Greek botanist Theophrastus as long ago as 200 B.C., but the first records of actual coal-mining in Europe do not date back before the twelfth century. By the seventeenth century, England, deforested and desperately short of wood for its navy, began to shift to the large-scale use of coal for fuel, a switch that laid the groundwork for the Industrial Revolution.

The shift was slow elsewhere. Even in 1800, wood supplied 94 per cent of the fuel needs in the young, forest-rich United States. In 1885, it supplied only 50 per cent of the fuel needs, and by the 1960s, only 3 per cent. The balance, moreover, has shifted beyond coal to oil and natural gas. In 1900, the energy supplied by coal in the United States was ten times that supplied by oil and gas together. Half a century later, coal supplied only one third the energy supplied by oil and gas. Coal, oil, and gas are 'fossil fuels', relics of plant life aeons old, and cannot be replaced once they are used up. With respect to coal and oil, man is living on his capital at an extravagant rate.

The oil, particularly, is going fast. The world is now burning a million barrels of oil each hour, and the rate of consumption is rising rapidly. Although well over a billion barrels remain in the earth, it is estimated that by 1980 oil production will reach its peak and begin to decline. Of course, additional oil can be formed by the combination of the more common coal with hydrogen under pressure. This process was first developed by the German chemist Friedrich Bergius in the 1920s, and he shared in the Nobel Prize for chemistry in 1931 as a result. The coal reserve is large indeed, perhaps as large as 7 billion tons, but not all of it is easy to mine. By the twenty-fifth century or sooner, coal may become an expensive commodity.

We can expect new finds. Perhaps surprises in the way of coal and oil await us in Australia, in the Sahara, even in Antarctica. Moreover, improvements in technology may make it economical to exploit thinner and deeper coal seams, to plunge more and more deeply for oil, and to extract oil from oil shale and from sub-sea reserves.

No doubt we shall also find ways to use our fuel more efficiently. The process of burning fuel to produce heat to convert water to steam to drive a generator to create electricity wastes a good deal of energy along the way. Most of these losses could be side-stepped if heat could be converted directly into electricity. The possibility of doing this appeared as long ago as 1823, when a German physicist, Thomas Johann Seebeck, observed that, if two different metals were joined in a closed circuit and if the junction of the two elements were heated, a compass needle in the vicinity would be deflected. This meant that the heat was producing an electric current in the circuit ('thermoelectricity'), but Seebeck misinterpreted his own work, and his discovery was not followed up.

With the coming of semi-conductor techniques, however, the old 'Seebeck effect' underwent a renaissance. Current thermo-

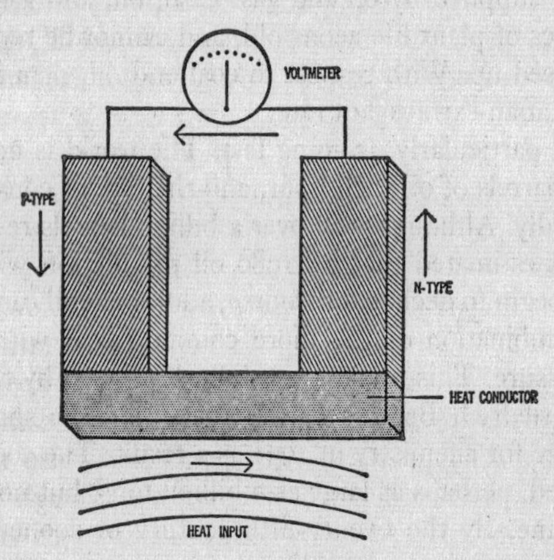

The thermoelectric cell. Heating the conductor causes electrons to flow towards the cold end of the n-type semi-conductor and from the cold to the warm region of the p-type. If a circuit is formed, current flows in the direction shown by the arrows. Thus heat is converted to electrical energy.

electric devices make use of semi-conductors. Heating one end of a semi-conductor creates an electric potential in the material: in a p-type semi-conductor the cold end becomes negative; in an n-type it becomes positive. Now if these two types of semi-conductor are joined in a U-shaped structure, with the n–p junction at the bottom of the U, heating the bottom will cause the upper end of the p branch to gain a negative charge and the upper end of the n branch to acquire a positive charge. As a result, current will flow from one end to the other, and it will be generated so long as the temperature difference is maintained. (In reverse, the use of a current can bring about a temperature drop, so that a thermoelectric device can also be used as a refrigerator.)

The thermoelectric cell, requiring no expensive generator or bulky steam engine, is portable and could be set up in isolated areas as a small-scale supplier of electricity. All it needs as an energy source is a kerosene heater. Such devices are reported to be used routinely in rural areas of the Soviet Union.

Notwithstanding all possible increases in the efficiency of using fuel and the likelihood of new finds of coal and oil, these sources of energy are definitely limited. The day will come, and not far in the future, when neither coal nor oil can serve as an important large-scale energy source.

And yet man's energy needs will continue and even be far larger than those of today. What can be done?

One possibility is to make increasing use of renewable energy sources: to live on the earth's energy income rather than its capital. Wood can be such a resource if forests are grown and harvested as a crop, though wood alone could not come anywhere near meeting all our energy needs. We could also make much more use of wind power and water power, though these again could never be more than subsidiary sources of energy. The same must be said about certain other potential sources of energy in the earth, such as tapping the heat of the interior (e.g., in hot springs) or harnessing the ocean tides.

Far more important, for the long run, is the possibility of directly tapping some of the vast energy pouring on the earth from the sun. This 'insolation' produces energy at a rate that is some 50,000 times as great as man's current rate of energy consumption. In this respect, one particularly promising device is the 'solar battery', which also makes use of semi-conductors.

As developed by the Bell Telephone Laboratories in 1954, the solar battery is a flat sandwich of n-type and p-type semiconductors. Sunlight striking the plate knocks some electrons out of place. The transfer is connected, as an ordinary battery would be, in an electric circuit. The freed electrons move towards the positive pole and holes move towards the negative pole, thus constituting a current. The solar battery can develop electric

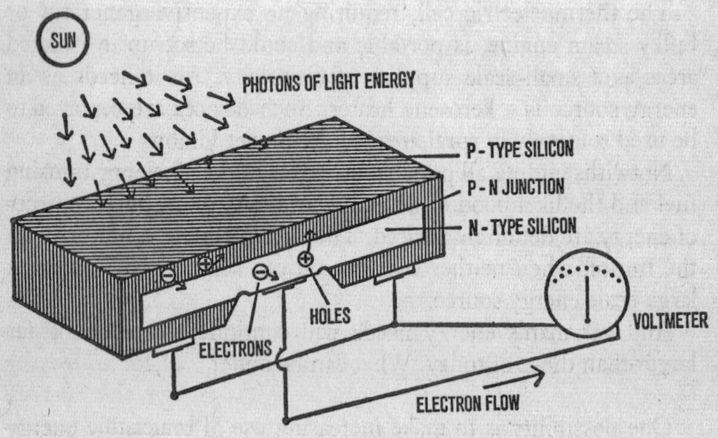

A solar battery cell. Sunlight striking the thin wafer frees electrons, thus forming electron-hole pairs. The p-n junction acts as a barrier, or electric field, separating electrons from holes. A potential difference therefore develops across the junction, and current then flows through the wire circuit.

potentials of up to half a volt, and up to 9 watts of power, from each square foot exposed to the sun. This is not much, but the beauty of the solar battery is that it has no liquids, no corrosive chemicals, no moving parts – it just keeps on generating electricity indefinitely merely by lying in the sun.

The artificial satellite Vanguard I, launched by the United States on 17 March 1958, was the first to be equipped with a solar battery to power its radio signals.

The amount of energy falling upon one acre of a generally sunny area of the earth is 9·4 million kilowatt-hours per year. If substantial areas in the earth's desert regions, such as Death Valley and the Sahara, were covered with solar batteries and electricity-storing devices, they could provide the world with its electricity needs for an indefinite time – for as long, in fact, as the human race is likely to endure, if it does not commit suicide.

But the tapping of solar energy, it seems, is not likely to be achieved on any great scale in this generation or the next. Fortunately we have an immense source of energy, here on the earth, which can tide us over for hundreds of years after we run out of inexpensive coal and oil. It is the energy in the atomic nucleus.

Nuclear energy is commonly called 'atomic energy', but that is a misnomer. Strictly speaking, atomic energy is the energy yielded by chemical reactions, such as the burning of coal and oil, because they involve the behaviour of the atom as a whole. The energy released by changes in the nucleus is of a totally different kind and vastly greater in magnitude.

Soon after the discovery of the neutron by Chadwick in 1932, physicists realized that they had a wonderful key for unlocking the atomic nucleus. Since it had no electric charge, the neutron could easily penetrate the charged nucleus. Physicists immediately began to bombard various nuclei with neutrons to see what nuclear reactions they could bring about; among the most ardent investigators with this new tool was Enrico Fermi of Italy. In the space of a few months, he had prepared new radioactive isotopes of thirty-seven different elements.

Fermi and his associates discovered that they got better results if they slowed down the neutrons by passing them through water or paraffin first. Bouncing off protons in the water or paraffin, the neutrons are slowed just as a billiard ball is by hitting other billiard balls. When a neutron is reduced to 'thermal' speed (the normal speed of motion of atoms), it has a greater chance of being absorbed by a nucleus, because it remains in the vicinity of the nucleus longer. Another way of looking at it is to consider that the wavelength of the wave associated with the neutron is longer, for the wavelength is inversely proportional to the momentum of the particle. As the neutron slows down, its wavelength increases. To put it metaphorically, the neutron grows fuzzier and takes up more volume. It therefore hits a nucleus more easily, just as a bowling ball has more chance of hitting a tenpin than a golf ball would have.

The probability that a given species of nucleus will capture a neutron is called its 'cross-section'. This term, metaphorically, pictures the nucleus as a target of a particular size. It is easier to hit the side of a barn with a tennis ball than it is to hit a foot-wide board at the same distance. The cross-sections of nuclei under neutron bombardment are reckoned in million trillionths of a square centimetre (10^{-24} square centimetre). That unit, in fact, was named a 'barn' by the American physicists M. G. Holloway and C. P. Baker in 1942. The name served to hide what was really going on in those hectic wartime days.

When a nucleus absorbs a neutron, its atomic number is unchanged (because the charge of the nucleus remains the same), but its mass number goes up by one unit. Hydrogen 1 becomes hydrogen 2, oxygen 17 becomes oxygen 18, and so on. The energy delivered to the nucleus by the neutron as it enters may 'excite' the nucleus – that is, increase its energy content. This surplus energy is then emitted as a gamma ray.

The new nucleus often is unstable. For example, when aluminium 27 takes in a neutron and becomes aluminium 28, one of the neutrons in the new nucleus soon changes to a proton (by emitting an electron). This increase in the positive charge of the nucleus transforms the aluminium (atomic number 13) to silicon (atomic number 14).

Because neutron bombardment is an easy way of converting an element to the next higher one, Fermi decided to bombard uranium to see if he could form an artificial element – number 93. In the products of the bombardment of uranium, he and his co-workers did find signs of new radioactive substances. They thought they had made element 93, and called it 'uranium X'. But how could the new element be identified positively? What sort of chemical properties should it have?

Well, element 93, it was thought, should fall under rhenium in the periodic table, so it ought to be chemically similar to rhenium. (Actually, though no one realized it at the time, element 93 belonged in a new rare-earth series, which meant that it would resemble uranium, not rhenium – see p. 246. Thus, the search for its identification got off on the wrong foot entirely.) If it were like rhenium, perhaps the tiny amount of 'element 93' created might be identified by mixing the products of the neutron bombardment with rhenium and then separating out the rhenium by chemical methods. The rhenium would act as a 'carrier', bringing out the chemically similar 'element 93' with it. If the rhenium proved to have radioactivity attached to it, this would indicate the presence of element 93.

The German physicist Otto Hahn and the Austrian physicist Lise Meitner, working together in Berlin, pursued this line of experiment. Element 93 failed to show up with rhenium. Hahn and Meitner then went on to try to find out whether the neutron bombardment had transformed uranium into other elements near it in the periodic table. At this point, in 1938, Germany occupied Austria, and Miss Meitner, who, until then, as an Austrian national, had been safe despite the fact that she was Jewish, was forced to flee from Hitler's Germany to the safety of Stockholm. Hahn continued his work with the German physicist Fritz Strassman.

Several months later Hahn and Strassman found that barium, when added to the bombarded uranium, carried off some radioactivity. They decided that this radioactivity must belong to radium, the element below barium in the periodic table. The conclusion was, then, that the neutron bombardment of uranium changed some of it to radium.

But this radium turned out to be peculiar stuff. Try as they would, Hahn and Strassman could not separate it from the barium. In France, Irène Joliot-Curie and her co-worker P. Savitch undertook a similar task and also failed.

And then Meitner, the refugee in Scandinavia, boldly cut through the riddle and broadcast a thought that Hahn was voicing in private but was hesitating to publish. In a letter published in the British journal *Nature* in January of 1939, she suggested that the 'radium' could not be separated from the barium because no radium was there. The supposed radium was actually radioactive barium: it was barium that had been formed in the neutron bombardment of uranium. This radioactive barium decayed by emitting a beta particle and formed lanthanum. (Hahn and Strassman had found that ordinary lanthanum added to the products brought out some radioactivity, which they assigned to actinium; actually it was radioactive lanthanum.)

But how could barium be formed from uranium? Barium was only a middleweight atom. No known process of radioactive decay could transform a heavy element into one only about half its weight. Meitner made so bold as to suggest that the uranium nucleus had split in two. The absorption of a neutron had caused it to undergo what she termed 'fission'. The two elements into which it had split, she said, were barium and element 43, the element above rhenium in the periodic table. A nucleus of barium and one of element 43 (later named technetium) would make up a nucleus of uranium. What made it a particularly daring suggestion was that neutron bombardment supplied only 6 million electron-volts, and the main thought of the day concerning nuclear structure made it seem that hundreds of millions would be required.

Meitner's nephew, Otto Robert Frisch, hastened to Denmark to place the new theory before Bohr, even in advance of publication. Bohr had to face the surprising ease with which this would require the nucleus to split, but fortunately he was evolving the liquid-drop theory of nuclear structure, and it seemed to him that this would explain it. (In later years the liquid-drop theory, taking into account the matter of nuclear shells, was to explain even the fine details of nuclear fission and why the nucleus broke into unequal halves.)

In any case, theory or not, Bohr grasped the implications at once. He was just leaving to attend a conference on theoretical physics in Washington, and there he told physicists what he had heard in Denmark of the fission suggestion. In high excitement, the physicists went back to their laboratories to test the hypothesis, and within a month half a dozen experimental confirmations were announced. The Nobel Prize for chemistry went to Hahn in 1944 as a result.

And so began the work that led to the most terrible weapon of destruction ever devised.

The Nuclear Bomb

The fission reaction released an unusual amount of energy, vastly more than did ordinary radioactivity. But it was not solely the additional energy that made fission so portentous a phenomenon. More important was the fact that it released two or three neutrons. Within two months after the Meitner letter, the awesome possibility of a 'nuclear chain reaction' had occurred to a number of physicists.

The expression 'chain reaction' has acquired an exotic meaning, but actually it is a very common phenomenon. The burning of a piece of paper is a chain reaction. A match supplies the heat required to start it; once the burning has begun, this supplies the very agent, heat, needed to maintain and spread the flame. Burning brings about more burning on an ever-expanding scale.

That is exactly what happens in a nuclear chain reaction. One neutron fissions a uranium nucleus; this releases two neutrons that can produce two fissions that release four neutrons which can produce four fissions, and so on. The first atom to fission yields 200 Mev of energy; the next step yields 400 Mev, the next 800 Mev, the next 1,600 Mev, and so on. Since the successive stages take place at intervals of about a 50-billionth of a second, you see that within a tiny fraction of a second a staggering amount of energy will be released. (The actual average number of neutrons produced per fission is 2·47, so matters go even more quickly than this simplified calculation indicates.) The fission of one ounce of uranium produces as much energy as the burning of 90 tons of coal or of 2,000 gallons of fuel oil. Peacefully used, uranium fission could relieve all our immediate worries about vanishing fossil fuels and man's mounting consumption of energy.

But the discovery of fission came just before the world was plunged into an all-out war. The fissioning of an ounce of uranium, physicists estimated, would yield as much explosive power as 600 tons of TNT. The thought of the consequences of a war fought with such weapons was horrible, but the thought of a world in which Nazi Germany laid its hands on such an explosive before the Allies did was even more horrible.

The Hungarian-American physicist Leo Szilard, who had been thinking of nuclear chain reactions for years, foresaw the possible future with complete clarity. He and two other Hungarian-American physicists, Eugene Wigner and Edward Teller, prevailed on the gentle and pacific Einstein in the summer of 1939 to write a letter to President Franklin Delano Roosevelt, pointing out the potentialities of uranium fission and suggesting that every effort be made to develop such a weapon before the Nazis managed to do so.

The letter was written on 2 August 1939 and was delivered to the President on 11 October 1939. Between those dates, the Second World War had erupted in Europe. Physicists at Columbia University, under the supervision of Fermi, who had left Italy for America the previous year, worked to produce sustained fission in a large quantity of uranium.

Eventually, the government of the United States itself took action in the light of Einstein's letter. On 6 December 1941, President Roosevelt (taking a huge political risk in case of failure) authorized the organization of a giant project, under the deliberately noncommittal name of 'Manhattan Engineer District' for the purpose of devising an atom bomb. The next day, the Japanese attacked Pearl Harbor, and the United States was at war.

As was to be expected, practice did not by any means follow easily from theory. It took a bit of doing to arrange a uranium chain reaction. In the first place, you had to have a substantial amount of uranium, refined to extreme purity so that neutrons would not be wasted in absorption by impurities. Uranium is a rather common element in the earth's crust, averaging about 2 grams per ton of rock, which makes it 400 times as common as gold. But it is well spread out, and there are few places in the world where it occurs in rich ores or even in reasonable concentration. Furthermore, before 1939 uranium had had almost no uses, and no methods for its purification had been worked out. Less than an ounce of uranium metal had been produced in the United States.

The laboratories at Iowa State College, under the leadership of Spedding, went to work on the problem of purification by ion-exchange resins (see p. 264), and in 1942 began to produce reasonably pure uranium metal.

That, however, was only a first step. Now the uranium itself had to be broken down to separate out its more fissionable fraction. The isotope uranium 238 (U 238) has an even number of protons (92) and an even number of neutrons (146). Nuclei with

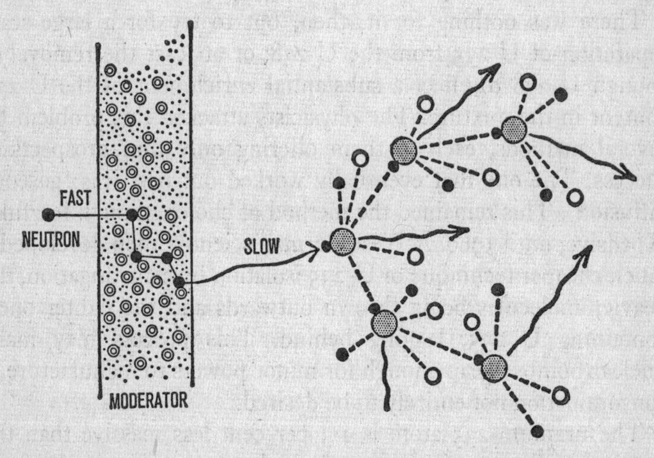

Nuclear chain reaction in uranium. The grey circles are uranium nuclei, the black dots neutrons, the wavy arrows gamma rays, and the small circles fission fragments.

even numbers of nucleons are more stable than those with odd numbers. The other isotope in natural uranium – uranium 235 – has an odd number of neutrons (143). Bohr had therefore predicted that it would fission more readily than uranium 238. In 1940, a research team under the leadership of the American physicist John Ray Dunning isolated a small quantity of uranium 235 and showed that Bohr's conjecture was true. U 238 fissions only when struck by fast neutrons of more than a certain energy, but U 235 would undergo fission upon absorbing neutrons of any energy, all the way down to simple thermal neutrons.

The trouble was that in purified natural uranium only one atom in 140 is U 235, the rest being U 238. This meant that most of the neutrons released by fission of U 235 would be captured by U-238 atoms without producing fission. Even if the uranium were bombarded with neutrons fast enough to split U 238, the neutrons released by the fissioning U 238 would not be energetic enough to carry on a chain reaction in the remaining atoms of this more common isotope. In other words, the presence of U 238 would cause the chain reaction to damp and die. It would be like trying to burn wet leaves.

There was nothing for it, then, but to try for a large-scale separation of U 235 from the U 238, or at least the removal of enough U 238 to effect a substantial enrichment of the U-235 content in the mixture. The physicists attacked this problem by several methods, each of them offering only thin prospects of success. The one that eventually worked out best was 'gaseous diffusion'. This remained the method of choice, though fearfully expensive, until 1960. A West German scientist then developed a much cheaper technique of U-235 isolation by centrifugation, the heavier molecules being thrown outwards and the lighter ones, containing U 235, lagging behind. This process may make nuclear bombs cheap enough for minor powers to manufacture, a consummation not entirely to be desired.

The uranium-235 atom is 1·3 per cent less massive than the uranium-238 atom. Consequently if the atoms were in the form of a gas, the U-235 atoms would move about slightly faster than the U-238 atoms. This meant they might be separated by reason of their faster diffusion through a series of filtering barriers. But first uranium had to be converted to a gas. About the only way to get it in this form was to combine it with fluorine and make uranium hexafluoride, a volatile liquid composed of one uranium atom and six fluorine atoms. In this compound a molecule containing U 235 would be less than 1 per cent lighter than one containing U 238; but that difference proved to be sufficient to make the method work.

The uranium hexafluoride gas was forced through porous barriers under pressure. At each barrier, the molecules containing U 235 got through a bit faster, on the average, and so with every passage through the successive barriers the advantage in favour of U 235 grew. To obtain sizeable amounts of almost pure uranium-235 hexafluoride required thousands of barriers, but well-enriched concentrations of U 235 could be achieved with a much smaller number of barriers.

By 1942, it was reasonably certain that the gaseous diffusion method (and one or two others) could produce 'enriched uranium' in quantity, and separation plants (costing a thousand million dollars each, and consuming as much electricity as all of New York City) were built at the secret city of Oak Ridge, Tennessee, sometimes called Dogpatch by irreverent scientists, after the mythical town of Al Capp's Li'l Abner.

Meanwhile, the physicists were calculating the 'critical size' that would be needed to maintain a chain reaction in a lump of enriched uranium. If the lump was small, too many neutrons would escape from its surface before being absorbed by U-235 atoms. To minimize this loss by leakage, the volume of the lump had to be large in proportion to its surface. At a certain 'critical size', enough neutrons would be intercepted by U-235 atoms to keep a chain reaction going.

The physicists also found a way to make efficient use of the available neutrons. 'Thermal' (i.e., slow) neutrons, as I have mentioned, are more readily absorbed by uranium 235 than are fast ones. The experimenters therefore used a 'moderator' to slow the neutrons from the rather high speeds they had on emerging from the fission reaction. Ordinary water would have been an excellent slowing agent, but unfortunately the nuclei of ordinary hydrogen hungrily snap up neutrons. Deuterium (hydrogen 2) fills the bill much better; it has practically no tendency to absorb neutrons. Consequently the fission experimenters became very interested in preparing supplies of heavy water.

Up to 1943, it was prepared by electrolysis for the most part. Ordinary water split into hydrogen and oxygen more readily than did heavy water, so that, if a large supply of water were electrolysed, the final bit of water was rich in heavy water and could be preserved. After 1943, careful distillation was the favoured method. Ordinary water had the lower boiling point, so that the last bit of unboiled water was rich in heavy water.

Heavy water was indeed valuable in the early 1940s. There is a thrilling story of how Joliot-Curie managed to smuggle France's supply of that liquid out of the country ahead of the invading Nazis in 1940. A hundred gallons of it, which had been prepared in Norway, did fall into the hands of the German Nazis. It was destroyed by a British commando raid in 1942.

Still, heavy water had drawbacks: it might boil away when the chain reaction got hot, and it would corrode the uranium. The scientists seeking to create a chain-reacting system in the Manhattan Project decided to use carbon, in the form of very pure graphite, as the moderator.

Another possible moderator, beryllium, had the disadvantage of toxicity. Indeed, the disease, berylliosis, was first recognized in the early 1940s in one of the physicists working on the atom bomb.

Now let us imagine a chain reaction. We start things off by sending a triggering stream of neutrons into the assembly of moderator and enriched uranium. A number of uranium-235 atoms undergo fission, releasing neutrons which go on to hit other uranium-235 atoms. They in turn fission and turn loose more neutrons. Some neutrons will be absorbed by atoms other than uranium 235; some will escape from the pile altogether. But if from each fission one neutron, and exactly one, takes effect in producing another fission, then the chain reaction will be self-sustaining. If the 'multiplication factor' is more than one, even very slightly more (e.g., 1·001), the chain reaction will rapidly build up to an explosion. This is good for bomb purposes but not for experimental purposes. Some device had to be worked out to control the rate of fissions. That could be done by sliding in rods of a substance such as cadmium, which has a high cross-section for neutron capture. The chain reaction develops so rapidly that the damping cadmium rods could not be slid in fast enough, were it not for the fortunate fact that the fissioning uranium atoms do not emit all their neutrons instantly. About one neutron in 150 is a 'delayed neutron', emitted a few minutes after fission, since it emerges, not directly from the fissioning atoms, but from the smaller atoms formed in fission. When the multiplication factor is only slightly above one, this delay is sufficient to give time for applying the controls.

In 1941, experiments were conducted with uranium-graphite mixtures, and enough information was gathered to lead physicists to decide that, even without enriched uranium, a chain reaction might be set up if only the lump of uranium were made large enough.

Physicists set out to build a uranium chain reactor of critical size at the University of Chicago. By that time some six tons of pure uranium were available; this amount was eked out with uranium oxide. Alternate layers of uranium and graphite were laid down one on the other, fifty-seven layers in all, with holes through them for insertion of the cadmium control rods. The structure was called a 'pile' – a noncommittal code name that did not give away its function. (During the First World War, the newly designed armoured vehicles on caterpillar treads were referred to as 'tanks' for the same purpose of secrecy. The name 'tank' stuck, but 'atomic pile' fortunately gave way eventually to the more descriptive name 'nuclear reactor'.)

The Chicago pile, built under the football stadium, measured 30 feet wide, 32 feet long, and 21½ feet high. It weighed 1,400 tons and contained 52 tons of uranium, as metal and oxide. (Using pure uranium 235, the critical size would have been, it is reported, no more than 9 ounces.) On 2 December 1942, the cadmium control rods were slowly pulled out. At 3.45 p.m. the multiplication factor reached one – a self-sustaining fission reaction was under way. At that moment mankind (without knowing it) entered the 'Atomic Age'.

The physicist in charge was Enrico Fermi, and Eugene Wigner presented him with a bottle of Chianti in celebration. Arthur Compton, who was at the site, made a long-distance telephone-call to James Bryant Conant at Harvard, announcing the success. 'The Italian navigator,' he said, 'has entered the new world.' Conant asked, 'How were the natives?' The answer came at once: 'Very friendly!'

It is curious that the first Italian navigator discovered one new world in 1492, and the second discovered another in 1942; those who value the mystic interplay of numbers make much of this coincidence.

Meanwhile another fissionable fuel had turned up. Uranium 238, upon absorbing a thermal neutron, forms uranium 239, which breaks down quickly to neptunium 239, which in turn breaks down almost as quickly to plutonium 239.

Now the plutonium-239 nucleus has an odd number of neutrons (145) and is more complex than uranium 235, so it should be highly unstable. It seemed a reasonable guess that plutonium, like uranium 235, might undergo fission with thermal neutrons. In 1941, this was confirmed experimentally. Still uncertain whether the preparation of uranium 235 would prove practical, the physicists decided to hedge their bets by trying to make plutonium in quantity.

Special reactors were built in 1943 at Oak Ridge and at Hanford, in the State of Washington, for the purpose of manufacturing plutonium. These reactors represented a great advance over the first pile in Chicago. For one thing, the new reactors were designed so that the uranium could be removed from the pile periodically. The plutonium produced could be separated from the uranium by chemical methods, and the fission products, some of them strong neutron absorbers, could be removed. In addition, the new reactors were water-cooled to prevent overheating. (The Chicago pile could operate only for short periods, because it was cooled merely by air.)

By 1945, enough purified uranium 235 and plutonium 239 were available for the construction of bombs. This portion of the task was undertaken at a third secret city, Los Alamos, New Mexico, under the leadership of the American physicist J. Robert Oppenheimer.

For bomb purposes it was desirable to make the nuclear chain reaction mount as rapidly as possible. This called for making the reaction go with fast neutrons, to shorten the intervals between fissions, so the moderator was omitted. The bomb was also enclosed in a massive casing to hold the uranium together long enough for a large proportion of it to fission.

Since a critical mass of fissionable material would explode spontaneously (sparked by stray neutrons from the air), the bomb fuel was divided into two or more sections. The triggering mechanism was an ordinary explosive (TNT?) which drove these sections together when the bomb was to be detonated. One arrangement was called 'the thin man' – a tube with two pieces of uranium 235 at its opposite ends. Another, 'the fat man', had the form of a ball in which a shell composed of fissionable material was 'imploded' towards the centre, making a dense critical mass held together momentarily by the force of the implosion and by a heavy outer casing called the 'tamper'. The tamper also served to reflect back neutrons into the fissioning mass and, therefore, to reduce the critical size.

To test such a device on a minor scale was impossible. The bomb had to be above critical size or nothing. Consequently, the first test was the explosion of a full-scale nuclear-fission bomb, usually called incorrectly, an 'atom bomb' or 'A-bomb'. At 5.30 a.m. on 16 July 1945, at Alamogordo, New Mexico, a bomb was exploded with truly horrifying effect; it had the explosive force of 20,000 tons of TNT. The physicist I. I. Rabi, on being asked later what he had witnessed, is reported to have said mournfully, 'I can't tell you, but don't expect to die a natural death.' (It is only fair to add that the gentleman so addressed by Rabi did die a natural death some years later.)

Two more fission bombs were prepared. One, a uranium bomb called 'Little Boy', 10 feet long by 2 feet wide and weighing 4½ tons, was dropped on Hiroshima on 6 August 1945; it was set off by radar echo. Days later, the second, a plutonium bomb, 11 feet by 5 feet, weighing 5 tons, and named 'Fat Man', was dropped on Nagasaki. Together, the two bombs had the explosive force of 35,000 tons of TNT. With the bombing of Hiroshima, the Atomic Age, already nearly three years old, broke on the consciousness of the world.

For four years after that, Americans lived under the delusion that there was an 'atom-bomb secret' which could be kept from other nations for ever if only security measures were made tight enough. Actually, the facts and theories of nuclear fission had been matters of public record since 1939, and the Soviet Union was fully engaged in research on the subject in 1940. If the Second World War had not occupied her lesser resources to a far greater extent than it occupied the greater resources of the uninvaded United States, the U.S.S.R. might have made an atomic bomb by 1945, as the United States did. As it was, the Soviet Union exploded her first atomic bomb on 22 September 1949, to the dismay and unnecessary amazement of most Americans. It had six times the power of the Hiroshima bomb and had an explosive effect equal to 210,000 tons of TNT.

On 3 October 1952, Great Britain became the third atomic power by exploding a test bomb of its own; on 13 February 1960, France joined the 'atomic club' as the fourth member, setting off a plutonium bomb in the Sahara. On 16 October 1964, the People's Republic of China (Communist China) announced the explosion of an atomic bomb, and thus became the fifth member.

The bomb became more versatile, too. In 1953, the United States, for the first time, fired a fission bomb from a cannon, rather than dropping it from a plane. Thus 'atomic artillery' (or 'tactical atomic weapons') was developed.

Meanwhile the fission bomb had been reduced to triviality. Man had succeeded in setting off another energetic nuclear reaction which made super-bombs possible.

In the fission of uranium, 0·1 per cent of the mass of the uranium atom is converted to energy. But in the fusion of hydrogen atoms to form helium, fully 0·5 per cent of their mass is converted to energy, as had first been pointed out in 1915 by the American chemist William Draper Harkins. At temperatures in the millions of degrees, the energy of protons is high enough to allow them to fuse. Thus two protons may unite and, after emitting a positron and a neutrino (a process which converts one of the protons to a neutron), become a deuterium nucleus. A deuterium nucleus may then fuse with a proton to form a tritium nucleus, which can fuse with still another proton to form helium 4. Or deuterium and tritium nuclei will combine in various ways to form helium 4.

Because such nuclear reactions take place only under the stimulus of high temperatures, they are referred to as 'thermonuclear reactions'. In the 1930s, the one place where the necessary temperatures were believed to exist was at the centre of stars. In 1938, the German-born physicist Hans Albrecht Bethe (who had left Hitler's Germany for the United States in 1935) proposed that fusion reactions were responsible for the energy that the stars radiated. It was the first completely satisfactory explanation of stellar energy since Helmholtz had raised the question nearly a century earlier.

Now the uranium fission bomb provided the necessary temperatures on the earth. It could serve as a match hot enough to ignite a fusion chain reaction in hydrogen. For a while it looked very doubtful that the reaction could actually be made to work in the form of a bomb. For one thing, the hydrogen fuel, in the form of a mixture of deuterium and tritium, had to be condensed to a dense mass, which meant that it had to be liquefied and kept at a temperature only a few degrees above absolute zero. In other words, what would be exploded would be a massive refrigerator. Furthermore, even assuming a hydrogen bomb could be made, what purpose would it serve? The fission bomb was already devastating enough to knock out cities; a hydrogen bomb would merely pile on destruction and wipe out whole civilian populations.

Nevertheless, despite the unappetizing prospects, the United States and the Soviet Union felt compelled to go on with it. The United States Atomic Energy Commission proceeded to produce some tritium fuel, set up a 65-ton fission–fusion contraption on a coral atoll in the Pacific, and on 1 November 1952 produced the first thermonuclear explosion (a 'hydrogen bomb' or 'H-bomb') on our planet. It fulfilled all the ominous predictions: the explosion yielded the equivalent of 10 million tons of TNT (10 'megatons') – 500 times the puny 20-'kiloton' energy of the Hiroshima bomb. The blast wiped out the atoll.

The Russians were not far behind; on 12 August 1953, they also produced a successful thermonuclear explosion, and it was light enough to be carried in a plane. The U.S.A. did not produce a portable one until early 1954. Where the U.S.A. developed the fusion bomb $7\frac{1}{2}$ years after the fission bomb, the Soviets took only 5 years.

Meanwhile a scheme for generating a thermonuclear chain reaction in a simpler way and packing it into a portable bomb had been conceived. The key to this reaction was the element lithium. When the isotope lithium 6 absorbs a neutron, it splits into nuclei of helium and tritium, giving forth 4·8 Mev of energy in the process. Suppose, then, that a compound of lithium and hydrogen (in the form of the heavy isotope deuterium) is used as the fuel. This compound is a solid, so there is no need for refrigeration to condense the fuel. A fission trigger would provide neutrons to split the lithium. And the heat of the explosion would cause the fusion of the deuterium present in the compound and of the tritium produced by the splitting of lithium. In other words, several energy-yielding reactions would take place: the splitting of lithium, the fusion of deuterium with deuterium, and the fusion of deuterium with tritium.

Now besides releasing tremendous energy, these reactions would also yield a great number of surplus neutrons. It occurred to the bomb builders: Why not use the neutrons to fission a mass of uranium? Even common uranium 238 could be fissioned with fast neutrons (though less readily than U 235). The heavy blast of fast neutrons from the fusion reactions might fission a considerable number of U-238 atoms. Suppose one built a bomb with a U-235 core (the igniting match), a surrounding explosive charge of lithium deuteride, and around all this a blanket of uranium 238 which would also serve as explosive. That would make a really big bomb. The U-238 blanket could be made almost as thick as one wished, because there is no critical size at which uranium 238 will undergo a chain reaction spontaneously. The result is sometimes called a 'U-bomb'.

The bomb was built; it was exploded at Bikini in the Marshall Islands on 1 March 1954; and it shook the world. The energy yield was around 15 megatons. Even more dramatic was a rain of radioactive particles that fell on twenty-three Japanese fishermen in a fishing boat named the *Lucky Dragon*. The radioactivity destroyed the cargo of fish, made the fishermen ill, eventually killed one, and did not exactly improve the health of the rest of the world.

Since 1954, fission–fusion–fission bombs have become items in the armaments of the United States, the Soviet Union, and Great Britain. In 1967, China became the fourth member of the fusion club, having made the transition from fission in only three years. The Soviet Union has exploded hydrogen bombs in the 50- to 100-megaton range and the United States is perfectly capable of building such bombs, or even larger ones, at short notice.

There are hints, too, that it may be possible to design a hydrogen bomb that would deliver a highly concentrated stream of neutrons, rather than heat. This would destroy life without doing much damage to property. Such a 'neutron bomb' or 'N-bomb' seems desirable to those who worry about property and hold life cheap.

Nuclear Power

The dramatic use of nuclear power in the form of unbelievably destructive bombs has done more to present the scientist in the role of an ogre than anything else that has occurred since the beginnings of science.

In a way this portrayal has its justifications, for no arguments or rationalizations can change the fact that scientists did indeed construct the atomic bomb, knowing from the beginning its destructive powers and that it would probably be put to use.

It is only fair to add that this was done under the stress of a great war against ruthless enemies and with an eye to the frightful possibility that a man as maniacal as Adolf Hitler might get such a bomb first. It must also be added that, on the whole, the scientists working on the bomb were deeply disturbed about it and that many opposed its use, while some even left the field of nuclear physics afterwards in what can only be described as remorse. Far fewer pangs of conscience were felt by most of the political and military leaders who made the actual decision to use the bombs.

Furthermore, we cannot and should not subordinate the fact that in releasing the energy of the atomic nucleus scientists put at man's disposal a power that can be used constructively as well as destructively. It is important to emphasize this in a world and at a time in which the threat of nuclear destruction has put science and scientists on the shamefaced defensive, and in a country like the United States, which has a rather strong Rousseauan tradition against book learning as a corrupter of the simple integrity of man in a state of nature.

Even the explosion of an atomic bomb need not be purely destructive. Like the lesser chemical explosives long used in mining and in the construction of dams and highways, nuclear explosives could be vastly helpful in construction projects. All kinds of dreams of this sort have been advanced: excavating harbours, digging canals, breaking up underground rock-formations, preparing heat reservoirs for power – even the long-distance propulsion of spaceships. In the sixties, however, the furore for such far-out hopes died down. The prospects of the danger of radioactive contamination, or of unlooked-for expense, or both, served as dampers.

Yet one constructive use of nuclear power that was realized lay in the kind of chain reaction that was born under the football stadium at the University of Chicago. A controlled nuclear reactor can develop huge quantities of heat, which, of course, can be drawn off by a 'coolant', such as water or even molten metal, to produce electricity or heat a building.

Experimental nuclear reactors that produced electricity were built in Great Britain and the United States within a few years after the war. The United States now has a fleet of nuclear-powered submarines, the first of which, the U.S.S. *Nautilus* (having cost 50 million dollars), was launched in January 1954.

This vessel, as important for its day as Fulton's *Clermont* was in its time, introduced engines with a virtually unlimited source of power that permits submarines to remain underwater for indefinitely long periods, whereas ordinary submarines must surface frequently to recharge their batteries by means of diesel generators that require air for their working. Furthermore, where ordinary submarines travel at a speed of eight knots, a nuclear submarine travels at twenty knots or more.

The first *Nautilus* reactor core lasted for 62,500 miles; included among those miles was a dramatic demonstration. The *Nautilus* made an underwater crossing of the Arctic Ocean in 1958. This trip demonstrated that the ocean depth at the North Pole was 13,410 feet (2½ miles), far deeper than had been thought previously. A second, larger nuclear submarine, the U.S.S. *Triton*, circumnavigated the globe underwater along Magellan's route in eighty-four days, between February and May of 1960.

The Soviet Union also possesses nuclear submarines, and, in December 1957, it launched the first nuclear-powered surface vessel, the *Lenin*, an ice-breaker. Shortly before that, the United States had laid the keel for a nuclear-powered surface vessel, and, in July 1959, the U.S.S. *Long Beach* (a cruiser) and the *Savannah* (a merchant ship) were launched. The *Long Beach* is powered by two nuclear reactors.

Less than ten years after the launching of the first nuclear vessels, the United States had sixty-one nuclear submarines and four nuclear surface ships operating, being built, or authorized for future building. And yet, except for submarines, enthusiasm for nuclear propulsion also waned. In 1967, the *Savannah* was retired after two years of life. It took $3 million a year to run her, and this was considered too expensive.

But it is not the military alone who must be served. The first nuclear reactor built for the production of electric power for civilian use was put into action in the Soviet Union in June of 1954. It was a small one, with a capacity of not more than 5,000 kilowatts. By October 1956, Great Britain had its Calder Hall plant in operation, with a capacity of more than 50,000 kilowatts. The United States was third in the field. On 26 May 1958, Westing-

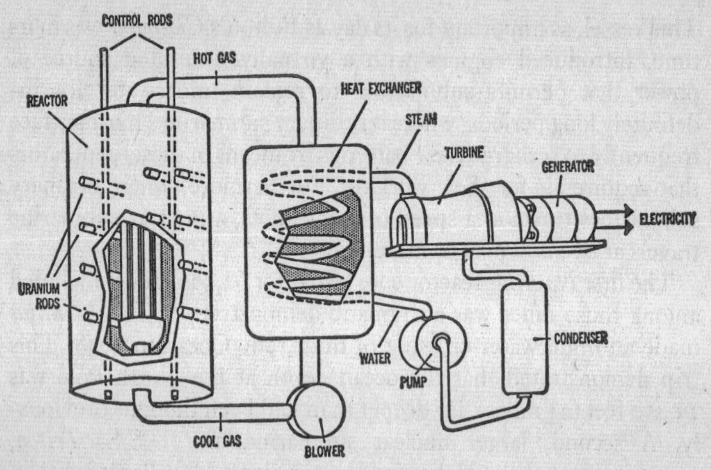

A nuclear power plant of the gas-cooled type, shown in a schematic design. The reactor's heat here is transferred to a gas, which may be a vaporized metal circulating through it, and the heat is then used to convert water to steam.

house completed a small nuclear reactor for the production of civilian electric power at Shippingport, Pennsylvania, with a capacity of 60,000 kilowatts. Other reactors quickly followed both in the United States and elsewhere.

Within little more than a decade, there were nuclear reactors in a dozen countries, and nearly half the supply of civilian electricity in the United States was being supplied by fissioning nuclei. Even outer space was invaded, for a satellite powered by a small reactor was launched on 3 April 1965. And yet the problem of radioactive contamination is a serious one. When the 1970s opened, public opposition to the continued proliferation of nuclear power plants was becoming louder.

If fission eventually replaces coal and oil as the world's chief source of energy, how long will the new fuel last? Not very long, if we have to depend entirely on the scarce fissionable material uranium 235. But, fortunately, man can create other fissionable fuels with uranium 235 as a starter.

We have seen that plutonium is one of these man-made fuels.

Suppose we build a small reactor with enriched uranium fuel and omit the moderator, so that fast neutrons will stream into a surrounding jacket of natural uranium. These neutrons will convert uranium 238 in the jacket into plutonium. If we arrange things so that few neutrons are wasted, from each fission of a uranium-235 atom in the core we may get more than one plutonium atom manufactured in the jacket. In other words, we will breed more fuel than we consume.

The first such 'breeder reactor' was built under the guidance of the Canadian-American physicist Walter H. Zinn at Arco, Idaho, in 1951. It was called EBR-1 (Experimental Breeder Reactor-1). Besides proving the workability of the breeding principle, it produced electricity. It was retired as obsolescent (so fast is progress in this field) in 1964.

Breeding could multiply the fuel supply from uranium many times, because all of the common isotope of uranium, uranium 238, would become potential fuel.

The element thorium, made up entirely of thorium 232, is another potential fissionable fuel. Upon absorbing fast neutrons, it is changed to the artificial isotope thorium 233, which soon decays to uranium 233. Now uranium 233 is fissionable by slow neutrons, and will maintain a self-sustaining chain reaction. Thus thorium can be added to the fuel supply, and thorium appears to be about five times as abundant as uranium in the earth. In fact, it has been estimated that the top hundred yards of the earth's crust contains an average of 12,000 tons of uranium and thorium per square mile. Naturally, not all of this material is easily available.

All in all, the total amount of power conceivably available from the uranium and thorium supplies of the earth is about twenty times that available from the coal and oil we have left.

Radioactivity

The arrival of the Atomic Age brought to man a hazard almost entirely new to his experience. The unlocking of the nucleus released floods of nuclear radiations. To be sure, life on the earth had always been exposed to natural radioactivity and cosmic rays. But man's concentration of naturally radioactive substances, such as radium, which ordinarily exist as greatly diluted traces in the earth's crust, vastly compounded the danger. Some early workers with X-rays and radium even received lethal doses: both Marie Curie and her daughter Irène Joliot-Curie died of leukaemia from their exposures, and there is the famous case of the watch-dial painters in the 1920s who died as the result of pointing their radium-tipped brushes with their lips.

The fact that the general incidence of leukaemia has doubled in the last two decades may be due, partly, to the increasing use of X-rays for numerous purposes. The incidence of leukaemia in doctors, who are likely to be so exposed, is twice that of the general public. In radiologists, who are medical specialists in the use of X-rays, the incidence is ten times greater. It is no wonder that attempts are being made to substitute other techniques, such as those making use of ultrasonic sound, for X-rays. The coming of fission added new force to the danger. Whether in bombs or in power reactors, it unleashes radioactivity on a scale that could make the entire atmosphere, the oceans, and everything we eat, drink, or breathe dangerous to human life. Fission has introduced a form of pollution that will tax man's ingenuity to control.

When the uranium or plutonium atom splits, its 'fission products' take various forms. The fragments may include isotopes of barium, or technetium, or any of a number of other possibilities. All told, some 200 different radioactive fission products have been identified. These are troublesome in nuclear technology, for some strongly absorb neutrons and place a damper on the fission reaction. For this reason, the fuel in a reactor must be removed and purified every once in a while.

In addition, these fission fragments are all dangerous to life in varying degrees, depending on the energy and nature of the radiation. Alpha particles taken into the body, for instance, are more dangerous than beta particles. The rate of decay also is important: a nuclide that breaks down rapidly will bombard the receiver with more radiation per second or per hour than one that breaks down slowly.

The rate of breakdown of a radioactive nuclide is something that can be spoken of only when large numbers of the nuclide are involved. An individual nucleus may break down at any time – the next instant or a thousand million years hence or any time in between – and there is no way of predicting when it will. Each radioactive species, however, has an average rate of breakdown, so if a large number of atoms is involved, it is possible to predict with great accuracy what proportion of them will break down in any unit of time. For instance, let us say that experiment shows that, in a given sample of an atom we shall call X, the atoms are breaking down at the rate of one out of two per year. At the end of a year, 500 of every 1,000 original X atoms in the sample would be left as X atoms; at the end of two years, 250; at the end of three years, 125; and so on. The time it takes for half of the original atoms to break down is called that particular atom's 'half-life' (an expression introduced by Rutherford in 1904); consequently, the half-life of atom X is one year. Every radioactive nuclide has its own characteristic half-life, which never changes under ordinary conditions. (The only kind of outside influence that can change it is bombardment of the nucleus with a particle or the extremely high temperature in the interior of a star – in other words, a violent event capable of attacking the nucleus *per se*.)

The half-life of uranium 238 is 4,500 million years. It is not surprising, therefore, that there is still uranium 238 left in the universe, despite the decay of uranium atoms. A simple calculation will show that it will take a period more than six times as long as the half-life to reduce a particular quantity of a radioactive nuclide to 1 per cent of its original quantity. Even 30,000 million years from now there will still be two pounds of uranium left from each ton of it now in the earth's crust.

Although the isotopes of an element are practically identical chemically, they may differ greatly in their nuclear properties. Uranium 235, for instance, breaks down six times as fast as uranium 238; its half-life is only 710 million years. It can be reasoned, therefore, that in aeons gone by, uranium was much richer in uranium 235 than it is today. Six thousand million years ago, for instance, uranium 235 would have made up about 70 per cent of natural uranium. Mankind is not, however, just catching the tail end of the uranium 235. Even if he had been delayed another million years in discovering fission, the earth would still have 99·9 per cent as much uranium 235 then as it has now.

Clearly any nuclide with a half-life of less than 100 million years would have declined to the vanishing point in the long lifetime of the universe. This explains why we cannot find more than traces of plutonium today. The longest-lived plutonium isotope, plutonium 244, has a half-life of only 70 million years.

The uranium, thorium, and other long-lived radioactive elements thinly spread through the rocks and soil produce small quantities of radiation, which is always present in the air about us. Man is even slightly radioactive himself, for all living tissue contains traces of a comparatively rare, unstable isotope of potassium (potassium 40), which has a half-life of 1,300 million years. (Potassium 40, as it breaks down, produces some argon 40, and this probably accounts for the fact that argon 40 is by far the most common inert-gas nuclide existing on earth. Potassium–argon ratios have been used to test the age of meteorites.)

The various naturally occurring radioactive nuclides make up what is called 'background radiation' (to which cosmic rays also contribute). The constant exposure to natural radiation probably has played a part in evolution by producing mutations and may be partly responsible for the affliction of cancer. But living organisms have lived with it for millions of years. Nuclear radiation has become a serious hazard only in our own time, first as man began to experiment with radium, and then with the coming of fission and nuclear reactors.

By the time the atomic-energy project began, physicists had learned from painful experience how dangerous nuclear radiation was. The workers in the project were therefore surrounded with elaborate safety precautions. The 'hot' fission products and other radioactive materials were placed behind thick shielding walls, and looked at only through lead glass. Instruments were devised to handle the materials by remote control. Each person was required to wear strips of photographic film or other detecting devices to 'monitor' his accumulated exposure. Extensive animal experiments were carried out to estimate the 'maximum permissible exposure'. (Mammals are more sensitive to radiation than are other forms of life, but man is averagely resistant, for a mammal.)

Despite everything, accidents happened, and a few nuclear physicists died of 'radiation sickness' from massive doses. Yet there are risks in every occupation, even the safest; the nuclear-energy workers have actually fared better than most, thanks to increasing knowledge of what the hazards are and care in avoiding them.

But a world full of nuclear power reactors, spawning fission products by the ton and the thousands of tons, will be a different story. How will all that deadly material be disposed of?

A great deal of it is short-lived radioactivity which fades away to harmlessness within a matter of weeks or months; it can be stored for that time and then dumped. Most dangerous are the nuclides with half-lives of one to thirty years. They are short-lived enough to produce intense radiation, yet long-lived enough to be hazardous for generations. A nuclide with a thirty-year half-life will take two centuries to lose 99 per cent of its activity.

Fission products can be put to good use. As sources of energy, they can power small devices or instruments. The particles emitted by the radioactive isotope are absorbed and their energy converted to heat. This in turn produces electricity in thermocouples. Batteries that produce electricity in this fashion are radio-isotope power generators, usually referred to as SNAP ('Systems for Nuclear Auxiliary Power') or, more dramatically, as 'atomic batteries'. They can be as light as four pounds, generate up to sixty watts, and last for years. SNAP batteries have been used in satellites; in Transit 4A and Transit 4B, for instance, which were put in orbit by the United States in 1961 to serve, ultimately, as navigational aids.

The isotope most commonly used in SNAP batteries is strontium 90, which will soon be mentioned in another connection. Isotopes of plutonium and curium are also used in some varieties.

Radionuclides also have large potential uses in medicine (e.g., for treatment of cancer), in killing bacteria and preserving food, and in many fields of industry, including chemical manufacturing. For instance, the Hercules Powder Company has designed a reactor to use radiation in the production of the anti-freeze ethylene glycol.

Yet when all is said and done, no conceivable uses could employ more than a small part of the vast quantities of fission products that power reactors will discharge. This represents an important difficulty in connection with nuclear power generally. The more obvious danger of explosions due to a sudden uncontrolled fission reaction (a 'nuclear excursion', as it is called) has always been in the minds of planners. It is to their credit that this has almost never happened, although one such case did indeed kill three men in Idaho in 1961 and spread radioactive contamination over the station. The matter of fission products, however, is far more difficult to handle. It is estimated that every 200,000 kilowatts of nuclear-produced electricity will involve the production of a pound and a half of fission products per day. What to do with it? Already the United States has stored millions of gallons of radioactive liquid underground and it is estimated that by A.D. 2000 as much as half a million gallons of radioactive liquid will require disposal each day! Both the United States and Great Britain have dumped concrete containers of fission products at sea. There have been proposals to drop the radioactive wastes in oceanic abysses, to store them in old salt mines, to incarcerate them in molten glass, and bury the solidified material. But there is always the nervous thought that in one way or another the radioactivity will escape in time and contaminate the soil or the seas. One particularly haunting nightmare is the possibility that a nuclear-powered ship might be wrecked and spill its accumulated fission products into the ocean. The sinking of the American nuclear submarine U.S.S. *Thresher* in the North Atlantic on 10 April 1963 lent new substance to this fear, although in this case such contamination apparently did not take place.

If radioactive pollution by peaceful nuclear energy is a potential danger, at least it will be kept under control, and probably successfully, by every possible means. But there is a pollution which has already spread over the world and which, indeed, in a nuclear war might be broadcast deliberately. This is the fall-out from atomic bombs.

Fall-out is produced by all nuclear bombs, even those not fired in anger. Because fall-out is carried round the world by the winds and brought to earth by rainfall, it is virtually impossible for any nation to explode a nuclear bomb in the atmosphere without detection. In the event of a nuclear war, fall-out in the long run might produce more casualties and do more damage to living things in the world at large than the fire and blast of the bombs themselves would wreak on the countries attacked.

Fall-out is divided into three types: 'local', 'tropospheric', and 'stratospheric'. Local fall-out results from ground explosions in which radioactive isotopes are adsorbed on particles of soil and settle out quickly within a hundred miles of the blast. Air blasts of fission bombs in the kiloton range send fission products into the troposphere. These settle out in about a month, being carried some thousands of miles eastwards by the winds in that interval of time.

The huge output of fission products from the thermonuclear super-bombs is carried into the stratosphere. Such stratospheric fall-out takes a year or more to settle and is distributed over a whole hemisphere, falling eventually on the attacker as well as the attacked.

The intensity of the fall-out from the first super-bomb, exploded in the Pacific on 1 March 1954, caught scientists by surprise. They had not expected the fall-out from a fusion bomb to be so 'dirty'. Seven thousand square miles were seriously contaminated, an area nearly the size of Massachusetts. But the reason became clear when they learned that the fusion core was supplemented with a blanket of uranium 238 that was fissioned by the neutrons. Not only did this multiply the force of the explosion, but it gave rise to a vastly greater cloud of fission products than a simple fission bomb of the Hiroshima type.

The fall-out from the bomb tests to date has added only a small amount of radioactivity to the earth's background radiation. But even a small rise above the natural level may increase the incidence of cancer, cause genetic damage, and shorten the average life expectancy slightly. The most conservative estimators of the hazards agree that, by increasing the mutation rate (see Chapter 12 for a discussion of mutations), fall-out is storing up a certain amount of trouble for future generations.

One of the fission products is particularly dangerous for human life. This is strontium 90 (half-life, twenty-eight years), the isotope so useful in SNAP generators. Strontium 90 falling on the soil and water is taken up by plants and thereafter incorporated into the bodies of those animals (including man) that feed directly or indirectly on the plants. Its peculiar danger lies in the fact that strontium, because of its chemical similarity to calcium, goes to the bones and lodges there for a long time. The minerals in bone have a slow 'turnover'; that is, they are not replaced nearly as rapidly as are the substances in the soft tissues. For that reason strontium 90, once absorbed, may remain in the body for a major part of a person's lifetime.

Strontium 90 is a brand-new substance in our environment; it did not exist on the earth in any detectable quantity until man fissioned the uranium atom. But today, within less than a generation, some strontium 90 has become incorporated in the bones of every human being on earth and, indeed, in all vertebrates. Considerable quantities of it are still floating in the stratosphere, sooner or later to add to the concentration in our bones.

The strontium-90 concentration is measured in 'strontium units' (S.U.). One S.U. is one micromicrocurie of strontium 90 per gram of calcium in the body. A 'curie' is a unit of radiation (named in honour of the Curies, of course) originally meant to be equivalent to that produced by a gram of radium in equilibrium with its breakdown product, radon. It is now more generally accepted as meaning 37,000 million disintegrations per second. A micromicrocurie is a billionth of a curie, or 2·12 disintegrations per minute. A strontium unit would therefore mean 2·12 disintegrations per minute per grain of calcium present in the body.

The concentration of strontium 90 in the human skeleton varies greatly from place to place and among individuals. Some persons have been found to have as much as seventy-five times the average amount. Children average at least four times as high a concentration as adults, because of the higher turnover of material in their growing bones. Estimates of the averages themselves vary, because they are based mainly on estimates of the amounts of strontium 90 found in the diet. (Incidentally, milk is not a particularly hazardous food, from this point of view, because

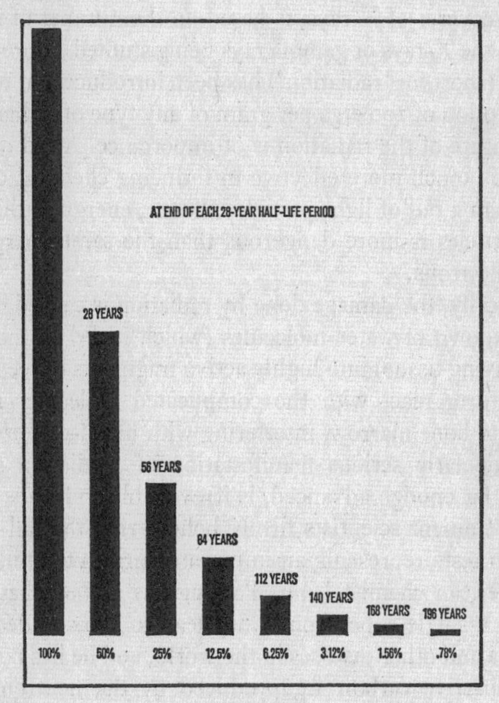

Decay of strontium 90 over approximately 200 years.

calcium obtained from vegetables has more strontium 90 associated with it. The cow's 'filtration system' eliminates some of the strontium it gets in its plant fodder.) The estimates of the average strontium-90 concentration in the bones of people in the United States in 1959 ranged from less than one strontium unit to well over five strontium units. (The 'maximum permissible' was established by the International Commission on Radiation Protection at 67 S.U.) But the averages mean little, particularly since strontium 90 may collect in 'hot spots' in the bones and reach a high enough level there to initiate leukaemia or cancer.

The importance of radiation effects has, among other things, resulted in the adoption of a number of types of unit designed to measure these effects. One such, the 'roentgen', named in honour of the discoverer of X-rays, is based on the number of ions produced by the X-rays or gamma rays being studied. More recently, the 'rad' (short for 'radiation') has been introduced. It represents the absorption of 100 ergs per gram of any type of radiation.

The nature of the radiation is of importance. A rad of massive particles is much more effective in inducing chemical change in tissues than a rad of light particles; hence, energy in the form of alpha particles is more dangerous than the same energy in the form of electrons.

Chemically, the damage done by radiation is caused chiefly by the breakdown of water molecules (which make up most of the mass of living tissue) into highly active fragments ('free radicals') that, in turn, react with the complicated molecules in tissue. Damage to bone marrow, interfering with blood-cell production, is a particularly serious manifestation of 'radiation sickness', which, if far enough advanced, is irreversible and leads to death.

Many eminent scientists firmly believe that the fall-out from the bomb tests represents an important peril to the human race. The American chemist Linus Pauling has argued that the fall-out from a single super-bomb may lead to 100,000 deaths from leukaemia and other diseases in the world, and he has pointed out that radioactive carbon 14, produced by the neutrons from a nuclear explosion, constitutes a serious genetic danger. He has, for this reason, been extremely active in pushing for cessation of testing of nuclear bombs; he endorses all movements designed to lessen the danger of war and to encourage disarmament. On the other hand, some scientists, including the Hungarian-American physicist Edward Teller, minimize the seriousness of the fall-out hazard.

The sympathy of the world generally lies with Pauling, as might be indicated by the fact that he was awarded the Nobel Peace Prize in 1963. (Nine years earlier, Pauling had won a Nobel Prize in chemistry; he thus joins Marie Curie as the only members of the select group who have been awarded two Nobel Prizes.)

In the fall of 1958, the United States, the U.S.S.R., and Great Britain suspended bomb testing by a gentleman's agreement (which, however, did not prevent France from exploding her first atomic bomb in the spring of 1960). For three years, things looked rosy; the concentration of strontium 90 reached a peak and levelled off about 1960 at a point well below what is estimated to be the maximum consistent with safety. Even so, some 25 million curies of strontium 90 and cesium 137 (another dangerous fission product) had been delivered into the atmosphere during the thirteen years of nuclear testing when some 150 bombs of all varieties were exploded. Only two of these were exploded in anger, but the results were dire indeed.

In 1961, without warning, the Soviet Union ended the moratorium and began testing again. Since the U.S.S.R. exploded thermonuclear bombs of unprecedented power, the United States felt that it was forced to begin testing again. World public opinion, sharpened and concentrated by the relief of the moratorium, reacted with great indignation.

On 10 October 1963, therefore, the three chief nuclear powers signed a partial test-ban treaty (*not* a mere gentleman's agreement) in which nuclear-bomb explosions in the atmosphere, in space, or underwater were banned. Only underground explosions were permitted, since these did not produce fall-out. This has been the most hopeful move in the direction of human survival since the opening of the Atomic Age. The chief danger now, assuming the test-ban treaty is observed by all signatories, is that France and the People's Republic of China (the newest members of the atomic club) have refused to sign the treaty so far.

Nuclear Fusion

For more than twenty years, nuclear physicists have had in the back of their minds a dream even more attractive than turning fission to constructive uses. It is the dream of harnessing fusion energy. Fusion, after all, is the engine that makes our world go round: the fusion reactions in the sun are the ultimate source of all our forms of energy and of life itself. If somehow we could reproduce and control such reactions on the earth, all our energy problems would be solved. Our fuel supply would be as big as the ocean, for the fuel would be hydrogen.

Oddly enough, this would not be the first time hydrogen will have been used as a fuel. Not long after hydrogen was discovered and its properties studied, it gained a place as a chemical fuel. The American scientist Robert Hare devised an oxyhydrogen torch in 1801, and the hot flame of hydrogen burning in oxygen has served industry ever since. Now, however, as a nuclear fuel, a much more glittering possibility lay before it.

Fusion power would be immensely more convenient than fission power. Pound for pound, a fusion reactor would deliver about five to ten times as much power as a fission reactor. A pound of hydrogen, on fusion, could produce 35 million kilowatt-hours of energy. Furthermore, fusion produces no radioactive ashes. Finally, a fusion reaction, in the event of any conceivable malfunction, could only collapse and go out, whereas a fission reaction might conceivably (though not very probably) go out of control and into a full explosion.

Of the three isotopes of hydrogen, hydrogen 1 is the most common, but it is also the one most difficult to force into fusion. It is the particular fuel of the sun, but the sun has it by the billions of cubic miles, together with an enormous gravity field to hold it together and central temperatures in the many millions of degrees. Only a tiny percentage of the hydrogen within the sun is fusing at any given moment, but given the vast mass present, even a tiny percentage is enough.

Hydrogen 3 is the easiest to bring to fusion, but it exists in such tiny quantities and can be made only at so fearful an expenditure of energy that it is hopeless to think of it as a practical fuel all by itself.

That leaves hydrogen 2, which is easier to handle than hydrogen 1 and much more common than hydrogen 3. In all the hydrogen of the world, only one atom out of 6,000 is deuterium, but that is enough. There is 35 billion tons of deuterium in the ocean, enough to supply man with ample energy for all the foreseeable future.

Yet there are problems. That might seem surprising, since fusion bombs exist. If we can make hydrogen fuse, why can't we make a reactor as well as a bomb? Ah, but to make a fusion bomb, we need to use a fission-bomb igniter and then let it go. To make a fusion reactor, we need a gentler igniter, obviously, and we must then keep the reaction going at a constant, controlled – and non-explosive – rate.

The first problem is the less difficult. Heavy currents of electricity, high-energy sound-waves, laser beams, and so on, can all produce temperatures in the millions of degrees very briefly. There is no doubt that the required temperature will be reached.

Maintaining the temperature while keeping the (it is to be

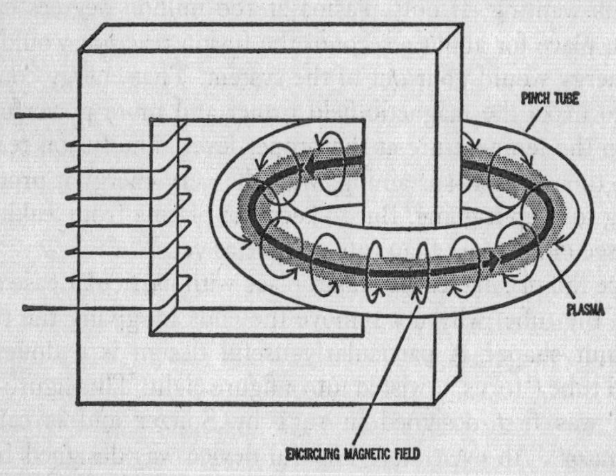

Magnetic bottle designed to hold a hot gas of hydrogen nuclei (a plasma). The ring is called a 'torus'.

hoped) fusing hydrogen in place is something else. Obviously no material container can hold a gas at anything like a temperature of over 100 million degrees. Either the container would vaporize or the gas would cool. The first step towards a solution is to reduce the density of the gas to far below normal pressure; this cuts down the heat content, though the energy of the particles remains high. The second step is a concept of great ingenuity. A gas at very high temperature has all the electrons stripped off its atoms; it is a 'plasma' (a term introduced by Irving Langmuir in the early 1930s) made up of electrons and bare nuclei. Since it consists entirely of charged particles, why not use a strong magnetic field, taking the place of a material container, to hold it? The fact that magnetic fields could restrain charged particles and 'pinch' a stream of them together had been known since 1907, when it was named the 'pinch effect'. The 'magnetic bottle' idea was tried and it worked – but only for the briefest instant. The wisps of plasma pinched in the bottle immediately writhed like a snake, broke up, and died out.

Another approach is to have a magnetic field stronger at the ends of the tube so that plasma is pushed back and kept from leaking. This is also found wanting. It doesn't seem as though much is wanting. If only plasma at 100 million degrees can be held in place for about a second, the fusion reaction would start and energy would pour out of the system. That energy could be used to make the magnetic field firmer and more powerful and to keep the temperature at the proper level. The fusion reaction would then be self-sustaining, with the very energy it produced serving to keep it going. But to keep the plasma from leaking for just a second is more than can be done as yet.

Since the plasma leakage takes place with particular ease at the end of the tube, why not remove the ends by giving the tube a doughnut shape? A particularly useful design is a doughnut-shaped tube ('torus') twisted into a figure eight. This figure-eight device was first designed in 1951 by Spitzer and is called a 'stellarator'. An even more hopeful device was designed by the Soviet physicist Lev Andreevich Artsimovich. It is called 'Toroidal Kamera Magnetic', a name which is abbreviated to 'Tokamak'.

The Tokamak only works with very rarefied gases, but the Soviets, using hydrogen at one millionth atmospheric density, have achieved a temperature of 100 million degrees for one hundredth of a second. Hydrogen so rare must be held in place for longer than a second, but if the Soviets can make hydrogen 2 just ten times denser and then hold on for a second, they might be able to make it.

American physicists are also working with Tokamaks and in addition with a device called 'Scyllac', which is designed to hold denser gas and therefore require a smaller containment period.

For nearly twenty years physicists have been inching towards fusion power. Progress has been slow, but as yet no signs of a definite dead-end have appeared.

Meanwhile, practical applications of fusion research are to be found. Plasma torches emitting jets at temperatures up to 50,000°C in absolute silence can far outdo ordinary chemical torches. And it is suggested that the plasma torch is the ultimate waste-disposal unit. In its flame everything, *everything*, would be broken down to its constituent elements, and all the elements would be available for recycling and for conversion into useful materials again.

Appendix

Mathematics in Science

Gravitation

As I explained in Chapter 1, Galileo initiated science in its modern sense by introducing the concept of reasoning back from observation and experiment to basic principles. In doing so, he also introduced the essential technique of measuring natural phenomena accurately and abandoned the practice of merely describing them in general terms. In short, he turned from the qualitative description of the universe by the Greek thinkers to a quantitative description.

Although science depends so much on mathematical relationships and manipulations, and could not exist in the Galilean sense without it, I have nevertheless written this book non-mathematically, and have done so deliberately. Mathematics, after all, is a highly specialized tool. To have discussed the developments in science in mathematical terms would have required a prohibitive amount of space, as well as a sophisticated knowledge of mathematics on the part of the reader. But in this appendix I would like to present an example or two of the way in which simple mathematics has been fruitfully applied to science. How better to begin than with Galileo himself?

Galileo (like Leonardo da Vinci nearly a century earlier) suspected that falling objects steadily increased their velocity as they fell. He set out to measure exactly by how much and in what manner the velocity increased.

The measurement was anything but easy for Galileo, with the tools he had at his disposal in 1600. To measure a velocity requires the measurement of time. We speak of velocities of 60 miles *an hour*, of 13 feet *a second*. But there were no clocks in Galileo's time that could do more than strike the hour at approximately equal intervals.

Galileo resorted to a crude water clock. He let water trickle slowly from a small spout, assuming, hopefully, that it dripped at a constant rate. This water he caught in a cup, and by the weight of water caught during the interval in which an event took place, Galileo measured the elapsed time. (He also used his pulse beat for the purpose on occasion.)

One difficulty was, however, that a falling object dropped so rapidly that Galileo could not collect enough water, in the interval of falling, to weigh accurately. What he did, then, was to 'dilute' the pull of gravity by having a brass ball roll down a groove in an inclined plane. The more nearly horizontal the plane, the more slowly the ball moved. Thus Galileo was able to study falling bodies in whatever degree of 'slow motion' he pleased.

Galileo found that a ball rolling on a perfectly horizontal plane moved at constant speed. (This supposes a lack of friction, a condition which could be assumed within the limits of Galileo's crude measurements.) Now a body moving on a horizontal track is moving at right angles to the force of gravity. Under such conditions, the body's velocity is not affected by gravity either way. A ball resting on a horizontal plane remains at rest, as anyone can observe. A ball set to moving on a horizontal plane moves at a constant velocity, as Galileo observed.

Mathematically, then, it can be stated that the velocity v of a body, *in the absence of any external force*, is constant k, or:

$$v = k$$

If k is equal to any number other than zero, the ball is moving at constant velocity. If k is equal to zero, the ball is at rest; thus, rest is a 'special case' of constant velocity.

Nearly a century later, when Newton systematized the discoveries of Galileo in connection with falling bodies, this finding became the First Law of Motion (also called the 'principle of inertia'). This law can be stated: Every body persists in a state of rest or of uniform motion in a straight line unless compelled by external force to change that state.

When a ball rolls down an inclined plane, however, it is under the continuous pull of gravity. Its velocity then, Galileo found, was not constant but increased with time. Galileo's measurements showed that the velocity increased in proportion to the lapse of time t.

In other words, when a body was under the action of constant external force, its velocity, starting at rest, could be expressed as:

$$v = kt$$

What was the value of k?

That, it was easy to find by experiment, depended on the slope of the inclined plane. The more nearly vertical the plane, the more quickly the rolling ball gained velocity and the higher the value of k. The maximum gain in speed would come when the plane was vertical – in other words, when the ball dropped freely under the undiluted pull of gravity. The symbol g (for 'gravity') is used where the undiluted force of gravity is acting, so that the velocity of a ball in free fall, starting from rest, was:

$$v = gt$$

Let us consider the inclined plane in more detail. In the diagram:

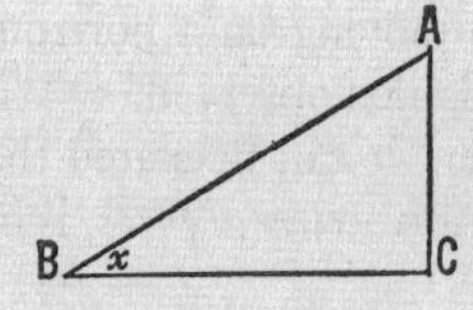

the length of the inclined plane is AB, while its height at the upper end is AC. The ratio of AC to AB is the sine of the angle x, usually abbreviated as sin x.

The value of this ratio – that is, of sin x – can be obtained approximately by constructing triangles with particular angles and actually measuring the height and length involved. Or it can be calculated by mathematical techniques to any degree of precision and the results can be embodied in a table. By using such a table, we can find, for instance, that sin 10° is approximately equal to 0·17365, that sin 45° is approximately equal to 0·70711, and so on.

There are two important special cases. Suppose that the 'inclined' plane is precisely horizontal. Angle x is then zero and as the height of the inclined plane is zero, the ratio of its height to its length is also zero. In other words, $\sin 0° = 0$. When the 'inclined' plane is precisely vertical, the angle it forms with the ground is a right angle, or 90°. Its height is then exactly equal to its length, so that the ratio of one to the other is just 1. Consequently, $\sin 90° = 1$.

Now let us return to the equation showing that the velocity of a ball rolling down an inclined plane is proportional to time:

$$v = kt$$

It can be shown by experiment that the value of k changes with the sine of the angle so that:

$$k = k' \sin x$$

(where k' is used to indicate a constant that is different from k).

(As a matter of fact, the role of the sine in connection with the inclined plane was worked out somewhat before Galileo's time by Simon Stevinus, who also performed the famous experiment of dropping different masses from a height, an experiment traditionally, but wrongly, ascribed to Galileo. Still, if Galileo was not the very first to experiment and measure, he was the first to impress the scientific world, indelibly, with the necessity to experiment and measure, and that is glory enough.)

In the case of a completely vertical inclined plane, $\sin x$ becomes sin 90°, which is 1, so that in free fall

$$k = k'$$

It follows that k' is the value of k in free fall under the undiluted pull of gravity, which we have already agreed to symbolize as g. We can substitute g for k' and, for any inclined plane:

$$k = g \sin x.$$

The equation for the velocity of a body rolling down an inclined plane is therefore:

$$v = (g \sin x)t$$

On a horizontal plane with $\sin x = 0° = 0$, the equation for velocity becomes:

$$v = 0$$

This is another way of saying that a ball on a horizontal plane, starting from rest, will remain motionless regardless of the passage of time. An object at rest tends to remain at rest, and so on. That is part of the First Law of Motion, and it follows from the inclined plane equation of velocity.

Suppose that a ball does not start from rest but has an initial motion before it begins to fall. Suppose, in other words, you have a ball moving along a horizontal plane at 5 feet per second, and it suddenly finds itself at the upper end of an inclined plane and starts rolling downwards.

Experiment shows that its velocity thereafter is 5 feet per second greater, at every moment, than it would have been if it had started rolling down the plane from rest. In other words, the equation for the motion of a ball down an inclined plane can be expressed more completely as follows:

$$v = (g \sin x)t + u$$

where u is the original starting velocity. If an object starts at rest, then u is equal to 0 and the equation becomes as we had it before:

$$v = (g \sin x)t$$

If we next consider an object with some initial velocity on a horizontal plane, so that angle x is 0°, the equation becomes:

$$v = (g \sin 0°) + u$$

or, since sin 0° is 0:

$$v = u$$

Thus the velocity of such an object remains its initial velocity, regardless of the lapse of time. That is the rest of the First Law of Motion, again derived from observed motion on an inclined plane.

The rate at which velocity changes is called 'acceleration'. If, for instance, the velocity (in feet per second) of a ball rolling down an inclined plane is, at the end of successive seconds, 4, 8, 12, 16 . . . then the acceleration is 4 feet per second per second.

In a free fall, if we use the equation:

$$v = gt$$

each second of fall brings an increase in velocity of g feet per second. Therefore, g represents the acceleration due to gravity.

The value of g can be determined from inclined-plane experiments. By transposing the inclined-plane equation, we get:

$$g = v/(t \sin x)$$

Since v, t, and x can all be measured, g can be calculated, and it turns out to be equal to 32 feet per second per second at the earth's surface. In free fall under normal gravity at earth's surface, then, the velocity of fall is related to time thus:

$$v = 32t$$

This is the solution to Galileo's original problem, namely, determining the rate of fall of a falling body and the manner in which that rate changes.

The next question is: How far does a body fall in a given time? From the equation relating the velocity to time, it is possible to relate distance to time by the process in calculus called 'integration'. It is not necessary to go into that, however, because the equation can be worked out by experiment, and, in essence, Galileo did this.

He found that a ball rolling down an inclined plane covered a distance proportional to the square of the time. In other words, doubling the time increased the distance fourfold, tripling it increased the distance ninefold, and so on.

For a freely falling body, the equation relating distance d and time is:

$$d = \frac{1}{2}gt^2$$

or, since g is equal to 32:

$$d = 16t^2$$

Next, suppose that instead of dropping from rest, an object is thrown horizontally from a position high in the air. Its motion would then be a compound of two motions – a horizontal one and a vertical one.

The horizontal motion, involving no force other than the single original impulse (if we disregard wind, air resistance, and so on), is one of constant velocity, in accordance with the First Law of Motion, and the distance the object covers horizontally is proportional to the time elapsed. The vertical motion, however, covers a distance, as I have just explained, that is proportional to the square of the time elapsed. Prior to Galileo, it had been vaguely believed that a projectile such as a cannon ball travelled in a straight line until the impulse that drove it was somehow exhausted, after which it fell straight down. Galileo, however, made the great advance of *combining* the two motions.

The combination of these two motions (proportional to time horizontally, and proportional to the square of the time vertically) produces a curve called a parabola. If a body is thrown, not horizontally, but upwards or downwards, the curve of motion is still a parabola.

Such curves of motion, or trajectories, apply, of course, to a projectile such as a cannon ball. The mathematical analysis of trajectories, stemming from Galileo's work, made it possible to calculate where a cannon ball would fall when fired with a given propulsive force and a given angle of elevation of the cannon. Although men had been throwing objects for fun, to get food, to attack, and to defend, for uncounted thousands of years, it was only due to Galileo that for the first time, thanks to experiment and measurement, there was a science of 'ballistics'. As it happened, then, the very first achievement of modern science proved to have a direct and immediate military application.

It also had an important application in theory. The mathematical analysis of combinations of more than one motion answered several objections to the Copernican theory. It showed an object thrown upwards would not be left behind by the moving earth, since the object would have two motions, one imparted to it by the impulse of throwing, and one that it shared along with the moving earth. It also made it reasonable to expect the earth to have two motions at once, rotation about its axis and revolution about the sun – a situation that some of the non-Copernicans insisted was unthinkable.

Isaac Newton extended the Galilean concepts of motion to the heavens and showed that the same set of laws of motion applied to the heavens and the earth alike.

He began by considering that the moon might be falling towards the earth in response to the earth's gravity but never struck the earth's surface because of the horizontal component of its motion. A projectile fired horizontally, as I said, follows a parabolically curved path downwards to intersection with earth's surface. But the earth's surface curves downwards, too, since the earth is a sphere. A projectile given a sufficiently rapid horizontal motion might curve downwards no faster than the earth's surface and would therefore eternally circle the earth.

Now the moon's elliptical motion around the earth can be split into horizontal and vertical components. The vertical component is such that in the space of a second the moon falls a trifle more than 1/20 of an inch towards the earth. In that time it also moves about 3,300 feet in the horizontal direction, just far enough to compensate for the fall and carry it around the earth's curvature.

The question was whether this 1/20 inch fall of the moon was caused by the same gravitational attraction that caused an apple, falling from a tree, to drop 16 feet in the first second of its fall.

Newton visualized the earth's gravitational force as spreading out in all directions like a vast, expanding sphere. The surface area A of a sphere is proportional to the square of its radius r:

$$A = 4\pi r^2$$

He therefore reasoned that the gravitational force, spreading out over the spherical area, must weaken as the square of the radius. The intensity of light and of sound weakened as the square of the distance from the source – why not the force of gravity as well?

The distance from the earth's centre to an apple on its surface is roughly 4,000 miles. The distance from the earth's centre to the moon is roughly 240,000 miles. Since the distance to the moon was 60 times greater than to the apple, the force of the earth's gravity at the moon must be 60^2, or 3,600, times weaker than at the apple. Divide 16 feet by 3,600, and you come out with roughly 1/20 of an inch. It seemed clear to Newton that the moon did indeed move in the grip of the earth's gravity.

Newton was persuaded further to consider 'mass' in relation to gravity. Ordinarily, we measure mass as weight. But weight is only the result of the attraction of the earth's gravitational force. If there were no gravity, an object would be weightless; nevertheless, it would still contain the same amount of matter. Mass, therefore, is independent of weight and should be capable of measurement by a means not involving weight.

Suppose you tried to pull an object on a perfectly frictionless surface in a direction horizontal to the earth's surface, so that there was no resistance from gravity. It would take effort to set the body in motion and to accelerate its motion, because of the body's inertia.

If you measured the applied force accurately, say by pulling on a spring balance attached to the object, you would find that the force f required to bring about a given acceleration a would be directly proportional to the mass m. If you doubled the mass, it would take double the force. For a given mass, the force required would be directly proportional to the acceleration desired. Mathematically, this is expressed in the equation:

$$f = ma$$

The equation is known as Newton's Second Law of Motion.

Now, as Galileo had found, the pull of the earth's gravity accelerates all bodies, heavy or light, at precisely the same rate. (Air resistance may slow the fall of very light bodies, but in a vacuum a feather will fall as rapidly as a lump of lead, as can easily be demonstrated.) If the Second Law of Motion is to hold, one must conclude that the earth's gravitational pull on a heavy body must be greater than on a light body, in order to produce the same acceleration. To accelerate a mass that is eight times as great as another, for instance, takes eight times as much force. It follows that the earth's gravitational pull on any body must be exactly proportional to the mass of that body. (That, in fact, is why mass on the earth's surface can be measured quite accurately as weight.)

Newton evolved a Third Law of Motion, too: 'For every action there is an equal and opposite reaction.' This applies to force. In other words, if the earth pulls at the moon with a certain force, then the moon pulls on the earth with an equal force. If the moon were suddenly doubled in mass, the earth's gravitational force upon it would also be doubled, in accordance with the Second Law; of course, the moon's gravitational force on the earth would then have to be doubled in accordance with the Third Law.

Similarly, if it were the earth rather than the moon that doubled in mass, it would be the moon's gravitational force on the earth that would double, according to the Second Law, and the earth's gravitational force on the moon that would double, in accordance with the Third.

If both the earth and the moon were to double in mass, there would be a doubled doubling, each body doubling its gravitational force twice, for a fourfold increase all told.

Newton could only conclude by this sort of reasoning that the gravitational force between any two bodies in the universe was directly proportional to the product of the masses of the bodies. And, of course, as he had decided earlier, it is inversely proportional to the square of the distance (centre to centre) between the bodies. This is Newton's Law of Universal Gravitation.

If we let f represent the gravitational force, m_1 and m_2 the masses of the two bodies concerned, and d the distance between them, then the law can be stated:

$$f = \frac{Gm_1m_2}{d^2}$$

G is the 'gravitational constant', the determination of which made it possible to 'weigh the earth' (see p. 119). It was Newton's surmise that G had a fixed value throughout the universe. As time went on, it was found that new planets, undiscovered in Newton's time, tempered their motions to the requirements of Newton's law; even double stars incredibly far away danced in time to Newton's analysis of the universe.

All this came from the new quantitative view of the universe pioneered by Galileo. As you see, much of the mathematics involved was really very simple. Those parts of it I have quoted here are secondary-school algebra.

In fact, all that was needed to introduce one of the greatest intellectual revolutions of all time was:

1. A simple set of observations any schoolboy studying physics might make with a little guidance.

2. A simple set of mathematical generalizations.

3. The transcendent genius of Galileo and Newton, who had the insight and originality to make these observations and generalizations for the first time.

Relativity

The laws of motion as worked out by Galileo and Newton depended on the assumption that such a thing as absolute motion existed – that is, motion with reference to something at rest. But everything that we know of in the universe is in motion: the earth, the sun, the Galaxy, the systems of galaxies. Where in the universe, then, can we find absolute rest against which to measure absolute motion?

It was this line of thought that led to the Michelson–Morley experiment, which in turn led to a scientific revolution as great, in some respects, as that initiated by Galileo (see p. 370). Here, too, the basic mathematics is rather simple.

The experiment was an attempt to detect the absolute motion of the earth againt an 'aether' that was supposed to fill all space and to be at rest. The reasoning behind the experiment was as follows.

Suppose that a beam of light is sent out in the direction in which the earth is travelling through the aether, and that at a certain distance in that direction there is a fixed mirror which reflects the light back to the source. Let us symbolize the velocity of light as c, the velocity of the earth through the aether as v, and the distance of the mirror as d. The light starts with the velocity $c+v$: its own velocity plus the earth's velocity. (It is travelling with a tail wind, so to speak.) The time it takes to reach the mirror is d divided by $(c+v)$.

On the return trip, however, the situation is reversed. The reflected light now is bucking the head wind of the earth's velocity, and its net velocity is $c-v$. The time it takes to return to the source is d divided by $(c-v)$.

The total time for the round trip is:

$$\frac{d}{c+v}+\frac{d}{c-v}$$

Combining the terms algebraically, we get:

$$\frac{d(c-v)+d(c+v)}{(c+v)(c-v)}=\frac{dc-dv+dc+dv}{c^2-v^2}=\frac{2dc}{c^2-v^2}$$

Now suppose that the light beam is sent out to a mirror at the

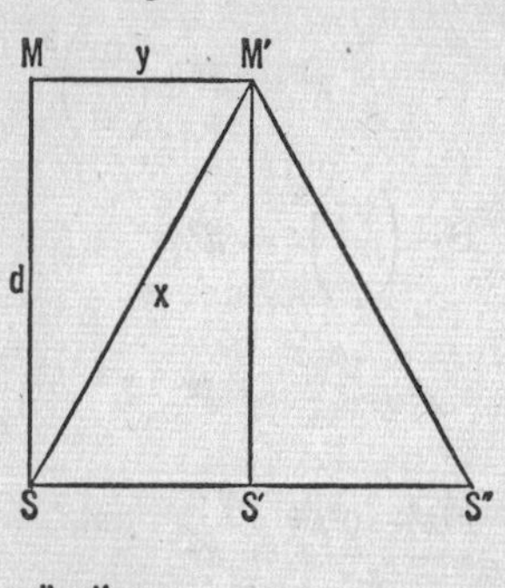

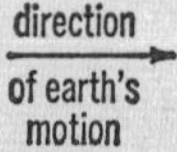

same distance in a direction at right angles to the earth's motion through the aether.

The beam of light is aimed from S (the source) to M (the mirror) over the distance d. However, during the time it takes the light to reach the mirror, the earth's motion has carried the mirror from M to M', so that the actual path travelled by the light beam is from S to M'. This distance we call x, and the distance from M to M' we call y.

While the light is moving the distance x at its velocity c, the mirror is moving the distance y at the velocity of the earth's motion v. Since both the light and the mirror arrive at M' simultaneously, the distances travelled must be exactly proportional to the respective velocities. Therefore:

$$\frac{y}{x} = \frac{v}{c}$$

or

$$y = \frac{vx}{c}$$

Now we can solve for the value of x by use of the Pythagorean theorem, which states that the sum of the squares of the sides of a right triangle is equal to the square of the hypotenuse. In the right triangle SMM', then, substituting vx/c for y:

$$x^2 = d^2 + \left(\frac{vx}{c}\right)^2$$

$$x^2 - \left(\frac{vx}{c}\right)^2 = d^2$$

$$x^2 - \frac{v^2x^2}{c^2} = d^2$$

$$\frac{c^2x^2 - v^2x^2}{c^2} = d^2$$

$$(c^2 - v^2)x^2 = d^2c^2$$

$$x^2 = \frac{d^2c^2}{c^2 - v^2}$$

$$x = \frac{dc}{\sqrt{c^2 - v^2}}$$

The light is reflected from the mirror at M' to the source, which meanwhile has travelled on to S'. Since the distance $S'S''$ is equal to SS', the distance $M'S''$ is equal to x. The total path travelled by the light beam is therefore $2x$, or $2dc/\sqrt{c^2 - v^2}$.

The time taken by the light beam to cover this distance at its velocity c is:

$$\frac{2dc}{\sqrt{c^2 - v^2}} \div c = \frac{2d}{\sqrt{c^2 - v^2}}$$

How does this compare with the time that light takes for the round trip in the direction of the earth's motion? Let us divide the time in the parallel case $(2dc/(c^2 - v^2))$ by the time in the perpendicular case $(2d/\sqrt{c^2 - v^2})$:

$$\frac{2dc}{c^2 - v^2} \div \frac{2d}{\sqrt{c^2 - v^2}} = \frac{2dc}{c^2 - v^2} \times \frac{\sqrt{c^2 - v^2}}{2d} = \frac{c\sqrt{c^2 - v^2}}{c^2 - v^2}$$

Now any number divided by its square root gives the same square root as a quotient, that is, $x/\sqrt{x} = \sqrt{x}$. Conversely, $\sqrt{x}/x = 1/\sqrt{x}$. So the last equation simplifies to:

$$\frac{c}{\sqrt{c^2 - v^2}}$$

This expression can be further simplified if we multiply both the numerator and the denominator by $\sqrt{1/c^2}$ (which is equal to $1/c$).

$$\frac{c\sqrt{1/c^2}}{\sqrt{c^2 - v^2}\sqrt{1/c^2}} = \frac{c/c}{\sqrt{c^2/c^2 - v^2/c^2}} = \frac{1}{\sqrt{1 - v^2/c^2}}$$

And there you are. That is the ratio of the time that light should take to travel in the direction of the earth's motion as compared with the time it should take in the direction perpendicular to the earth's motion. For any value of v greater than zero, the expression $1/\sqrt{1-v^2/c^2}$ is greater than 1. Therefore, if the earth is moving through a motionless aether, it should take longer for light to travel in the direction of the earth's motion than in the perpendicular direction. (In fact, the parallel motion should take the maximum time and the perpendicular motion the minimum time.)

Michelson and Morley set up their experiment to try to detect the directional difference in the travel time of light. By trying their beam of light in all directions, and measuring the time of return by their incredibly delicate interferometer, they felt they ought to get differences in apparent velocity. The direction in which they found the velocity of light to be at a minimum should be parallel to the earth's absolute motion, and the direction in which the velocity would be at a maximum should be perpendicular to the earth's motion. From the difference in velocity, the amount (as well as the direction) of the earth's absolute motion could be calculated.

They found no differences at all in the velocity of light with changing direction! To put it another way, the velocity of light was always equal to c, regardless of the motion of the source – a clear contradiction of the Newtonian laws of motion. In attempting to measure the absolute motion of the earth, Michelson and Morley had thus managed to cast a doubt not only on the existence of the aether, but on the whole concept of absolute rest and absolute motion, and upon the very basis of the Newtonian system of the universe (see Chapter 7).

The Irish physicist G. F. FitzGerald conceived a way to save the situation. He suggested that all objects decreased in length in the direction in which they were moving by an amount equal to $\sqrt{1-v^2/c^2}$. Thus:

$$L' = L\sqrt{1-v^2/c^2}$$

where L' is the length of a moving body in the direction of its motion and L is what the length would be if it were at rest.

The foreshortening fraction $\sqrt{1-v^2/c^2}$, FitzGerald showed, would just cancel the ratio $1/\sqrt{1-v^2/c^2}$, which related the maximum and minimum velocities of light in the Michelson-Morley experiment. The ratio would become unity, and the velocity of light would seem to our foreshortened instruments and sense organs to be equal in all directions, regardless of the movement of the source of light through the aether.

Under ordinary conditions the amount of foreshortening is very small. Even if a body were moving at one tenth the velocity of light, or 18,628 miles per second, its length would be foreshortened only slightly, according to the FitzGerald equations. Taking the velocity of light as 1, the equation says:

$$L' = L\sqrt{\left(1-\frac{0{\cdot}1}{1}\right)^2}$$

$$L' = L\sqrt{1-0{\cdot}01}$$

$$L' = L\sqrt{0{\cdot}99}$$

Thus L' turns out to be approximately equal to $0{\cdot}995L$, a foreshortening of about half of 1 per cent.

For moving bodies, velocities such as this occur only in the realm of the sub-atomic particles. The foreshortening of an aeroplane travelling at 2,000 miles per hour is infinitesimal, as you can calculate for yourself.

At what velocity will an object be foreshortened to half its rest length? With L' equal to one half L, the FitzGerald equation is:

$$L/2 = L\sqrt{1-v^2/c^2}$$

or, dividing by L:

$$1/2 = \sqrt{1-v^2/c^2}$$

Squaring both sides of the equation:

$$1/4 = 1-v^2/c^2$$

$$v^2/c^2 = 3/4$$

$$v = \sqrt{3c/4} = 0{\cdot}866c$$

Since the velocity of light in a vacuum is 186,282 miles per second, the velocity at which an object is foreshortened to half its length is 0·866 times 186,282, or roughly 161,300 miles per second.

If a body moves at the speed of light, so that v equals c, the FitzGerald equation becomes:

$$L' = L\sqrt{1-c^2/c^2} = L\sqrt{0} = 0$$

At the speed of light, then, length in the direction of motion becomes zero. It would seem, therefore, that no velocity faster than that of light is possible.

In the decade after FitzGerald had advanced his equation, the electron was discovered, and scientists began to examine the properties of tiny charged particles. Lorentz worked out a theory that the mass of a particle with a given charge was inversely proportional to its radius. In other words, the smaller the volume into which a particle crowded its charge, the greater its mass.

Now if a particle is foreshortened because of its motion, its radius in the direction of motion is reduced in accordance with the FitzGerald equation. Substituting the symbols R and R' for L and L', we write the equation:

$$R' = R\sqrt{1-v^2/c^2}$$

$$R'/R = \sqrt{1-v^2/c^2}$$

The mass of a particle is inversely proportional to its radius. Therefore:

$$\frac{R'}{R} = \frac{M}{M'}$$

where M is the mass of the particle at rest and M' is its mass when in motion.

Substituting M/M' for R'/R in the preceding equation, we have:

$$\frac{M}{M'} = \sqrt{1-v^2/c^2}$$

$$M' = \frac{M}{\sqrt{1-v^2/c^2}}$$

The Lorentz equation can be handled just as the FitzGerald equation was. It shows, for instance, that for a particle moving at a velocity of 18,628 miles per second (one tenth the speed of light) the mass M' would appear to be 0·5 per cent higher than the rest mass M. At a velocity of 161,300 miles per second the apparent mass of the particle would be twice the rest mass.

Finally, for a particle moving at a velocity equal to that of light, so that v is equal to c, the Lorentz equation becomes:

$$M' = \frac{M}{\sqrt{1 - c^2/c^2}} = \frac{M}{0}$$

Now as the denominator of any fraction with a fixed numerator becomes smaller and smaller ('approaches zero'), the value of the fraction itself becomes larger and larger without limit. In other words, from the equation above, it would seem that the mass of any object travelling at a velocity approaching that of light becomes infinitely large. Again, the velocity of light would seem to be the maximum possible.

All this led Einstein to recast the laws of motion and of gravitation. He considered a universe, in other words, in which the results of the Michelson–Morley experiments were to be expected.

Yet even so we are not quite through. Please note that the Lorentz equation assumes some value for M that is greater than zero. This is true for most of the particles with which we are familiar and for all bodies from atoms to stars, that are made up of such particles. There are, however, neutrinos and antineutrinos for which M, the mass at rest, or 'rest-mass', is equal to zero. This is also true of photons.

Such particles travel at the speed of light in a vacuum, provided they are indeed in a vacuum. The moment they are formed they begin to move at such a velocity without any measurable period of acceleration.

We might wonder how it is possible to speak of the 'rest-mass' of a photon or a neutrino, if they are never at rest but can only exist while travelling (in the absence of interfering matter) at a constant speed of 186,280 miles per second. The physicists O. M. Bilaniuk and E. C. G. Sudarshan have therefore suggested that M

be spoken of as 'proper mass'. For a particle with mass greater than zero, the proper mass is equal to the mass measured when the particle is at rest relative to the instruments and observer making the measurement. For a particle with mass equal to zero, the proper mass is obtained by indirect reasoning. Bilaniuk and Sudarshan also suggest that all particles with a proper mass of zero be called 'luxons' (from the Latin word for 'light') because they travel at light-speed, while particles with a proper mass greater than zero be called 'tardyons' because they travel at less than light-speed, or at 'subluminal velocities'.

In 1962, Bilaniuk and Sudarshan began to speculate on the consequences of faster-than-light velocities ('superluminal velocities'). Any particle travelling with faster-than-light velocities would have an imaginary mass. That is, the mass would be some ordinary value multiplied by the square root of -1.

Suppose, for instance, a particle were going at twice the speed of light, so that in the Lorentz equation $v = 2c$. In that case:

$$M' = \frac{M}{\sqrt{1-(2c)^2/c^2}} = \frac{M}{\sqrt{1-4c^2/c^2}} = \frac{M}{\sqrt{-3}}$$

This works out to the fact that its mass while in motion would be some proper mass (M) divided by $\sqrt{-3}$. But $\sqrt{-3}$ is equal to $3\times\sqrt{-1}$ and therefore to $1{\cdot}74\sqrt{-1}$. The proper mass M is therefore equal to $M'\times 1{\cdot}74\times\sqrt{-1}$. Since any quantity that includes $\sqrt{-1}$ is called imaginary, we conclude that particles at superluminal velocities must have imaginary proper masses.

Ordinary particles in our ordinary universe always have masses that are zero or positive. An imaginary mass can have no imaginable significance in our universe. Does this mean that faster-than-light particles cannot exist?

Not necessarily. Allowing the existence of imaginary proper-masses, we can make such faster-than-light particles fit all the equations of Einstein's Special Theory of Relativity. Such particles, however, display an apparently paradoxical property: the more slowly they go, the more energy they contain. This is the precise reverse of the situation in our universe and is perhaps the significance of the imaginary mass. A particle with an imaginary mass speeds up when it meets resistance and slows down when it is pushed ahead by a force. As its energy declines, it moves faster and faster, until when it has zero energy it is moving at infinite speed. As its energy increases, it moves slower and slower until, as its energy approaches the infinite, it slows down to approach the speed of light.

Such faster-than-light particles have been given the name of 'tachyons' from the Greek word for 'speed', by the American physicist Gerald Feinberg.

We may imagine, then, the existence of two kinds of universes. One, our own, is the tardyon-universe, in which all particles go at subluminal velocities and may accelerate to nearly the speed of light as their energy increases. The other is the tachyon-universe, in which all particles go at superluminal velocities and may decelerate to nearly the speed of light as their energy increases. Between is the infinitely narrow 'luxon wall' in which there are particles that go at exactly luminal velocities. The luxon wall can be considered as being held by both universes in common.

If a tachyon is energetic enough and therefore moving slowly enough, it might have sufficient energy and remain in one spot for a long enough period of time to give off a detectable burst of photons. (Tachyons would leave a wake of photons even in a vacuum as a kind of Cerenkov radiation.) Scientists are watching for those bursts, but the chance of happening to have an instrument in just the precise place where one of those (possibly very infrequent) bursts appears for a billionth of a second or less is not very great.

There are those physicists who maintain that 'anything that is not forbidden is compulsory'. In other words, any phenomenon that does not actually break a conservation law *must* at some time or another take place; or, if tachyons do not actually violate special relativity, they *must* exist. Nevertheless, even physicists most convinced of this as a kind of necessary 'neatness' about the universe would be rather pleased (and perhaps relieved) to obtain some evidence for the non-forbidden tachyons. So far, they have not been able to.

One consequence of the Lorentz equation was worked out by Einstein to produce what has become perhaps the most famous scientific equation of all time.

The Lorentz equation can be written in the form:

$$M' = M(1-v^2/c^2)^{-1/2}$$

since in algebraic notation $1/\sqrt{x}$ can be written $x^{-1/2}$. This puts the equation into a form which can be expanded (that is, converted into a series of terms) by a formula discovered by, of all people, Newton. The formula is the binomial theorem.

The number of terms into which the Lorentz equation can be expanded is infinite, but since each term is smaller than the one before, if you take only the first two terms you are approximately correct, the sum of all the remaining terms being small enough to be neglected. The expansion becomes:

$$(1-v^2/c^2)^{-1/2} = 1+\frac{\frac{1}{2}v^2}{c^2} \ldots.$$

Substituting that in the Lorentz equation, we get:

$$M' = M\left(1+\frac{\frac{1}{2}v^2}{c^2}\right) = M+\frac{\frac{1}{2}Mv^2}{c^2}$$

Now in classical physics the expression $\frac{1}{2}Mv^2$ represents the energy of a moving body. If we let the symbol e stand for energy, the equation above becomes:

$$M' = M+e/c^2$$

or

$$M'-M = e/c^2$$

The increase in mass due to motion ($M'-M$) can be represented as m, so:

$$m = e/c^2$$

or

$$e = mc^2$$

It was this equation that for the first time indicated mass to be a form of energy. Einstein went on to show that the equation applied to all mass, not merely to the increase in mass due to motion.

Here again, most of the mathematics involved is only at the secondary-school level. Yet it presented the world with the beginnings of a view of the universe greater and broader even than that of Newton, and also pointed the way to concrete consequences. It pointed the way to the nuclear reactor and the atom bomb, for instance.

Bibliography

A guide to physical science would be incomplete without a guide to more reading. I am setting down here a brief selection of books. The list is a miscellaneous one and does not pretend to be a comprehensive collection of the best modern books about physical science, but I have read most or all of each of them myself and can highly recommend all of them, even my own.

Chapter 1 – *What is Science?*

BERNAL, J. D., *Science in History*, Watts, London, 1965.
CLAGETT, MARSHALL, *Greek Science in Antiquity*, Abelard-Schuman, London, 1957.
CROMBIE, A. C., *Medieval and Early Modern Science* (2 vols.), Penguin Books, Harmondsworth, 1969.
DAMPIER, SIR WILLIAM CECIL, *A History of Science*, Cambridge University Press, Cambridge, 1948.
DREYER, J. L. E., *A History of Astronomy from Thales to Kepler*, Dover Publications, New York, 1953.
FORBES, R. J., and DIJKSTERHUIS, E. J., *A History of Science and Technology* (2 vols.), Penguin Books, Harmondsworth, 1963.
MASON, S. F., *A History of Science: Main Currents of Scientific Thought*, Routledge & Kegan Paul, London, 1953.
TATON, R. (editor), *A General History of Science* (4 vols.), Thames & Hudson, London, 1963–6.

Chapter 2 – *The Universe*

ALTER, D., CLEMINSHAW, C. H., and PHILLIPS, J. G., *Pictorial Astronomy* (3rd rev. edn), Museum Press, London, 1964.

ASIMOV, ISAAC, *The Universe*, Allen Lane The Penguin Press, London, 1967.
BONNOR, WILLIAM, *The Mystery of the Expanding Universe*, Eyre & Spottiswoode, London, 1965.
BURBIDGE, G., and BURBIDGE, M., *Quasi-Stellar Objects*, Freeman, Folkestone, 1968.
FLAMMARION, G. C. *et al.*, *The Flammarion Book of Astronomy*, Allen & Unwin, London, 1964.
GAMOW, GEORGE, *The Creation of the Universe*, Macmillan, London, 1961.
HOYLE, FRED, *Astronomy*, Macdonald, London, 1962.
HOYLE, FRED, *Frontiers of Astronomy*, Mercury Books, London, 1961.
JOHNSON, MARTIN, *Astronomy of Stellar Energy and Decay*, Faber & Faber, London, 1950.
KRUSE, W., and DIECKVOSS, W., *The Stars*, University of Michigan Press, Ann Arbor, 1957.
LEY, WILLY, *Watchers of the Skies*, Sidgwick & Jackson, London, 1964.
MCLAUGHLIN, DEAN B., *Introduction to Astronomy*, Houghton Mifflin, Boston, 1961.
ÖPIK, ERNST J., *The Oscillating Universe*, New American Library, New York, 1960.
SCIAMI, D. W., *The Unity of the Universe*, Faber & Faber, London, 1959.
SHKLOVSKII, I. S. and SAGAN, CARL, *Intelligent Life in the Universe*, Holden-Day, San Francisco, 1966.
SMITH, F. GRAHAM, *Radio Astronomy*, Penguin Books, Harmondsworth, 1960.
STRUVE, OTTO, and ZEBERGS, VELTA, *Astronomy of the 20th Century*, Macmillan, New York and London, 1962.
WHIPPLE, FRED L., *Earth, Moon and Planets* (rev. edn), Penguin Books, Harmondsworth, 1971.
WHITHROW, G. J., *The Structure and Evolution of the Universe*, Harper & Brothers, New York, 1959.

Chapter 3 – *The Earth*

ADAMS, FRANK DAWSON, *The Birth and Development of the Geological Sciences*, Dover Publications, New York, 1938.
BURTON, MAURICE, *Life in the Deep*, Phoenix House, London, 1966.
GAMOW, GEORGE, *A Planet Called Earth*, Macmillan, London, 1965.
GILLULY, J., WATERS, A. C., and WOODFORD, A. O., *Principles of Geology*, W. H. Freeman, San Francisco and London, 1968.
GUTENBERG, B. (editor), *Internal Constitution of the Earth*, Dover Publications, New York, 1951.
HURLEY, PATRICK M., *How Old is the Earth?*, Heinemann, London, 1960.
KUENEN, P. H., *Realms of Water*, Cleaver-Hume Press, London, 1955.

MASON, BRIAN, *Principles of Geochemistry*, Chapman & Hall, London, 1952.
MOORE, RUTH, *The Earth We Live On*, Cape, London, 1957.
SCIENTIFIC AMERICAN (editors), *The Planet Earth*, Simon & Schuster, New York, 1957.

Chapter 4 – *The Atmosphere*

BATES, D. R. (editor), *The Earth and Its Atmosphere*, Basic Books, New York, 1957.
GLASSTONE, SAMUEL, *Sourcebook on the Space Sciences*, Van Nostrand, New York, 1965.
LEY, WILLY, *Rockets, Missiles, and Space Travel*, Chapman & Hall, London, 1957.
LOEBSACK, THEO, *Earth's Envelope*, Collins, London, 1959.
NEWELL, HOMER E., JR, *Window in the Sky*, McGraw-Hill, New York, 1959.
NININGER, H. H., *Out of the Sky*, Dover Publications, New York, 1952.
ORR, CLYDE, JR, *Between Earth and Space*, Collier Books, New York, 1961.

Chapter 5 – *The Elements*

ALEXANDER, W., and STREET, A., *Metals in the Service of Man*, Penguin Books, Harmondsworth, 1972.
ASIMOV, ISAAC, *A Short History of Chemistry*, Doubleday, New York, 1965.
ASIMOV, ISAAC, *The Noble Gases*, Basic Books, New York, 1966.
DAVIS, HELEN MILES, *The Chemical Elements*, Ballantine Books, Boston, 1950.
HOLDEN, ALAN, and SINGER, PHYLLIS, *Crystals and Crystal Growing*, Heinemann, London, 1961.
IHDE, AARON, J., *The Development of Modern Chemistry*, Harper & Row, New York, 1964.
JAFFE, BERNARD, *Chemistry Creates a New World*, Thomas Y. Crowell, New York, 1957.
LEICESTER, HENRY M., *The Historical Background of Chemistry*, Chapman & Hall, London, 1956.
PAULING, LINUS, *College Chemistry* (3rd edn), W. H. Freeman, San Francisco and London, 1964.
SCIENTIFIC AMERICAN (editors), *New Chemistry*, Simon & Schuster, New York, 1957.
WEAVER, E. C., and FOSTER, L. S., *Chemistry for Our Times*, McGraw-Hill, New York and London, 1947.
WEEKS, MARY E., and LEICESTER, H. M., *Discovery of the Elements* (7th edn), Journal of Chemical Education, Easton, Pa., 1968.

Chapter 6 – *The Particles*

ALFVÉN, HANNES, *Worlds Antiworlds*, Freeman, San Francisco and London, 1966.
ASIMOV, ISAAC, *The Neutrino*, Dobson, London, 1966.
FORD, KENNETH W., *The World of Elementary Particles*, Blaisdell New York, 1963.
FRIEDLANDER, G., KENNEDY, J. W., and MILLER, J. M., *Nuclear and Radiochemistry* (2nd edn), Wiley, New York, 1964.
GAMOW, GEORGE, *Mr Tompkins Explores the Atom*, Cambridge University Press, Cambridge, 1965.
GARDNER, MARTIN, *The Ambidextrous Universe*, Penguin Books, Harmondsworth, 1970.
GLASSTONE, SAMUEL, *Sourcebook on Atomic Energy* (3rd edn), D. Van Nostrand Company, Princeton, N.J., 1967.
HUGHES, DONALD J., *The Neutron Story*, Heinemann, London, 1960.
MASSEY, SIR HARRIE, *The New Age in Physics*, Elek, London, 1966.
PARK, DAVID, *Contemporary Physics*, Harcourt, Brace & World, New York, 1964.
SHAMOS, M. H. and MURPHY, G. M. (editors), *Recent Advances in Science*, New York University Press, New York, 1956.

Chapter 7 – *The Waves*

BENT, H. A., *The Second Law*, Oxford University Press, 1965.
BERGMANN, P. G., *The Riddle of Gravitation*, Murray, London, 1969.
BLACK, N. H., and LITTLE, E. P., *An Introductory Course in College Physics*, Macmillan, 1957.
EDDINGTON, SIR ARTHUR S., *The Nature of the Physical World*, Collins, London, 1958.
EINSTEIN, ALBERT, and INFELD, LEOPOLD, *The Evolution of Physics*, Cambridge University Press, Cambridge, 1938.
FREEMAN, IRA M., *Physics Made Simple*, Made Simple Books, New York, 1954.
GARDNER, MARTIN, *Relativity for the Million*, Macmillan, New York and London, 1962.
HOFFMAN, BANESH, *The Strange Story of the Quantum*, Penguin Books, Harmondsworth, 1963.
ROSSI, BRUNO, *Cosmic Rays*, Allen & Unwin, London, 1966.
SHAMOS, MORRIS H., *Great Experiments in Physics*, Henry Holt, New York, 1959.

Chapter 8 – *The Machine*

BITTER, FRANCIS, *Magnets*, Heinemann, London, 1960.
DE CAMP, L. SPRAGUE, *The Ancient Engineers*, Souvenir Press, London, 1963.
KOCK, W. E., *Lasers and Holography*, Doubleday, New York, 1969.
LARSEN, EGON, *Transport*, Phoenix House, London, 1959.
LEE, E. W., *Magnetism*, Penguin Books, Harmondsworth, 1963.
LENGYEL, BELA A., *Lasers*, John Wiley, New York, 1962.
NEAL, HARRY EDWARD, *Communication*, Phoenix House, London, 1963.
PIERCE, JOHN R., *Electrons, Waves and Messages*, Doubleday, New York, 1956.
PIERCE, JOHN R., *Symbols, Signals and Noise*, Harper, New York, 1961.
SINGER, CHARLES, HOLMYARD, E. J., and HALL, A. R. (editors), *A History of Technology* (5 vols.), Clarendon Press, Oxford, 1954.
TAYLOR, F. SHERWOOD, *A History of Industrial Chemistry*, Heinemann, London, 1957.
UPTON, MONROE, *Electronics for Everyone*, Faber & Faber, London, 1956.
USHER, ABBOTT PAYSON, *A History of Mechanical Inventions*, Beacon Press, Boston, 1959.
WARSCHAUER, DOUGLAS M., *Semiconductors and Transistors*, McGraw-Hill, New York, 1959.

Chapter 9 – *The Reactor*

ALEXANDER, PETER, *Atomic Radiation and Life*, Penguin Books, Harmondsworth, 1957.
BISHOP, AMASA S., *Project Sherwood*, Addison-Wesley, Reading, Mass., 1958.
FOWLER, JOHN M., *Survival: A Study of Superbombs, Strontium 90, and Fallout*, MacGibbon & Kee, London, 1960.
JUKES, JOHN, *Man-Made Sun*, Abelard-Schuman, New York and London, 1959.
JUNGK, ROBERT, *Brighter Than a Thousand Suns*, Penguin Books, Harmondsworth, 1958.
PURCELL, JOHN, *The Best-Kept Secret*, Vanguard Press, New York, 1963.
RIEDMAN, SARAH R., *Men and Women behind the Atom*, Abelard-Schuman, New York and London, 1958.
SCIENTIFIC AMERICAN (editors), *Atomic Power*, Simon & Schuster, New York, 1955.
WILSON, ROBERT R., and LITTAUER, R., *Accelerators*, Heinemann, London, 1962.

Appendix – *Mathematics in Science*

COURANT, RICHARD and ROBBINS, HERBERT, *What Is Mathematics?*, Oxford University Press, London, 1941.
DANTZIG, TOBIAS, *Number, the Language of Science*, Allen & Unwin, London, 1962.
FELIX, LUCIENNE, *The Modern Aspects of Mathematics*, Basic Books, New York, 1960.
FREUND, JOHN E., *A Modern Introduction to Mathematics*, Prentice-Hall, New York, 1956.
KLINE, MORRIS, *Mathematics and the Physical World*, Murray, London, 1960.
KLINE, MORRIS, *Mathematics in Western Culture*, Oxford University Press, 1953.
NEWMAN, JAMES R., *The World of Mathematics* (4 vols.), Allen & Unwin, 1960.
STEIN, SHERMAN K., *Mathematics, the Man-Made Universe*, W. H. Freeman, San Francisco and London, 1963.
VALENS, EVANS G., *The Number of Things*, Dutton, New York, 1964.

General

ASIMOV, ISAAC, *Asimov's Bibliographical Encyclopedia of Science and Technology*, Allen & Unwin, London, 1966.
ASIMOV, ISAAC, *Life and Energy*, Dobson, London, 1963.
ASIMOV, ISAAC, *Understanding Physics* (3 vols.), Walker, New York, 1966.
ASIMOV, ISAAC, *The Words of Science*, Houghton Mifflin, Boston, 1959.
CABLE, E. J. *et al.*, *The Physical Sciences*, Prentice-Hall, New York, 1959.
GAMOW, GEORGE, *Matter, Earth, and Sky*, Macmillan, London, 1959.
HUTCHINGS, EDWARD, JR (editor), *Frontiers in Science*, Allen & Unwin, London, 1960.
SHAPLEY, HARLOW, RAPPORT, SAMUEL, and WRIGHT, HELEN (editors), *A Treasury of Science* (3rd edn), Angus & Robertson, London, 1954.
SLABAUGH, W. H. and BUTLER, A. B., *College Physical Science*, Prentice-Hall, New York, 1958.
WATSON, JANE WERNER, *The World of Science*, Adprint, London, 1959.

Name Index

Subject Index

More about Penguins and Pelicans

Penguinews, which appears every month, contains details of all the new books issued by Penguins as they are published. From time to time it is supplemented by *Penguins in Print*, which is a complete list of all titles available. (There are some five thousand of these.)

A specimen copy of *Penguinews* will be sent to you free on request. For a year's issues (including the complete lists) please send 50p if you live in the British Isles, or 75p if you live elsewhere. Just write to Dept. EP, Penguin Books Ltd, Harmondsworth, Middlesex, enclosing a cheque or postal order, and your name will be added to the mailing list.

In the U.S.A.: For a complete list of books available from Penguin in the United States write to Dept. CS, Penguin Books Inc., 7110 Ambassador Road, Baltimore, Maryland 21207.

In Canada: For a complete list of books available from Penguin in Canada write to Penguin Books Canada Ltd, 41 Steelcase Road West, Markham, Ontario.

The Chemistry of Life

Steven Rose

The molecular structure of a protein (insulin) was described in detail for the first time in 1956: today such procedures are routine. Not only has the pace of biochemistry accelerated in recent years: with the perfection of the electron microscope and the development of cybernetics, the science has also widened and grown more complex.

The Chemistry of Life outlines the scope and achievement of a science which began as the study of the chemical constituents of living matter. Dealing successively with the chemical analysis of the living animal cell, the conversions induced between chemicals by the enzymes acting as catalysts, and the self-regulating nature of cells, Dr Rose explains how the design of particular cells influences their functions within the living organism as a whole.

Biochemistry is a difficult subject. But it is presented here as simply as accuracy will permit by a young research biochemist who conveys much of the adventure of discovery implicit in a science which may one day answer the eternal question: 'What is life?'

Atoms and the Universe

G. O. Jones, J. Rotblat, G. J. Whitrow

Third Revised Edition

Everything observed by the physicist or astronomer – on earth or in the sky – is packed with atomic matter and atomic energy. During the last two decades our knowledge about the universe has progressed by leaps and bounds, and there are signs that reasonably consistent theories about the universe are emerging.

In *Atoms and the Universe* two physicists and an astronomer survey the whole, vast field of physical science and modern astronomy, giving a complete and up-to-date guide to the structure of matter and the age and origins of the universe. They cover, among many subjects, the new accelerators developed by atomic physicists; the possibility of controlled thermonuclear reactors; the latest findings from artificial satellites and rockets; the speculations caused by quasi-stellar radio sources and exploding nebulae, and the important new work on the origin of elements and the evolution of stars.

This edition has been extensively revised and rewritten. Considered by many to be the outstanding work in its field, *Atoms and the Universe* will appeal not only to scientists, but to the layman with an intelligent interest in the nature of things.